Circularly Polarized Dielectric Resonator Antennas

For a complete listing of titles in the
Artech House Antennas and Propagation Library
turn to the back of this book.

Circularly Polarized Dielectric Resonator Antennas

Raghvendra Kumar Chaudhary

Rajkishor Kumar

Rakesh Chowdhury

ARTECH HOUSE

BOSTON | LONDON

artechhouse.com

Library of Congress Cataloging-in-Publication Data
A catalog record for this book is available from the U.S. Library of Congress.

British Library Cataloguing in Publication Data
A catalogue record for this book is available from the British Library.

Cover design by Charlene Stevens

ISBN 13: 978-1-63081-817-3

685 Canton Street
Norwood, MA 02062

10 9 8 7 6 5 4 3 2 1

Contents

1

Introductory Remarks

This chapter begins with an introductory description of different polarization states focusing on circular polarization (CP) and highlighting its benefit over linear polarization. Next, the historical developments of the dielectric resonator antenna (DRA) from energy storage elements to radiating elements are detailed along with some of the basic advantages that they offer. After discussion of CP and DRAs, a short overview of the generation of circularly polarized waves in DRAs is also incorporated in this chapter. Finally, this chapter concludes with a brief overview of the contents of the remaining chapters.

1.1 Introduction to Circular Polarization

Polarization [1, 2] is a property of electromagnetic (EM) waves that captures the geometrical orientation of the electric field vibrations in a plane. In general, an antenna polarization is of an elliptical nature (i.e., the tip traces out an ellipse in which the electric field vector rotates at a constant rate when time goes on). There are two parameters of an electric field that can be linked during wave propagation: its magnitude and direction. On the basis of this property, the elliptical locus of a net electric field transforms into linear and circular shapes. When the magnitude changes and the direction remains the same with respect to time, it is called linear polarization, and when the magnitude remains constant but the direction changes with respect to time, it is called circular polarization. There are various inherent properties offered by CP waves that make them superior to linearly polarized (LP) antennas, which are listed below.

1. Circularly polarized waves are more resistant to signal degradation in adverse, atmospheric, and bad weather conditions. The linearly polarized waves are more affected in these conditions as compared to CP signals. This is because the energy present in the CP signal vibrates on all planes (0° to 360°) and there is more probability of successful signal reception even if the signal is degraded in any particular plane.
2. For linearly polarized antenna systems, the transmitting and receiving of antenna alignments must be the same for maximum power transfer (horizontal-horizontal and vertical-vertical). In circular polarization, this condition of antenna alignment is not required at all. Even if receiver antenna alignment is disturbed, it will still be able to receive the full signal from other planes. It is worth mentioning that if an ideal CP signal is received by a linearly polarized antenna, the loss is 3 dB regardless of the orientation of the receiving antenna. Therefore, CP signals are most useful when antenna alignment is not known in a communication system.
3. During propagation, radio frequency (RF) signals go through several reflections, multipaths, or absorption of signal (different materials absorb the signal from different planes) depending on the material they interact with. Linear polarized signals can resolve these problems in only one plane. If the surface in contact does not reflect the signal precisely in the same plane, that signal strength will be lost. Since circularly polarized antennas send and receive in all planes, the signal strength is not lost, but is transferred to a different plane and is still utilized. As a result, a circularly polarized antenna gives you a higher probability of a successful link because it is transmitting on all planes and is much better at penetrating and bending around obstructions.
4. Circularly polarized signals resolve the problem of the Faraday effect in the upper atmosphere. It is mainly a rotation of the plane of polarization that is proportional to the component of the magnetic field present in the direction of propagation [3]. For linear polarization, this would cause polarization mismatch of the signal. CP antennas are immune to this effect, as every electric field component is equal in magnitude and rotated by the same angle.

To conclude from the above discussion, CP waves are found to be more superior than LP antennas in some applications.

1.2 Brief Overview of Different Polarization States

In the previous section, benefits of circular polarization were presented to introduce CP in this book. Now, the equation of Locus traced out by the net electric field is derived from where the special cases of elliptical polarization come into the picture, and then different polarization states will be discussed. Generally, the time-harmonic E-field will have two perpendicular components and , as shown in Figure 1.1.

E_{yo} = magnitude of y-component of E-field, E_{xo} = magnitude of x-component of E-field.

$\varnothing_y$= Phase angle of y-component of E-field, $\varnothing_x$ = Phase angle of x-component of E-field.

At fixed location (say $z = 0$), the net E-field vector can be written as

$$E_{net} = E_{xo}\cos(wt + \varnothing_x)\hat{x} + E_{yo}\cos(wt + \varnothing_y)\hat{y}$$

Now, y- component can be expanded as

$$E_y = E_{yo}\left\{\cos(wt)\cos(\varnothing_y) - \sin(wt)\sin(\varnothing_y)\right\}$$

$$\frac{E_y}{E_{yo}} = \cos(wt)\cos(\varnothing_y) - \sin(wt)\sin(\varnothing_y) \tag{1.1}$$

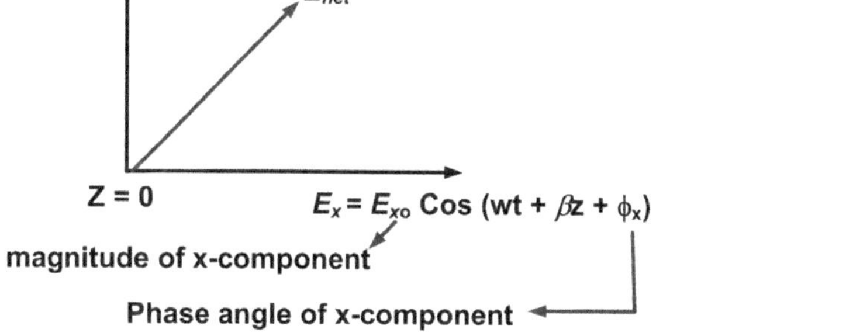

Figure 1.1 Decomposition of net electric field into its perpendicular components.

Similarly, *x*-component can be expanded as

$$E_x = E_{xo}\left\{\cos(wt)\cos(\varnothing_x) - \sin(wt)\sin(\varnothing_x)\right\}$$

$$\frac{E_x}{E_{xo}} = \cos(wt)\cos(\varnothing_x) - \sin(wt)\sin(\varnothing_x) \tag{1.2}$$

Multiplying (1.1) by $\sin\varnothing_x$ and (1.2) by $\sin\varnothing_y$

$$\frac{E_y}{E_{yo}}\sin(\varnothing_x) = \cos(wt)\cos(\varnothing_y)\sin(\varnothing_x) - \sin(wt)\sin(\varnothing_y)\sin(\varnothing_x) \tag{1.3}$$

$$\frac{E_x}{E_{xo}}\sin(\varnothing_y) = \cos(wt)\cos(\varnothing_x)\sin(\varnothing_y) - \sin(wt)\sin(\varnothing_x)\sin(\varnothing_y) \tag{1.4}$$

Subtracting (1.4) and (1.3) will get

$$\frac{E_x}{E_{xo}}\sin(\varnothing_y) - \frac{E_y}{E_{yo}}\sin(\varnothing_x) = \cos(wt)\sin(\varnothing_y - \varnothing_x) \tag{1.5}$$

Again, following the above steps in a similar manner after multiplying (1.1) by $\cos\varnothing_x$ and (1.2) by $\cos\varnothing_y$ and performing the subtraction operation in a similar manner, will get

$$\frac{E_x}{E_{xo}}\cos(\varnothing_y) - \frac{E_y}{E_{yo}}\cos(\varnothing_x) = \sin(wt)\sin(\varnothing_y - \varnothing_x) \tag{1.6}$$

Let $(\varnothing_y - \varnothing_x) = \delta$ (phase difference between E_x and E_y component as assumed before).

Squaring (1.5) and (1.6), and adding them, will get

$$\left(\frac{E_x}{E_{xo}}\sin(\varnothing_y) - \frac{E_y}{E_{yo}}\sin(\varnothing_x)\right)^2 + \left(\frac{E_x}{E_{xo}}\cos(\varnothing_y) - \frac{E_y}{E_{yo}}\cos(\varnothing_x)\right)^2 = \sin\delta^2 \tag{1.7}$$

Now expanding (1.7) and rearranging the equation, will get

$$\left(\frac{E_x}{E_{x0}}\right)^2 + \left(\frac{E_y}{E_{y0}}\right)^2 - 2\frac{E_x E_y}{E_{x0} E_{y0}}\cos\delta = sin^2\delta \tag{1.8}$$

So, from (1.8), it can be seen that it is the equation of an ellipse. Therefore, we conclude that in general the tip of the net electric field vector traces out an ellipse and this is called elliptical polarization. Linear and circular polarization are just special cases of elliptical polarization. The axial ratio (AR) is a far-field parameter that captures the polarizability of the field vector and is defined as the ratio of the major to the minor axis of the ellipse traced out by the net E-field vector. Mathematically, it is given by

$$Axial\ Ratio(dB) = 20\log\left(\frac{E_{major}}{E_{minor}}\right)$$

Where, E_{major} and E_{minor} are the magnitudes of maximum and minimum E-field. When both axes are of equal magnitude, then $AR = 0$ dB, which is the case for circular polarization and AR is infinite in the case of linear polarization. In practice, AR ≤ 3 dB is considered an acceptable value when CP radiation is needed. Now, two different cases are possible, which are discussed next.

Case I: Linear polarization: when $\delta = \varnothing_y - \varnothing_x = \eta\pi$, $\eta = 0, 1, 2, 3, \ldots$ (1.8) becomes.

$$\begin{aligned}
&\left(\frac{E_x}{E_{x0}}\right)^2 + \left(\frac{E_y}{E_{y0}}\right)^2 - 2\frac{E_x E_y}{E_{x0} E_{y0}} = 0 \\
&\left\{\frac{E_x}{E_{x0}} - \frac{E_y}{E_{y0}}\right\}^2 = 0 \\
&\frac{E_x}{E_{x0}} = \frac{E_y}{E_{y0}} \\
&E_y = \frac{E_{y0}}{E_{x0}} E_x
\end{aligned} \tag{1.9}$$

This (1.9) is the equation of a straight line and the associated polarization is called linear polarization. There are two types of linear polarization:

1. Horizontal polarization;
2. Vertical polarization.

The linear polarization is defined in the form of a time-harmonic wave and this wave shows the linearly polarized behavior at a given point in space if the electric field vector at that point is always oriented along the same straight line at every instant of time.

Case II: Circular polarization: Now, from (1.8), if $E_{y0} = E_{x0} = \mathrm{E}$ and $\delta = \pm(2n+1)\frac{\pi}{2}, n = 0,1,2,\ldots$ we get

$$\left(\frac{E_x}{E}\right)^2 + \left(\frac{E_y}{E}\right)^2 = 1 \tag{1.10}$$

This (1.10) is the equation of circular polarization, which shows that the net *E*-field traces a circle at a given point in space when time changes [4]. The sense of rotation of circular polarization is determined by the phase-leading field component toward the phase-lagging component. In the context of the above equations of wave orientation, circularly polarized waves can be generated in DRAs if following conditions are satisfied.

- It has two orthogonal linear field components;
- The two orthogonal linear field components have the same magnitude;
- These two linear field components have a time-phase difference of odd multiples of 90°.

Mathematically, circular polarization can be obtained when the two field components have the same magnitude and the time-phase difference between the two field components are odd multiples of $\pi/2$.

Figure 1.2 shows the conditions of two electric field components for different polarization states. This book is devoted to various techniques by which circular polarization can be synthesized from DRAs. Thus it is concluded that circular polarization is just a special case of elliptical polarization that can be achieved when required conditions are satisfied in DRAs.

1.3 History of Dielectric Resonators

The idea of getting electromagnetic radiation from dielectric bodies was first predicted by R. D. Richtmyer in 1939 [5] when he studied the resonant properties of the spherical, ring, and toroidal shaped dielectric materials. However, this idea did not gain any attention from the scientific community due to the unavailability of low loss dielectric materials. A standard dielectric resonator (DR) can be of any geometrical shape, the most popular being cylindrical, rect-

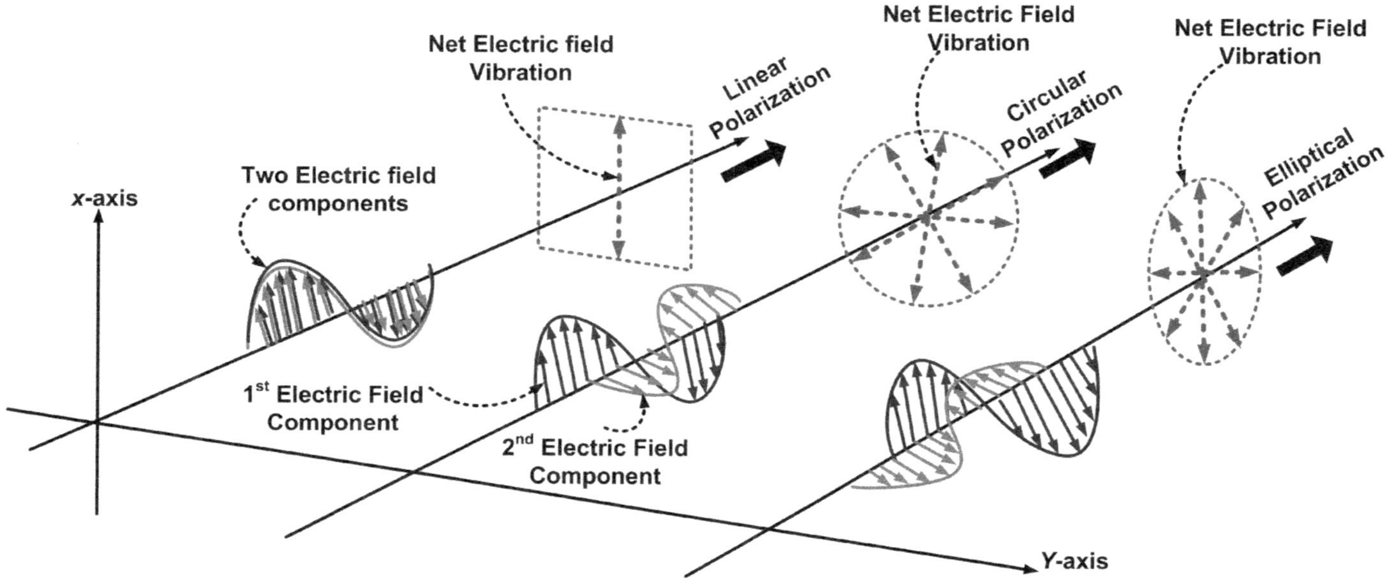

Figure 1.2 Orientation of electric field components for different states of polarization.

angular, and hemispherical ones, which are made of low loss and very high permittivity material [6,7]. In the early 1960s, due to the growth of the ceramic industry and accessibility of low loss dielectrics, this led to the usage of dielectric resonators as an alternative to metallic microwave resonators. Thus, initial research activity on dielectric resonators was focused on utilizing them for circuit applications such as oscillators and filters [8, 9]. In 1962, Okaya and Barash found that a portion of rutile crystals with low dielectric losses can provide high unloaded quality (*Q*) factors even higher than metal-walled cavity resonators [10]. In 1967, Gastine et al. reported the radiation *Q*-factor of isolated dielectric spheres [11]. Many research studies have been conducted on the investigation of electromagnetic mode field configuration and nomenclature in dielectric resonators of high dielectric constant [12–15]. Therefore, during this period, primary attention was given to dielectric resonators as circuit elements and various research was conducted for using dielectric resonator properties for oscillator and filter applications. For this purpose, DRs were surrounded by a metallic cavity to sustain high *Q*-factor and minimize any radiation leakage. However, if the metallic surrounding is removed and its dielectric constant is lowered, DRs can become very good radiators and this thought came into consideration among the scientific community as the idea of DRAs was developed. This was a transition period for dielectric resonators from circuit elements to radiating elements. The core idea that led to this transition was that electromagnetic fields could leak out of the cavity if the metallic enclosure is removed.

Thus, radiation can be expected from the same dielectric resonators by lowering its permittivity and removing the metallic surrounding. The systematic study of a dielectric resonator as antenna element began in the early 1980s and was first accomplished by Long, McAllister, and Shen [16–18] who examined the characteristics of cylindrical, hemispherical, and rectangular shapes and theoretically calculated their radiation pattern Their research revealed that dielectric resonator-based antennas could be a better alternative to traditional low gain metallic antennas. Since then, most of the research had been devoted to an examination of simple dielectric resonator shapes, possible resonant modes, and so on. During the early 1990s, research was devoted to applying numerical and analytical techniques to determine the input impedance, resonance frequency, and radiated *Q*-factor in different shapes of DRAs. Also, during the mid- to late 1990s, researchers' attention started turning to various feeding mechanisms for exciting dielectric resonators such as aperture coupling, probe feeding, and microstrip coupling [19]. Since the start of the new millennium, the research on DRAs has significantly increased into new research directions, such as designing of circularly polarized dielectric resonator antennas, array antennas, high gain antennas, and reconfigurable antennas. The timeline of DRA evolution from circuit element to the radiating element is shown in Figure 1.3.

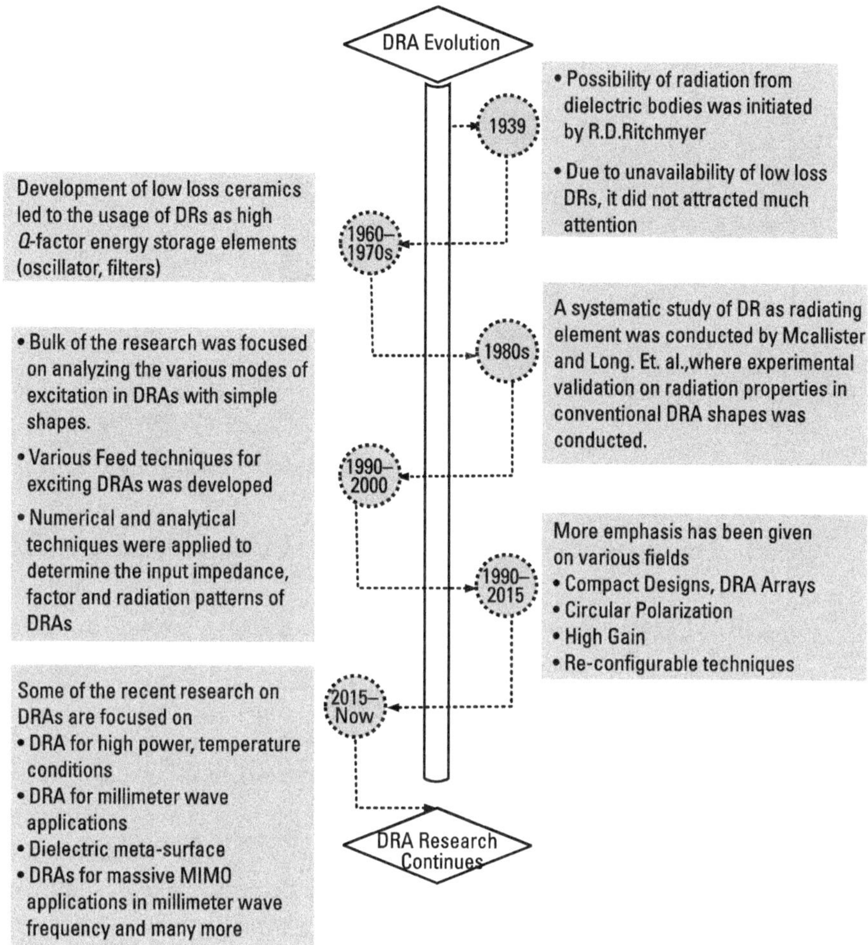

Figure 1.3 Diagram showing the evolution of dielectric resonator antennas.

1.4 Basic Features of Dielectric Resonator Antennas

DRAs have several unique features [20,21] that make them different from metallic antennas, which are shown in Figure 1.4. Some of the features are discussed below:

1. The size of a DRA is dependent on both the resonant frequency and dielectric constant of the materials.
2. The aspect ratio of the DRA affects the radiation *Q*-factor and resonant frequency.

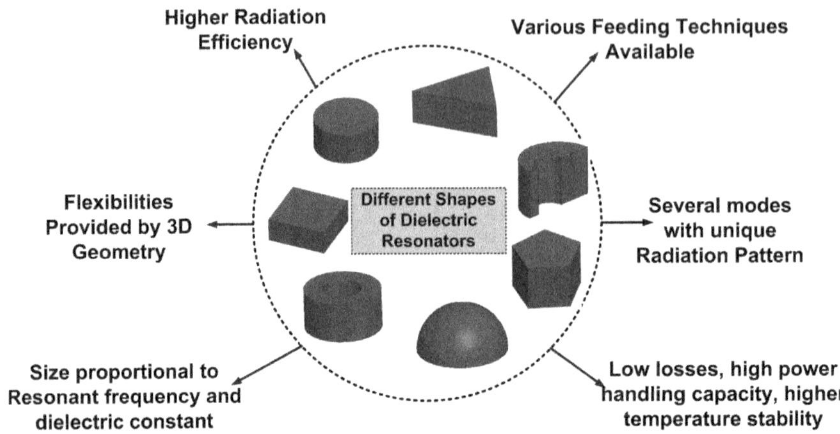

Figure 1.4 Diagram showing basic features of dielectric resonator antennas.

3. High radiation efficiency can be achieved by using dielectric materials with low loss.
4. DRAs possess more design flexibility than planar metallic antennas to control the physical size and impedance bandwidth by choosing a wide range of the dielectric constants.
5. DRAs provide a high degree of freedom and versatility to cover a wide range of frequencies of operations starting from 1 GHz to 100 GHz.
6. There are several feeding mechanisms, such as a coaxial probe, microstrip line, aperture coupling, conformal strip, and coplanar waveguide, used to efficiently excite the DRA. Because of the flexibility of the feeding mechanism, a DRA integrates various existing or current technologies.
7. DRAs also provides excitation of different modes with unique radiation patterns, such as broadside, end-fire, and omnidirectional, which are very important from an application point of view.

Besides all these advantages, DRAs also possess some disadvantages, which are listed below:

1. They have a high fabrication cost compared to planar antennas because of the selection of the high permittivity material. It is difficult to design a DR with different shapes, such as hexagonal and Koch, due to material rigidity.
2. It is difficult to get specific dielectric materials having desired dielectric constants.

1.5 Circular Polarization in Dielectric Resonator Antennas

DRAs have been investigated by numerous research communities by analyzing different possible ways of generation of circularly polarized radiation and improving their CP performance. There are two conditions for getting CP waves (i.e., generation of orthogonal components with equal amplitude (spatial orthogonality condition) and achieving phase difference of $\pm\pi/2$ between the orthogonal components (which is a time-domain condition)). In the case of DRAs, a CP wave can be generated by exciting two orthogonal components of the electric field in phase quadrature. Figure 1.5 shows the mechanism to generate a CP wave [22]. Generation of orthogonal E-field components is a primary requirement that requires two linearly polarized electric field components (let us say E1 and E2) separated by 90° in the space domain, as shown in Figure 1.5(a). Next, an essential requirement is establishing a phase quadrature relationship between both components, which in turn requires separation of

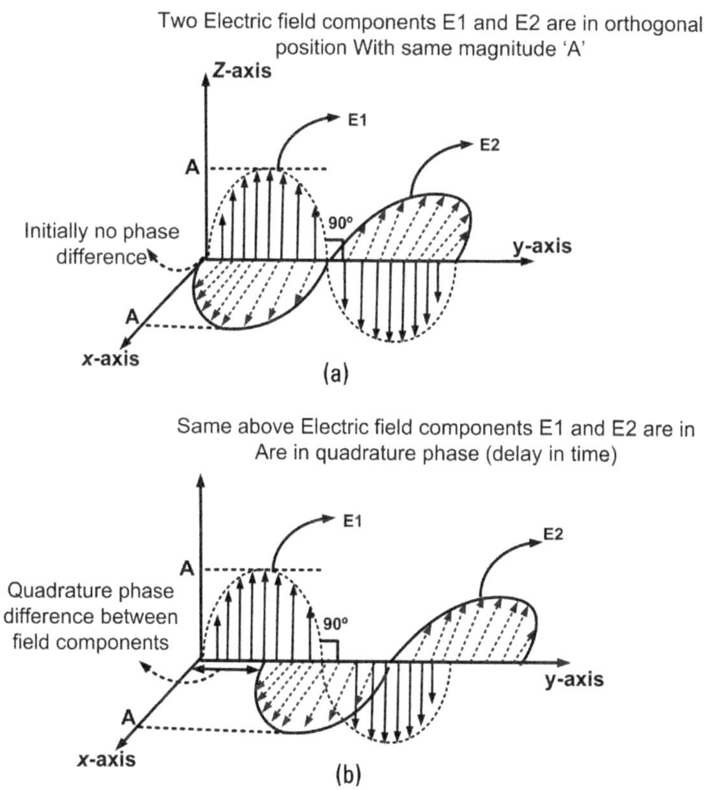

Figure 1.5 Generation of circularly polarized wave in DRAs. (a) Two electric field components separated in space (orthogonal signals), and (b) the same two orthogonal electric field components with time delay (to establish phase difference between them).

orthogonal components in time domain. Figure 1.5(b) shows the time delay between E_1 and E_2 that causes phase difference between them. By employing such conditions, the resultant electric field rotates in a circular manner at a particular point as time varies, resulting in circular polarization.

1.6 Agenda of the Book

The focus of this book is on various possible ways of designing and development of circularly polarized DRAs by using different feeding techniques and shape modification of DRAs. This book will provide the various aspects of CP generation techniques that will be helpful to readers who are searching for practical designing of DRAs. This book mainly discusses the feeding techniques, such as a coaxial probe, microstrip, conformal strip, aperture, and coplanar waveguide, as well as different shapes of DRA for improvement in CP performance. The organization of this book is shown in Figure 1.6.

Chapter 2 gives some background on dielectric resonator antennas focusing on conventional shapes, such as hemispherical, cylindrical, and rectangular. The knowledge of fundamental modes and coupling techniques to excite them

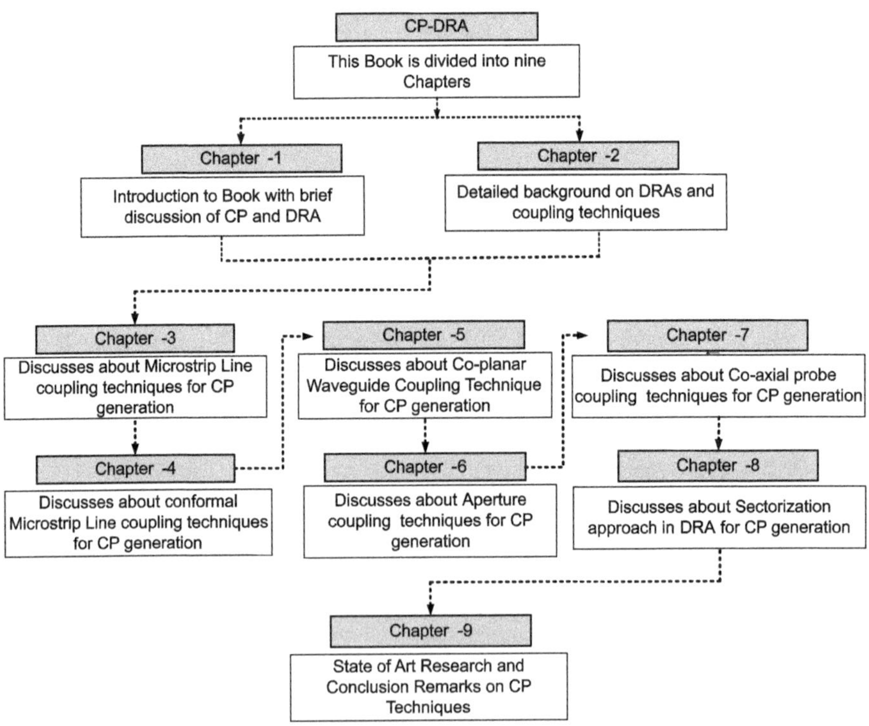

Figure 1.6 Summary of book organization.

is also briefed in this chapter to create a solid foundation for the remaining chapters. The next chapters are organized as follows: Chapter 3 discusses circular polarization techniques based on modified microstrip line techniques in a different configuration with a design example. In Chapters 4, 5, 6, and 7, the remaining feeding techniques—conformal microstrip line feeding, aperture feeding, coplanar feeding, and simple probe feed coupling—are discussed and examined, respectively, with design examples. In Chapter 8, the mechanism of CP generation is investigated by the sectorization approach in DRAs followed by conclusions in Chapter 9.

References

[1] Balanis, C. A., *Antenna Theory: Analysis and Design*, Hoboken, NJ: John Wiley & Sons, 2005.

[2] Stutzman, W. L., *Polarization in Electromagnetic Systems*, Norwood, MA: Artech House, 1992.

[3] Brookner, E., W. M. Hall, and R. H. Westlake, "Faraday Loss for L-band Radar and Communications Systems," *IEEE Transactions on Aerospace and Electronic Systems*, Vol. 21, No. 4, 1985, pp. 459–469.

[4] Kraus, J. D., *Antennas*, New York: McGraw-Hill, 1988.

[5] Richtmyer, R. D., "Dielectric Resonators," *Journal of Applied Physics*, Vol. 10, 1939, pp. 391–398.

[6] Luk, K. M., and K. W. Leung, *Dielectric Resonator Antennas*, Baldock, England: Research Studies Press Ltd, 2003.

[7] Petosa, A., *Dielectric Resonator Antenna Handbook*, Norwood, MA: Artech House, 2007.

[8] Cohn, S. B., "Microwave Bandpass Filters Containing High-Q Dielectric Resonators," *IEEE Transactions on Microwave Theory and Techniques*, Vol. 16, No. 4, pp. 218–227, 1968.

[9] Fiedziuszko, S. J., "Microwave Dielectric Resonators," *Microwave Journal*, Vol. 29, 1968, pp. 189–196.

[10] Okaya, A., and L. F. Barash, "The Dielectric Microwave Resonator," *Proceedings of the IRE*, Vol. 50, No. 10, 1962, pp. 2081–2092.

[11] Gastine, M., L. Courtois, and J. L. Dormann, "Electromagnetic Resonances of Free Dielectric Spheres," *IEEE Transactions on Microwave Theory and Techniques*, Vol. 15, No. 12, 1967, pp. 694–700.

[12] Kajfez, D., and P. Guillon, *Dielectric Resonators*, Norwood, MA: Artech House, 1986.

[13] Van Bladel, J., "On the Resonances of a Dielectric Resonator of Very High Permittivity," *IEEE Transaction on Microwave Theory and Techniques*, Vol. 23, 1975, pp. 199–208.

[14] Van Bladel, J., "The Excitation of Dielectric Resonators of Very High Permittivity," *IEEE Transaction on Microwave Theory and Techniques*, Vol. 23, 1975, pp. 208–215.

[15] Kajfez, D., A. W. Glisson, and J. James, "Computed Modal Field Distributions for Isolated Dielectric Resonators," *IEEE Transactions on Microwave Theory and Techniques*, Vol. 32, No. 12, 1984, pp. 1609–1616.

[16] Long, S. A., M. W. McAllister, and L. C. Shen, "The Resonant Cylindrical Dielectric Cavity Antenna," *IEEE Antennas and Wireless Propagation Letters*, Vol. 31, No. 3, 1983, pp. 406–412.

[17] McAllister, M. W., S. A. Long, and G. L. Conway, "Rectangular Dielectric Resonator Antenna," *IET Electronics Letters*, Vol. 19, No. 6, 1983, pp. 218-219.

[18] McAllister, M. W., and S. A. Long, "Resonant Hemispherical Dielectric Antenna," *IET Electronics Letters*, Vol. 20, No. 16, 1984, pp.657–659.

[19] Petosa, A., and A. Iittipiboon, "Dielectric Resonator Antennas: A Historical Review and the Current State of the Art," *IEEE Antennas and Propagation Magazine*, Vol. 52, No. 5, 2010, pp. 91–116.

[20] Luk, K. M., and K. W. Leung, *Dielectric Resonator Antenna*, Baldock, U.K.: Research Studies Press, 2003.

[21] Mongia, R. K., and P. Bhartia, "Dielectric Resonator Antennas-A Review and General Design Relations for Resonant Frequency and Bandwidth," *International Journal of Microwave and Millimeter-Wave Computer-Aided Engineering*, Vol. 4, No. 3, 1994, pp. 230–247.

[22] Gao, S., Q. Luo, and F. Zhu, "Circularly Polarized Antennas," West Sussex, U.K.: John Wiley & Sons, 2014

2

Dielectric Resonator Antenna Theory

This chapter provides a preliminary description of DRAs focusing on their historical developments from storage elements to radiating elements. There are three conventional shapes of dielectric resonators used as (1) hemispherical, (2) cylindrical, and (3) rectangular, which are investigated along with their description of modes. The following discussion will be built on a basic understanding of DRA functions that are required for the generation of circular polarization.

2.1 Dielectric Resonator as Circuit Element

Resonators are devices or circuit systems that exhibit electromagnetic resonance behavior at particular frequencies, called resonance frequency (f_r) [1]. At this frequency, the energy present in the system oscillates with larger amplitude than at other frequencies. Various categories of resonators used according to the frequency of operation are shown collectively in Figure 2.1.

A resonator is a crucial element of microwave-circuits, the main requirements of microwave-resonators are a low-loss or high-quality factor, and temperature stability to provide near-zero change in resonant frequency even when the ambient operating temperatures do vary [3]. The simplest resonant circuit is the RLC circuit, which consists of discrete components, such as resistor (R), an inductor (L), and a capacitor (C) in a series or parallel arrangement. This configuration acts as an electrical resonator at low frequencies in which the energy oscillates around the resonant frequency. Next, at higher frequencies, transmission lines are also used as resonators. At further higher frequencies, metallic cavity resonators are very popular and have been widely used. Energy is stored due to multiple reflections of the wave at the metallic boundary. The cavity is also

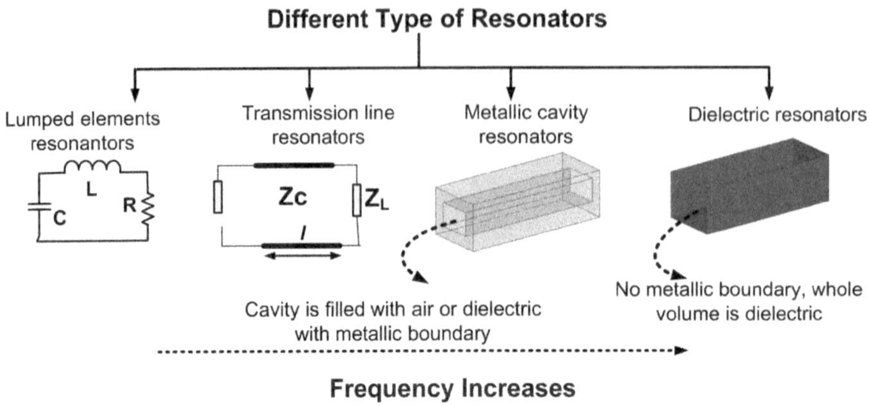

Figure 2.1 Various types of resonators used at a different range of frequencies.

sometimes occupied with dielectric materials. However, resonators with whole dielectric material were only possible after the finding of low-loss dielectric materials in the early 1960s. Various research was conducted on dielectric materials and it was found that they can also be used as energy storage elements with very high-quality factor (*Q*-factor) because it is characterized by its permittivity (ε_r). The larger the ε_r of the dielectric material, the more energy storage ability of the dielectric resonator there will be. This is because ε_r acts as a resistance to an electric field and does not allow it to easily leak out from the dielectric resonator. As discussed in the previous chapter, high permittivity dielectric-based resonators () were initially used as microwave devices such as microwave filters and oscillators because of their inherent advantageous properties like high-quality factor (*Q*-factor), which is typically in the range of 20–10,000, making it an excellent energy storage component. The degree to which the system exhibits this resonance behavior and how much energy is stored and lost is characterized by a quantity called quality factor. Thus, two parameters that are resonance frequency (f_o) and quality factor (*Q*-factor) describe any resonant device. *Q*-factor is defined as the ratio of maximum energy stored during one cycle to the average power lost or average energy dissipated per period. It is a figure of merit for evaluating the quality of a resonator.

$$Q = 2\pi \frac{\text{Maximum energy stored during one cycle}}{\text{average energy dissipated during one cycle}}$$

And mathematically it is given by

$$Q = \frac{\omega_o W}{P} \tag{2.1}$$

where, ω_o = resonant radian frequency, W = stored energy and P = power loss (dissipated) from the resonator per cycle [3]. The numerator of (2.1) captures the energy storage property, whereas the denominator deals with the energy dissipated or power leaked information from the resonator. Thus, quality factor becomes a fundamental factor that is an intermediate parameter, which links the antenna-like properties and resonator-like properties of a dielectric resonator.

In the case of dielectric resonators, various *Q*-factors are associated with them. Figure 2.2 shows the classification of *Q*-factors in a dielectric resonator, which are discussed as follows: while using dielectric resonators in a microwave system as a load, the overall *Q*-factor is called as loaded *Q*-factor denoted by Q_l. Loaded *Q*-factor can be calculated by incorporating two more *Q*-factors, which are

- Unloaded *Q*-factors (Q_o): accountable for internal losses;
- External *Q*-factor (Q_e): responsible for external losses.

These three *Q*-factors are related by (2.2)

$$Q_l = Q_o + Q_e \tag{2.2}$$

Unloaded *Q*-factor accounts for internal losses when it is not used as a load and is given by

$$Q_o = \frac{\omega_o W_o}{P_o} \tag{2.3}$$

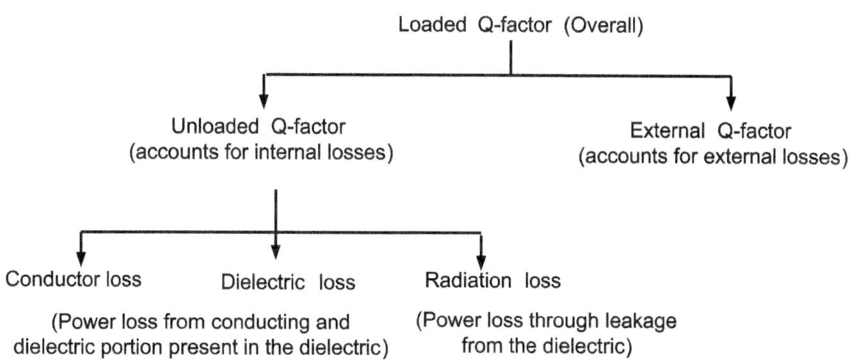

Figure 2.2 Different classification of quality factor in a dielectric resonator.

where P_o accounts for power losses from the resonator due to internal factors. Now, the power loss may be from the conductor (P_c) or dielectric fills (P_d) and some energy may be lost in the form of leakage power (P_r) from the boundary. Based on this, different types of Q-factors that contribute to Q_o are possible, which is given by (2.4–2.6)

$$Q_c = \frac{\omega_o W}{P_c} \tag{2.4}$$

$$Q_d = \frac{\omega_o W}{P_d} \tag{2.5}$$

$$Q_r = \frac{\omega_o W}{P_r} \tag{2.6}$$

where Q_c, Q_d, and Q_r represent conductor quality factor, dielectric quality factor, and radiation quality factor, respectively. Figure 2.2 shows the classification of Q-factors in a dielectric resonator. Total power lost in the resonator is due to three internal factors, which are

$$P_O = P_c + P_d + P_r \tag{2.7}$$

The unloaded Q-factor (Q_o) are related by its internal factors by (2.8) as given below

$$\frac{1}{Q_o} = \frac{1}{Q_c} + \frac{1}{Q_d} + \frac{1}{Q_r} \tag{2.8}$$

Now, it is evident that the lowest Q-factor value dominates other internal losses and accounts for the net unloaded Q-factor. Next, the external Q-factor is given as

$$Q_e = \frac{\omega_o W}{P_e} \tag{2.9}$$

Where, P_e represents power losses due to external load. Now, the overall Q_l is given by (2.10) found by rearranging the (2.2) (taking reciprocal) as

$$\frac{1}{Q_l} = \frac{1}{Q_o} + \frac{1}{Q_e} \tag{2.10}$$

To conclude, Q-factor is one of the deciding factors for the operation of dielectric resonator, because low Q-factors are required for radiating applications whereas high Q-factors are responsible for storage applications. Since quality factor is an important parameter that reflects its behavior, there are many Q-factors that are associated with a dielectric resonator. It is important to show which parameter dominates and contributes to radiation.

2.2 Dielectric Resonators as Antenna Element

The radiation mechanism of dielectric resonators can be simply understood by comparing it with metallic cavity resonators. In the case of metallic cavity resonators, the electromagnetic energy is stored by multiple reflections of the waves at the conducting boundary. When the incident and reflected waves are superimposed in phase, standing waves are formed. These standing waves inside the metallic cavity cannot be present or exist beyond the geometrical boundaries. Therefore, the power loss from leakage is zero, and power loss is insignificant, hence why the Q-factor is very high. However, in the case of DRs, electromagnetic energy is stored by multiple reflections of the waves at the dielectric boundary due to abrupt change in the permittivity at the dielectric-air interface. The boundary of the DR is not metallic, and instead, it is dielectric in nature. Thus, there is a possibility that fields could leak out, and there is a small power leakage from the resonator [5]. When DR is used as an energy storage device, its quality factor is kept very high by using a high value of permittivity with low dielectric losses. However, a small power leakage still exists. This radiation loss is further kept minimum by enclosing the DR in a metallic shielding as shown in Figure 2.3.

This radiation loss is a disadvantage for DRs acting as a resonator element, but this disadvantage is used for using the same DRs as an antenna element. It has been shown in the literature that by removing the shielding, lowering the permittivity of dielectric material, and proper selection of dielectric material, this radiation loss can be enhanced in an inefficient way and the Q-factor of lowest-order modes can be reduced significantly, as depicted in Figure 2.3. The proper selection of the feeding network and dimension of the resonator also plays an important role in determining the resonance frequency of the modes radiated. The impedance bandwidth of a particular mode is related to the unloaded Q-factor by (2.11)

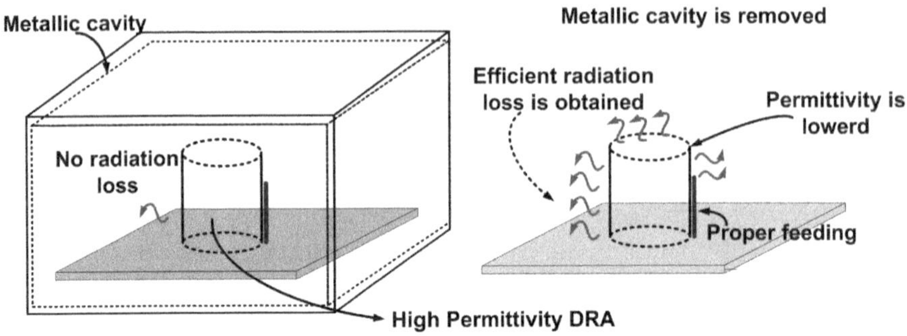

Figure 2.3 Picture showing the radiation loss from a dielectric resonator when metallic shielding is removed and permittivity of DR is lowered.

$$Bandwidth = \frac{s-1}{Q_o\sqrt{s}} \tag{2.11}$$

where s = maximum is the acceptable voltage standing wave ratio. For a dielectric resonator with low permittivity and absence of metallic shielding, it has an insignificant conductor and dielectric loss compared to its leaked power, and in this case, unloaded Q-factor is approximated by radiation Q-factor. In metallic antennas, the primary cause for the radiation is the acceleration and deceleration of charges that are free to move within the metal. However, in the case of dielectrics, there are no free charges present in the material. Earlier, Maxwell equations demonstrated that a source for radiation can be either a conductive current, displacement, or polarization current induced in the dielectrics. Thus, the acceleration and deceleration of polarization current induced in the DR due to external fields are the cause of radiation in a dielectric resonator. The bound charges in the dielectric medium polarize due to the external electric field, as shown in Figure 2.4.

Mathematically, it is challenging to calculate the responsible polarization currents for radiation, but it can be obtained from the known field distribution of any particular mode in the DRA by using surface equivalence theorem [7]. Once the polarization current equation is obtained for a particular mode, its far-field behavior can be predicted. Another very interesting explanation of radiation from dielectric resonators has been recently proposed by Sinha et al. [4]. In their research, it has been argued that electromagnetic radiation occurs under explicit symmetry breaking. In the case of a DRA, it behaves like an extended capacitor (dielectric resonator) and inductor system (from the ground plane and metallic feeding) in asymmetric excitation where one end is exposed to time-varying electric excitation and the other is floating in free space. This causes a swing of charges between its capacitive and inductive elements un-

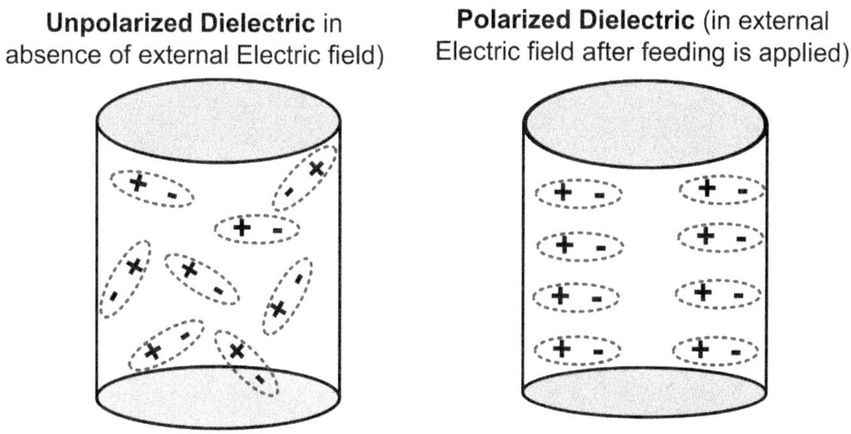

Figure 2.4 Polarization of dielectric in the presence of an external electric field.

der symmetry breaking and generation of electromagnetic waves, which are detached from the dielectric resonator.

The next important process is the study of wave detachment from the antennas. This can be explained very easily by comparing the wave detachment process with a dipole antenna. Figure 2.5 shows the top view of electric field distribution in a hemispherical DRA operating in its TM_{101} mode using an electromagnetic simulation software [6]. The initial counterclockwise fields may be considered as generated by polarization current induced in the dielectric, as shown in Figure 2.5 (a–d). In the next cycle, the fields turn in a clockwise direction, as shown in Figure 2.5 (e–i), where it can be expected that the direction of polarization current has changed from the previous direction. This is obvious because the charges are oscillating at a frequency equal to the applied frequency, and it is continuously changing its polarity. Thus, the fields associated with this change in polarity also changes. Now, when the new clockwise fields come into effect, the old counterclockwise field (due to the previous polarity of charges) has to detach them from the resonator forming closed loops and travels at the speed of light. In the next cycle again, new counterclockwise fields are formed as shown in Figure 2.5 (j–l), and the previous clockwise fields detach from the resonator. This process continues and electric field patterns are formed.

2.3 Mode Classification in Dielectric Resonators

The concept of electromagnetic modes in dielectric resonator was first developed through asymptotic theory by Van Bladel [8, 9]. According to the theory, the field analysis of isolated dielectric resonators of very high permittivity requires the electric and magnetic wall boundary conditions at the air-dielectric

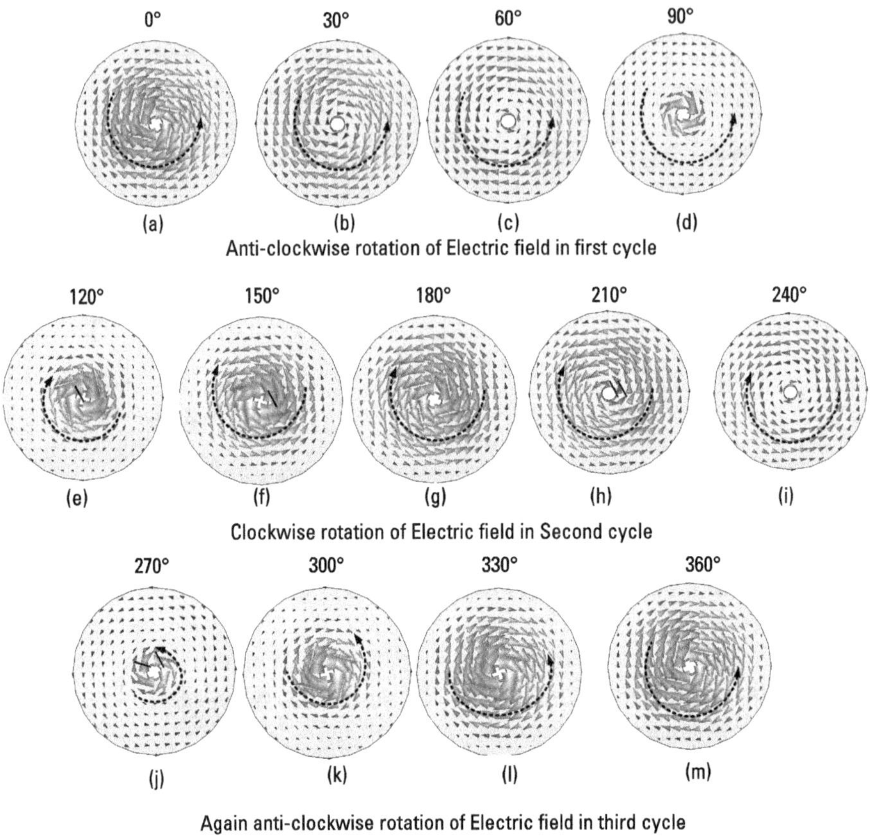

Figure 2.5 The variation of electric field distribution in hemispherical DRA working in TM101 mode (top view).

interface. The modes of an arbitrarily shaped dielectric resonator of very high permittivity are in general a nonconfined type and main radiation is contributed by the magnetic dipole term, hence it radiates like a magnetic dipole. There is an additional confined type of mode that is also possible that are supported by axis-symmetric resonator shapes (cylindrical and hemispherical). These modes radiate like an electric dipole. For both confined and nonconfined modes in DRs, the following conditions are satisfied at all the surfaces of the resonator:

$$\hat{n}.\vec{E} = 0 \tag{2.12}$$

$$\hat{n} \times \vec{H} = 0 \tag{2.13}$$

where $\vec{E}$, $\vec{H}$ and $\hat{n}$ denotes the electric-field intensity, magnetic field intensity and unit vector normal to the boundary of the resonator, respectively. The modes of a DR that satisfy both (2.12) and (2.13) are known as confined modes, whereas those which satisfy (2.12) are only known as nonconfined modes. The above categorization of modes in DRs was initially developed for high-permittivity DRs that were used for energy storage device in filters, oscillators, and so on. Since DRs came into consideration as a radiating element in 1980s, it has become popular in antennas (knows as DRAs), which usually employ low permittivity dielectric materials. On the basis of antenna-like behavior, the modes of a DR can be classified into radiating and nonradiating modes. Figure 2.6 depicts the possible radiating and nonradiating modes of the hemispherical, cylindrical, and rectangular DRAs. Radiating modes have a low-quality factor whereas nonradiating modes have a high-quality factor at an optimum value of permittivity.

2.4 Shapes of Dielectric Resonators

From a literature survey, the shapes of a dielectric resonator can be divided into two parts: conventional shapes, such as hemispherical, cylindrical, and rectangular [11–13], and modified or deformed shapes, such as ring cylindrical, triangular, half-split cylindrical, trapezoidal, and pentagonal, as shown in Figure 2.7 [14, 15].

Theoretical analyses for mode configurations and radiation characteristics of hemispherical, cylindrical, and rectangular dielectric resonators have been analyzed. That is why these shapes of dielectric resonators are known as conventional shapes. The advantages of traditional forms are that it allows design

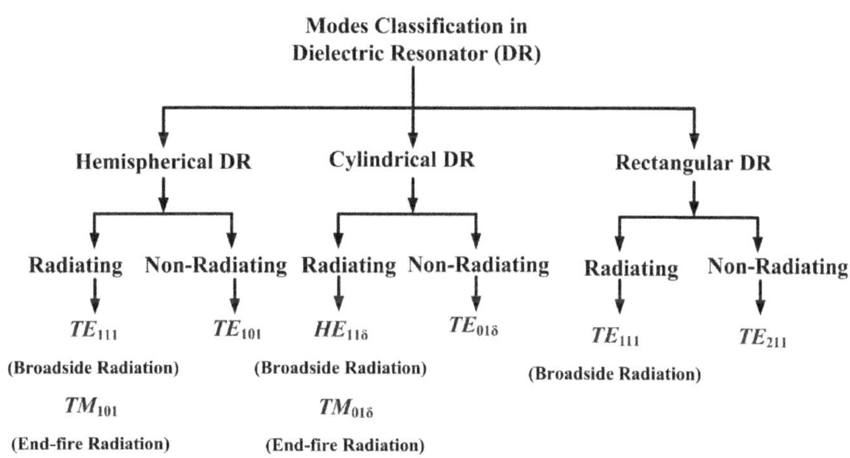

Figure 2.6 Classification of some useful radiating and nonradiating modes in conventional-shaped DRAs.

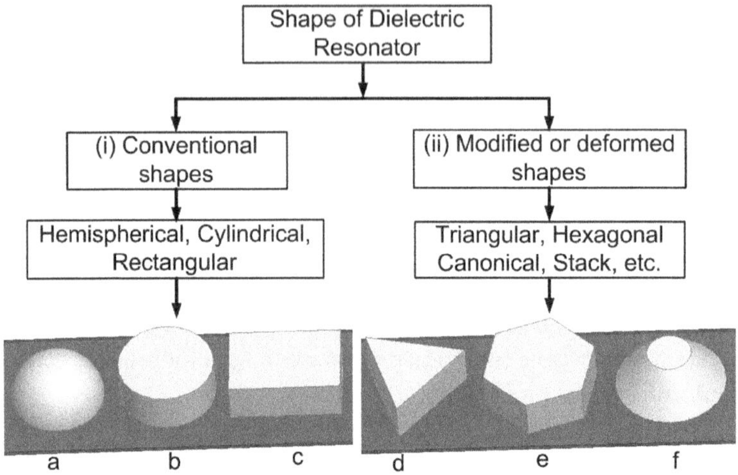

Figure 2.7 Conventional and modified shapes of the dielectric resonator: (a) hemispherical, (b) cylindrical, (c) rectangular, (d) triangular, (e) hexagonal, and (f) conical.

flexibility, modes, and offers different radiation characteristics along with field distributions.

The modified or deformed shapes of the dielectric resonator are generally utilized for reducing the effective dielectric constant so that it is useful for enhancing impedance bandwidth. Theoretical analyses of these deformed shapes, such as triangular, half-split cylindrical, trapezoidal, and pentagonal, are not possible. However, by considering some assumptions of the formulae given in the conventional shapes, the modes and radiation characteristics of these dielectric resonators could be calculated. The analytical method is very difficult to use for finding the characteristics and modes of conventional shapes of dielectric resonator antennas. Then, various techniques have been investigated based on the EM fields within the dielectric resonators. The EM fields help with finding the resonant frequencies and radiation characteristics of different modes of DRAs. Henceforth, the literature shows primarily two approximate techniques: perfect magnetic conductor or magnetic wall model, and dielectric waveguide model and effective dielectric constant model.

In the perfect magnetic conductor (PMC) model or magnetic wall model (MWM), the outer surface of the dielectric material is considered a perfect magnetic wall, hence normal electric and tangential magnetic fields vanish. However, antenna designers did not consider MWM because significant errors have been found while calculating the resonant frequencies of the dielectric resonator antennas. Hence, to remove this drawback, the dielectric waveguide model (DWM) was enforced. The DWM shown maximum error ~ 5% to 6% because the working principle is different from the MWM. In the DWM method, the outer surface of the dielectric is considered as a radiation wall instead of the magnetic wall compared with the MWM method [11]. Furthermore, to reduce the error of DWM, an effective dielectric constant (EDC) concept has been utilized. These methods were used to analyze and characterize the conventional shape of dielectric resonators [15]. The next sections provide a brief discussion of the traditional forms (i.e., hemispherical, cylindrical, and rectangular dielectric resonator antennas and their characteristics).

2.4.1 Hemispherical DRA

The geometry of a hemispherical dielectric resonator antenna has been extensively studied by most researchers, as shown in Figure 2.8. The hemispherical DRA primarily depends on the dielectric constant (ε_r) and radius (a), which is also called the degree of freedom provided by this geometry. For the radiation purpose, it is mounted on a ground plane and assumed to be a perfect electric conductor of an infinite extent. Hence, image theory can be applied for finding the characterization of modes in a hemispherical DRA. The hemispherical shape should be considered the dielectric sphere and the modes are decomposed into two fragments: transverse electric (TE), and transverse magnetic (TM). In TE mode, the radial component of the electric field is zero ($E_r = 0$) while the radial component of the magnetic field is also zero ($H_r = 0$) in TE mode.

The fundamental modes of the hemispherical DRA are TE_{111} and TM_{101} modes and are widely used in various applications. The radiation pattern of TE_{111} mode is similar to the short horizontal magnetic dipole, whereas the radiation pattern of TM_{101} mode is similar to the short electric monopole. The three subscripts of TE_{111} or TM_{101} modes represent field variations in r (radial), φ (azimuth), and θ (elevation) direction in spherical coordinates systems.

The resonant frequency and radiation quality factor of these fundamental modes (i.e., TE_{111} and TM_{101} modes can be determined by using transcendental equations) where J and H represent the Bessel function and Hankel function.

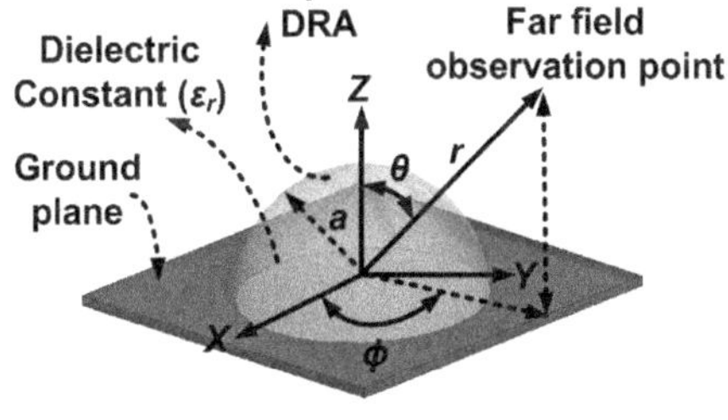

Figure 2.8 Configuration of hemispherical dielectric resonator antenna.

Therefore, the resonant frequency and radiation quality factor of TE_{111} and TM_{101} modes can be calculated by (2.14) and (2.15), respectively.

$$\frac{J_{1/2}\left(\sqrt{\varepsilon_r}k_0 a\right)}{J_{3/2}\left(\sqrt{\varepsilon_r}k_0 a\right)}=\frac{H_{1/2}^{(2)}\left(k_0 a\right)}{\sqrt{\varepsilon_r}H_{1/2}^{(2)}\left(k_0 a\right)} \tag{2.14}$$

$$\frac{1}{\sqrt{\varepsilon_r}k_0 a}-\frac{J_{1/2}\left(\sqrt{\varepsilon_r}k_0 a\right)}{J_{3/2}\left(\sqrt{\varepsilon_r}k_0 a\right)}=\frac{\sqrt{\varepsilon_r}}{k_0 a}-\sqrt{\varepsilon_r}\frac{H_{1/2}^{(2)}\left(k_0 a\right)}{H_{3/2}^{(2)}\left(k_0 a\right)} \tag{2.15}$$

where $J(x)$, $H^{(2)}(x)$, and k_0 are the first-order Bessel function, second-order Hankel function, and wave number in free-space. Equations (2.14) and (2.15) are solved for a given value of $\boldsymbol{a}$ and then for finding the value of k_0. The resonant frequency can be found from (2.16).

$$f_{GHz}=\frac{4.7713Re\left(k_0 a\right)}{a_{cm}} \tag{2.16}$$

where the radius (a) is in cm while the resonant frequency is in GHz, the radiation quality factor can be calculated using (1.17).

$$Q=\frac{Re\left(k_0 a\right)}{2Im\left(k_0 a\right)} \tag{2.17}$$

If the dielectric material is lossless, the lost power of the resonator is completely converted in the form of radiated power and then the Q-factor is equal to the radiating Q-factor (i.e., $Q = Q_{rad}$). To determine the fraction impedance bandwidth of hemispherical DRA, the radiation quality factor is utilized by using (2.18).

$$Bandwidth=\frac{\Delta f}{f_0}=\frac{VSWR-1}{\sqrt{s}Q} \tag{2.18}$$

where Δf and f_0 are the absolute bandwidth and resonant frequency, while VSWR stands for the voltage standing wave ratio. The distribution of electric and magnetic fields for the fundamental TE_{111} and TM_{101} modes of hemispherical DRA are shown in Figures 2.9 and 2.10, respectively.

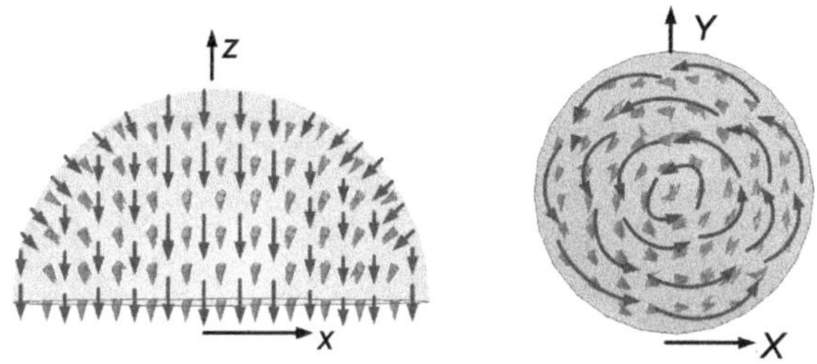

Figure 2.9 TM_{101} mode: eigenmode analysis done by high-frequency structure simulator (HFSS) to find electric and magnetic fields inside the hemispherical DRA: (a) E-field, side view in xz-plane, and (b) H-field, top view in xy-plane.

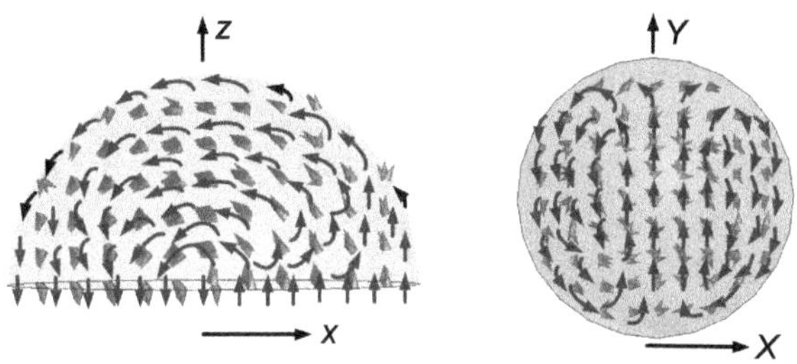

Figure 2.10 TE_{111} mode: eigenmode analysis done by HFSS to find electric and magnetic fields inside the hemispherical DRA: (a) E–field, side view in xz-plane, and (b) H–field, top view in xy-plane.

2.4.2 Cylindrical DRA

After the invention of the cylindrical shape of the dielectric resonator, a cylindrical DRA is the most popular shape among antenna engineers because it provides one more degree of freedom (i.e., height) compared to a hemispherical dielectric resonator. The geometry of the cylindrical dielectric resonator antenna depends on its height (h), radius (a), and dielectric constant (ε_r), respectively, as shown in Figure 2.11. Hence, the height plays the main role of a cylindrical dielectric resonator for choosing the different aspect ratios (a/h). The aspect ratios are also directly related to determining the radiation quality factor and values of k_0a for a given dielectric material [16].

The cylindrical DRA supports three different types of fundamental modes and these are named as TE ($E_z = 0$), TM ($H_z = 0$), and hybrid mode *HE*/*EH* (both E_z and H_z are presents). Based on the leading of electric or magnetic fields along the axial direction, hybrid mode can again be divided into two groups (i.e., HE (E_z is dominant) and EH (H_z is dominant) modes). The TE and TM modes are axial asymmetry; therefore, there is no azimuthal dependency, while hybrid mode has dependency on the azimuthal.

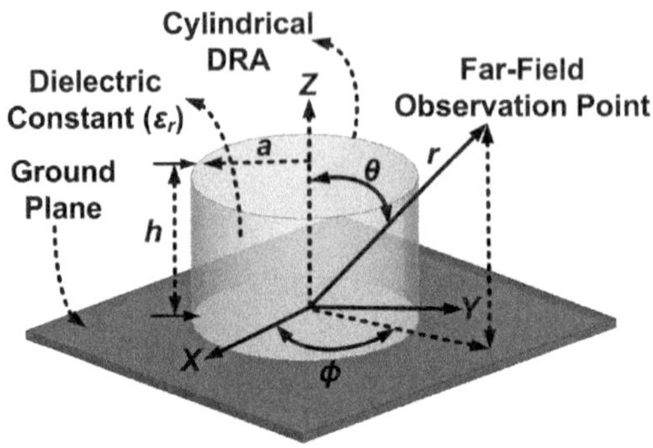

Figure 2.11 Configuration of cylindrical dielectric resonator antenna.

The three fundamental modes of cylindrical DRA are described by $TE_{01\delta}$, $TM_{01\delta,}$, and $HE_{11\delta}$, modes, respectively, where the subscripts represent the directions of azimuthal (φ), radial (r), and axial (z), which are referred to as field variations. If you know the variation of fields along with the three directions of a cylindrical DRA (i.e., azimuthal, radial, and z-direction) the three mode indices m, n, and p are added as a subscript of each mode. Therefore, the TE, TM, EH, and HE are denoted by $TE_{0np\ +\delta}$, $TM_{0np\ +\delta}$, $EH_{mnp\ +\delta,}$ and $HE_{mnp\ +\delta}$, respectively, where the value of δ ranges between 0 and 1.

The empirical formula has been given by Mongia et al. [17] for calculating the resonant frequencies and Q-factor of all the fundamental modes of a cylindrical DRA. The results obtained by using this empirical formula of all modes gave a slight variation in experimental results and it is found that the errors in resonant frequency and Q-factor are nearly 2.5% and 10%, respectively. The theoretical equation of resonant frequency and Q-factor and electric field distribution of all fundamental modes (i.e., TE, TM, EH, and HE) of a cylindrical DRA are discussed below.

- *$TE_{01\delta}$ mode:* The fundamental and lowest-order mode of a cylindrical DRA is $TE_{01\delta}$ mode. It has a far-field radiation pattern similar to the short vertical magnetic dipole, which is confirmed by the E- and H-field distributions inside the cylindrical DRA. It is observed in Figure 2.12 that the field lines are similar to the magnetic dipole, which is placed in a vertical position. The radiation pattern of the $TE_{01\delta}$ mode is omnidirectional with a null along the axis and radiates a maximum along an angle away from the broadside in the elevation plane, which is more suitable for broadside radiation. The pattern is very useful for many practical communication applications, mainly in satellites from locations of different latitudes on the earth. The theoretical formula for calculating the resonant frequency and Q-factor of $TE_{01\delta}$ mode is given by [17].

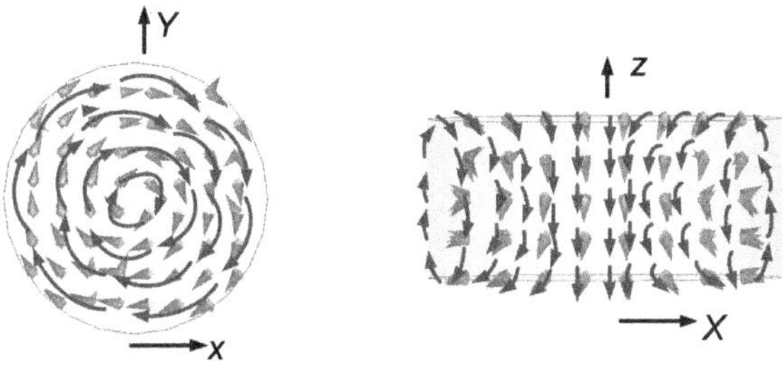

Figure 2.12 Eigenmode analysis done by HFSS to find E- and H-fields in cylindrical DRA for TE01δ mode: (a) TE01δ, E–field, and (b) TE01δ, H–field.

$$k_0 a = \frac{2.327}{\sqrt{\varepsilon_r + 1}} \left\{ 1 + 0.2123 \frac{a}{h} - 0.00898 \left(\frac{a}{h} \right)^2 \right\} \tag{2.19}$$

$$Q = 0.078192 \varepsilon^{1.27} \left\{ 1 + 17.31 \left(\frac{h}{a} \right) - 21.57 \left(\frac{h}{a} \right)^2 + 10.86 \left(\frac{h}{a} \right)^3 - 1.98 \left(\frac{h}{a} \right)^4 \right\} \tag{2.20}$$

- *$TM_{01\delta}$ mode:* Another important and fundamental mode of cylindrical DRA is $TM_{01\delta}$ mode, which has a far-field radiation pattern similar to short electric monopole. Figure 2.13 depicts the *E*- and *H*-field distributions inside the cylindrical DRA and these field lines are similar to vertical orientation electric dipole. It can be observed in Figure 2.13(a) that the mode radiates strongly toward the low elevation (i.e., null in broadside direction and suitable for end-fire radiation). Because of this mode pattern, it is widely used in vehicular applications. The resonant frequency and *Q*-factor of $TM_{01\delta}$ mode are calculated by the given empirical formula [17]:

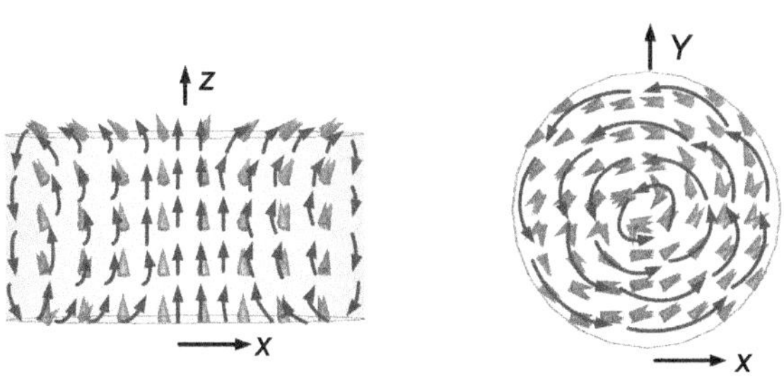

Figure 2.13 Eigenmode analysis done by HFSS to find E- and H-fields in cylindrical DRA for $TM_{01\delta}$ mode: (a) $TM_{01\delta}$, E–field, and (b) $TM_{01\delta}$, H–field.

$$k_0 a = \frac{\sqrt{3.83^2 + \left(\frac{\pi a}{2h}\right)^2}}{\sqrt{\varepsilon_r + 1}} \tag{2.21}$$

$$Q = 0.008721\varepsilon_r^{0.888413} e^{0.0397475\varepsilon_r} \left\{1 - \left(0.3 - 0.2\frac{h}{a}\right)\left(\frac{38-\varepsilon_r}{28}\right)\right\} \times \left\{9.498186\frac{a}{h} + 2058.33\left(\frac{a}{h}\right)^{4.322261} e^{-3.50099\left(\frac{a}{h}\right)}\right\} \tag{2.22}$$

- *$HEM_{11\delta}$ mode:* The most useful and fundamental mode of a cylindrical DRA used in the various applications is $HEM_{11\delta}$ mode. It has a far-field radiation pattern similar to a short horizontal magnetic dipole, and therefore radiates maximum in the broadside directions. Figure 2.14 illustrates the E-field distribution inside the cylindrical DRA for $HEM_{11\delta}$ mode. The resonant frequency and Q-factor of $HEM_{11\delta}$ mode are calculated by the given empirical formula [17]:

$$k_0 a = \frac{6.324}{\sqrt{\varepsilon_r + 2}} \left\{0.27 + 0.36\frac{a}{2h} + 0.02\left(\frac{a}{2h}\right)^2\right\} \tag{2.23}$$

$$Q = 0.01007\varepsilon_r^{1.3} \frac{a}{h} \left\{1 + 100 e^{-2.05\left(\frac{a}{2h} - \frac{1}{80}\left(\frac{a}{h}\right)^2\right)}\right\} \tag{2.24}$$

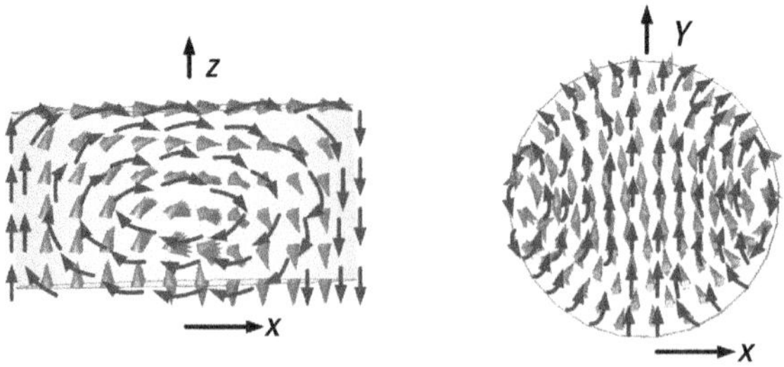

Figure 2.14 Eigenmode analysis have done by HFSS to find E- and H-fields in cylindrical DRA for $HEM_{11\delta}$ mode: (a) $HEM_{11\delta}$, E-field, and (b) $HEM_{11\delta}$, H-field.

2.4.3 Rectangular DRA

The structure of a rectangular dielectric resonator antenna depends on the width (w), length (l), height (h), and dielectric constant (ε_r), as shown in Figure 2.15. It was invented and briefly explained by S. A. Long and his colleagues in 1983 [11–13]. Thereafter, Mongia and Ittipiboon briefly investigated the modes and radiation characteristics of the rectangular DRA in 1997 [16, 18, 19].

The rectangular DRA offers one extra degree of freedom compared to the cylindrical DRA. Hence, the rectangular DRA has two degrees of freedom, which makes it the most versatile of the basic shapes. It offers a high extent of flexibility to achieve the desired bandwidth for a given resonant frequency and dielectric constant. This is possible because of the freedom to choose or control two aspects ratios w/h and w/l independently. The radiation Q-factor is also dependent on the aspect ratio, so it is another design flexibility provided by the rectangular DRA. No rigorous classification of modes of a rectangular DR can exist compared to the hemispherical or cylindrical DR's. In a hemispherical DR, the modes are TE or TM, while the modes of a cylindrical DR are of TE, TM, or a hybrid type. The concepts of modes in a dielectric resonator have been explained in Section 2.3. Okaya and Barash were the first to provide the approximate analysis of modes of the rectangular DRA. According to them, the modes of the rectangular DRA can be categorized into two types: (1) *TE(H)*, and (2) *TM(H)*, but the theory given by Okaya and Barash for all modes in a rectangular DRA is not observed experimentally. This occurs when a DWM is utilized to find the *TM(E)* modes in rectangular DRA and observed that it is not satisfied with (2.12); therefore, the appearance of *TM(E)* modes is doubtful. On the other hand, the lower-order *TE(H)* mode (i.e., $TE_{11\delta}$) is already proven and used in various circuits applications. It has also satisfied the conditions of nonconfined modes. The *TE* modes of a rectangular DRA can be any of three dimensions because it has three independent dimensions. As per the rectangular DRA coordinate system, the modes having lowest-order indexes are

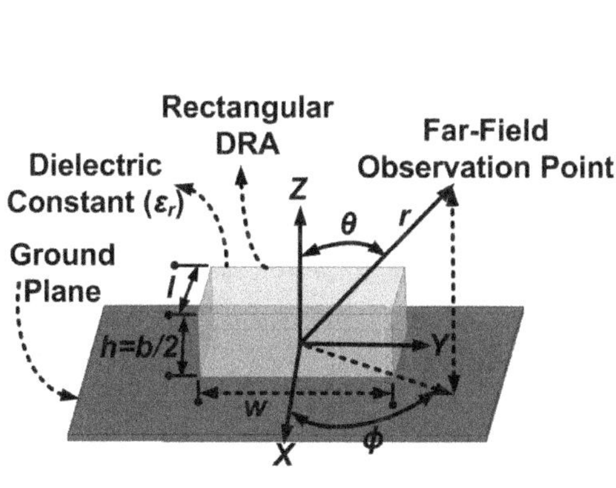

Figure 2.15 Geometry of rectangular dielectric resonator antenna.

TE^x_{111}, TE^y_{111}, and TE^z_{111}, which radiate like short magnetic dipoles in the x-, y-, and z-directions, respectively. The resonant frequencies of all these modes are dependent on the dimensions of the DRA. For example, if $w > l > b$, then the modes in the order of increasing resonant frequency are TE^z_{111}, TE^y_{111}, and TE^x_{111} , hence by selecting the proper dimensions of a DRA, the antenna designers avoided the generation of unwanted modes.

Here, the lowest-order mode (i.e., TE^x_{111}) is considered for derivation by using the dDWM to find the fields within the DRA, by utilizing these equations to find resonant frequency (f_0). The field components are given as [5, 16]:

$$H_x = A\frac{\left(k_y^2 + k_z^2\right)}{j\omega\mu_0} cos\left(k_x x\right) cos\left(k_y y\right) cos\left(k_z z\right) \tag{2.25}$$

$$H_y = A\frac{\left(k_y k_x\right)}{j\omega\mu_0} sin\left(k_x x\right) sin\left(k_y y\right) cos\left(k_z z\right) \tag{2.26}$$

$$H_y = A\frac{\left(k_z k_x\right)}{j\omega\mu_0} sin\left(k_x x\right) cos\left(k_y y\right) sin\left(k_z z\right) \tag{2.27}$$

$$E_x = 0 \tag{2.28}$$

$$E_y = A k_z \, cos\left(k_x x\right) cos\left(k_y y\right) sin\left(k_z z\right) \tag{2.29}$$

$$E_z = -A k_y \, cos\left(k_x x\right) sin\left(k_y y\right) cos\left(k_z z\right) \tag{2.30}$$

where A, k_x, $k_{y,}$, and k_z represent the random, constant wavenumbers along the x, y, and z directions, respectively, inside the dielectric resonator.

By solving above transcendental equation by utilizing the DWM, the following equations are obtained for the k_x wavenumber:

$$k_x tan\left(k_x l / 2\right) = \sqrt{\left(\varepsilon_r - 1\right) k_0^2 - k_x^2} \tag{2.31}$$

Applying magnetic wall conditions (i.e., $\vec{E}.\hat{n} = 0$) at the surfaces of the resonator ($|y| = w/2, |z| = b/2$) in the above transcendental equations then (2.32) is obtained for calculating the wavenumbers of k_y and k_z:

$$k_y = \frac{m\pi}{w} \text{ and } k_z = \frac{n\pi}{b} \tag{2.32}$$

where m and n are positive integers and consider 1 for the lowest-order mode. The wavenumbers can also satisfy the separation equations and is given by

$$k_x^2 + k_y^2 + k_z^2 = \varepsilon_r k_0^2 \tag{2.33}$$

where k_0 represents the free-space wavenumber ($k_0 = 2\pi/\lambda_0 = 2\pi f_0/c$) corresponds to the resonant frequency. The δ is the half-cycle variation of the field variation in the z-direction and is defined by

$$\delta = \frac{k_x}{\pi/d} \tag{2.34}$$

If the ratio of w / b as a function of l/b then the normalized frequency (F) is obtained by using (2.31) and is given by

$$F = \frac{2\pi w f_0 \sqrt{\varepsilon_r}}{c} \tag{2.35}$$

Or it can also rearrange and obtained by a more suitable (2.36) for finding the direct resonant frequency:

$$f = \frac{15F}{w \times \pi \times \sqrt{\varepsilon_r}} \tag{2.36}$$

Here, the unit of resonant frequency (f) and width (w) are GHz and cm.

The same procedure should be followed to find the resonant frequencies of the other two modes (i.e., TE_{111}^y and TE_{111}^z modes).

Figures 2.16 and 2.17 depict the eigenmode analysis of electric field distribution inside the rectangular DRA for some of the selected TE_{111} and TE_{121} modes, respectively.

2.5 Coupling Mechanism of Dielectric Resonator Antenna

To determine which modes are excited in the DRA, a selection of feed and its positions are very important. It also significantly affects the resonant frequency and Q-factor of the DRA. The dielectric resonator antennas have a significant

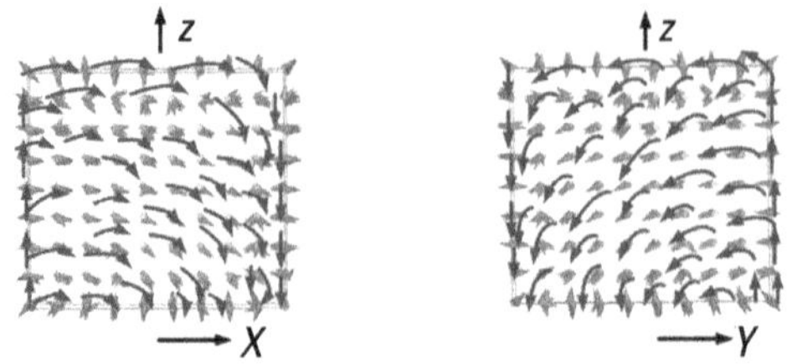

Figure 2.16 Eigenmode analysis to show E-fields inside the rectangular DRA for TE_{111} mode: (a) E-field, side view in xz-plane, and (b) E-field, side view in yz-plane.

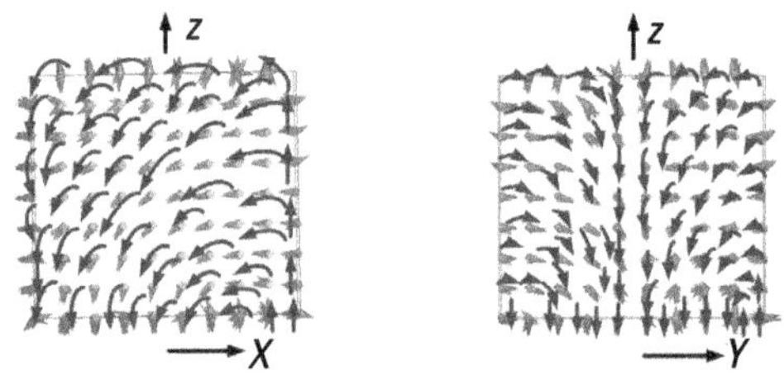

Figure 2.17 Eigenmode analysis to show E-fields inside the rectangular DRA for TE_{121} mode: (a) E-field, side view in xz-plane, and, (b) E-field, side view in yz-plane.

advantage over the conventional antenna, such as flexibility to choose different types of feeding or excitation mechanisms, as per its requirements to provide the maximum or minimum EM waves. These feedings include coaxial probe, microstrip feed, conformal strip, aperture coupling, coplanar waveguide fed, and dielectric image guide coupling. [3, 5].

Generally, two types of the source have been utilized to couple the DRA, either electric or magnetic source current, and these are given as

For an electric source (J_S),

$$\div = \int_V \vec{E}_{DRA}\cdot\vec{J}_S)\mathrm{dV} \tag{2.37}$$

For a magnetic source (M_S),

$$\chi = \int_V \vec{H}_{DRA}\cdot\vec{M}_S)dV \tag{2.38}$$

where χ, V, E_{DRA}, and H_{DRA} are the coupling factor, the volume occupied by the source, electric and magnetic fields. From (2.37) and (2.38), it is observed that to achieve strong coupling by using an electric (like a probe) or magnetic (like an aperture) current source, the source must be located in the area of strong electric or magnetic fields within the DRA.

2.5.1 Coaxial Probe Coupling

Initially, coaxial probe coupling is frequently used for exciting the dielectric resonator antennas because of its simplicity and provided 50Ω directly to the antenna without any matching network, as shown in Figure 2.18 [5, 11–13]. In this, a metallic probe is utilized for couples the electromagnetic energy to the DRA and flexibility to use inside or outside in the DRA to find the strong coupling, as shown in Figures 2.18 and 2.19. The radiation characteristics, modes, and antenna parameters also depend on the location and dimensions of the metallic probe, which are represented in Figure 2.18. For example, in cylindrical DRA, the center probe excites the $TM_{01\delta}$ mode (see Figure 2.20) while an adjacent probe excites the $HEM_{11\delta}$ mode (see Figure 2.21). Where as in a rectangular DRA, the location of the adjacent probe excites the $TE_{11\delta}$ mode, as shown in Figure 2.22.

2.5.2 Microstrip Coupling

The microstrip feed line is most suitable to avoid unwanted coupling and modes. It provides better matching, which is easily obtained by adjusting the position of the DRA with respect to the length and width of the microstrip feed line, as shown in Figures 2.23 and 2.24 [5]. The mode profiles of two different shapes of DRs have been seen by using the microstrip feed coupling. It behaves like a horizontal magnetic dipole, and hence excites the $TE_{01\delta}$ mode in a rectangular DRA, while $HEM_{11\delta}$ mode in a cylindrical DRA, as shown in Figures 2.25 and 2.26.

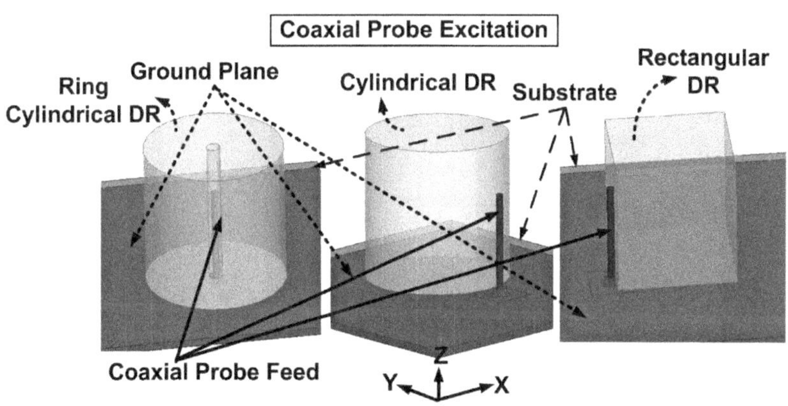

Figure 2.18 Excitation of a ring, cylindrical, and rectangular DRA by using coaxial probe coupling: (a) ring cylindrical DRA, (b) cylindrical DRA, and (c) rectangular DRA.

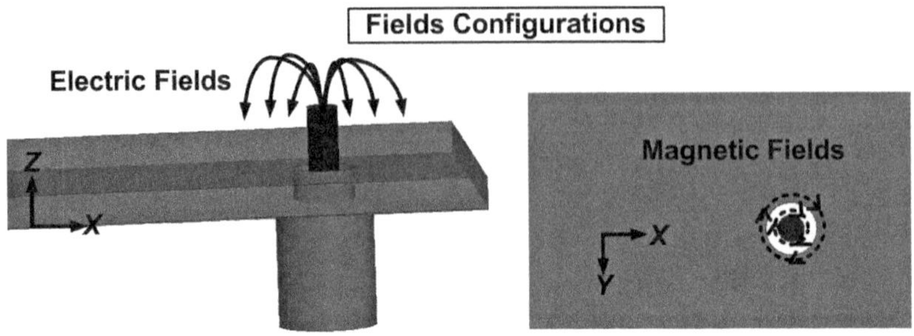

Figure 2.19 Electric and magnetic fields profile of coaxial probe coupling: (a) electric fields, and (b) magnetic fields.

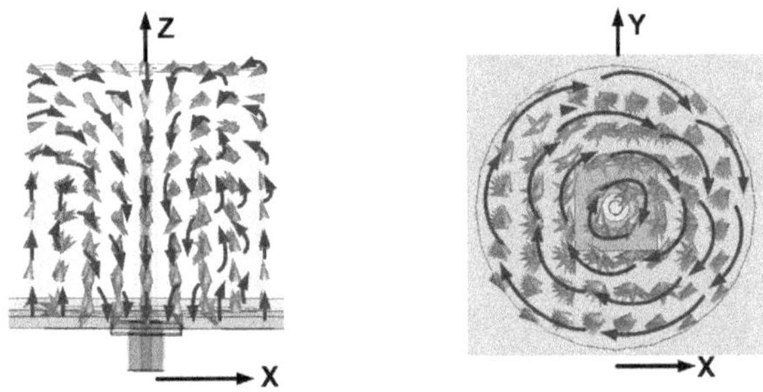

Figure 2.20 Distributions of electric and magnetic fields inside a cylindrical DRA done by HFSS to confirm the TM01δ mode: (a) E–field, side view in xz-plane, and (b) H–field, top view in xy-plane.

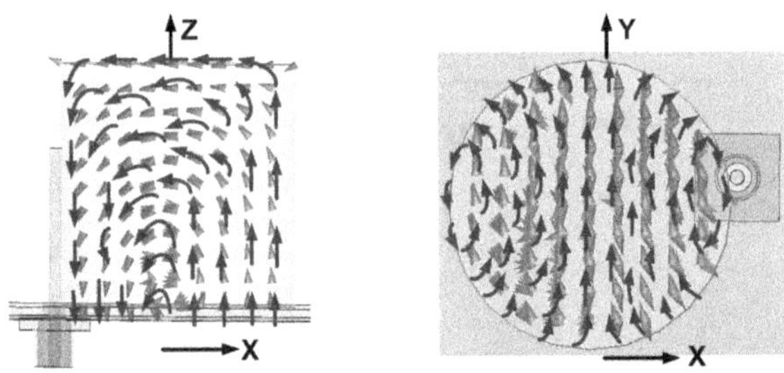

Figure 2.21 Distributions of electric and magnetic fields inside a cylindrical DRA done by HFSS to confirm the $HEM_{11\delta}$ mode: (a) E–field, side view in xz-plane, and (b) H–field, top view in xy-plane.

2.5.3 Conformal Coupling

The conformal strip is a combination of a coaxial probe and microstrip feed line. It has been used to excite the dielectric resonator antenna in place of the coaxial probe as well as microstrip feed line because drilling a hole in the di-

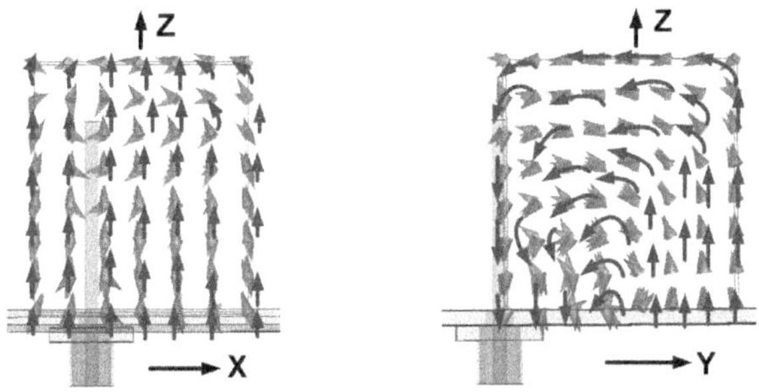

Figure 2.22 Distributions of electric fields inside a rectangular DRA done by HFSS to confirm the $TE_{01\delta}$ mode: (a) E–field, side view in xz-plane, and (b) E–field, side view in yz-plane.

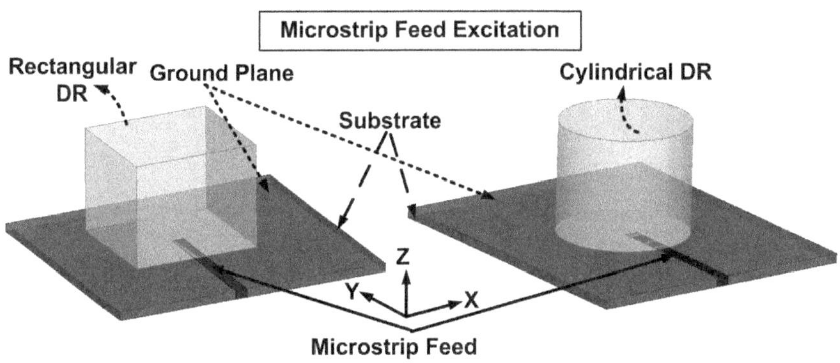

Figure 2.23 Excitation of rectangular and cylindrical DRA by using microstrip coupling: (a) rectangular DRA, and (b) cylindrical DRA.

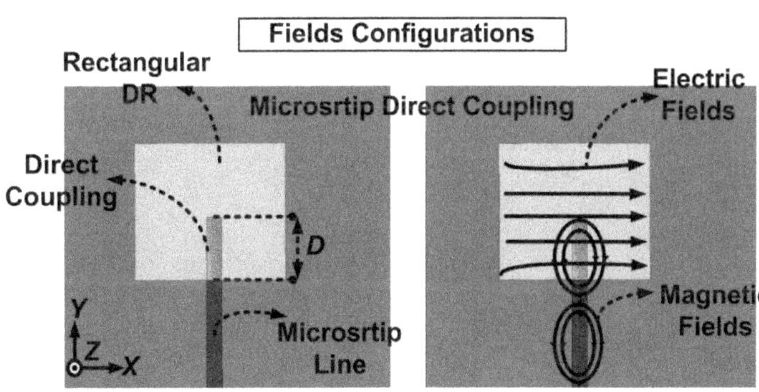

Figure 2.24 Magnetic field profile of direct microstrip line coupling to a rectangular DRA: (a) direct coupling, top view in xy-plane, and (b) magnetic fields in microstrip line and rectangular DRA.

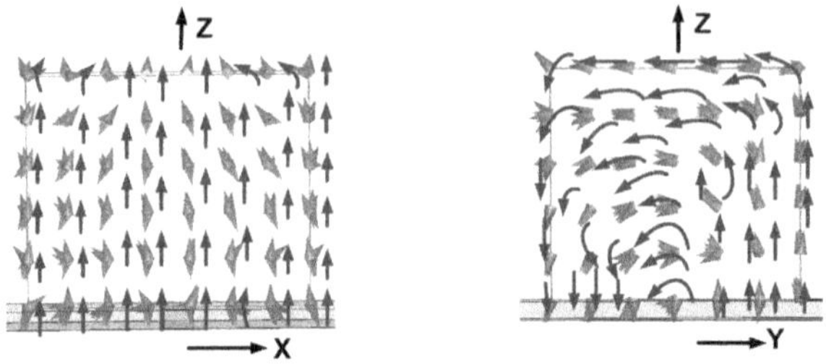

Figure 2.25 Distributions of electric fields inside a rectangular DRA done by HFSS to confirm the $TE_{01\delta}$ mode: (a) E-field, side view in xz-plane, and (b) E-field, side view in yz-plane.

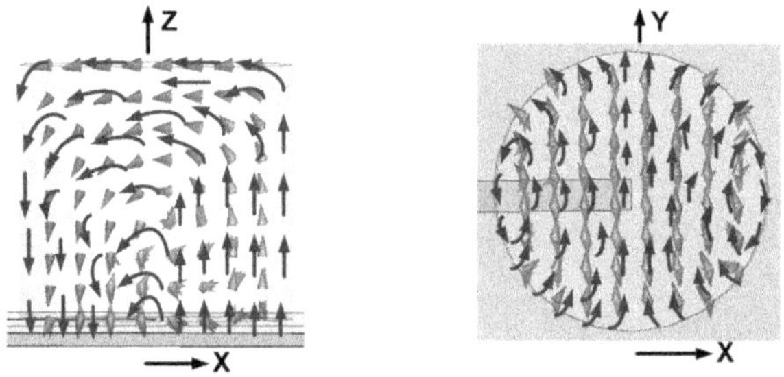

Figure 2.26 Distributions of electric and magnetic fields inside a cylindrical DRA to confirm the $HEM_{11\delta}$ mode: (a) E–field, side view in xz-plane, and (b) H–field, top view in xy-plane.

electric resonator is very hard and should be avoided, as shown in Figure 2.27 [20]. The electric and magnetic field lines for the conformal strip line are shown in Figure 2.28. For more clarity in coaxial probe coupling, it is very difficult to have a hole that perfectly fits the size of the coaxial probe, and hence an air gap exists between the probe and DRA, which causes the measured results to

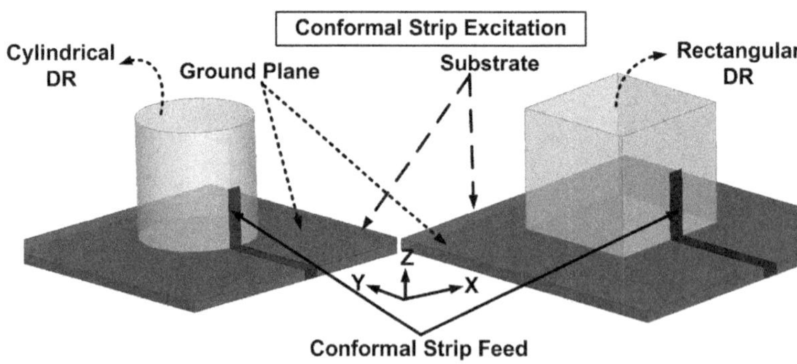

Figure 2.27 Excitation of a cylindrical and rectangular DRA by using conformal strip coupling: (a) cylindrical DRA, and (b) rectangular DRA.

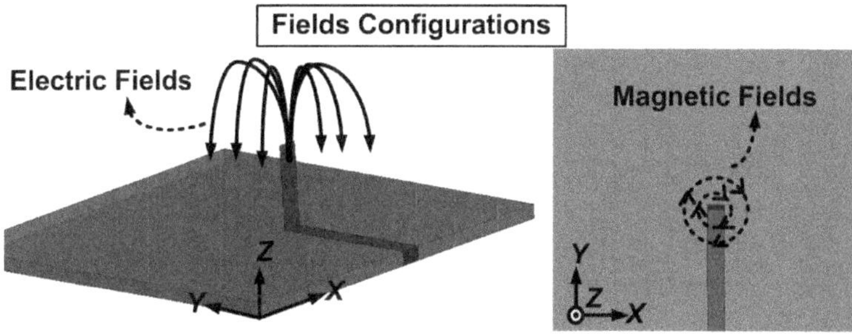

Figure 2.28 Electric and magnetic field profiles of conformal strip coupling: (a) electric fields, and (b) magnetic fields.

deviate from theoretical analysis. To avoid this, a conformal strip is very essential. The major advantage of the conformal strip is that it allows the complete electric current to flow on the surface of DRA, and hence it is more suitable and efficient in electromagnetic energy coupling rather than the coaxial probe. The distribution of electric and magnetic fields of two different shapes of DRs have been shown in the Figures 2.29 and 2.30 by using conformal feeding techniques. The mode profiles of the cylindrical DR and rectangular DR have been given and it was found that if a cylindrical DRA is used it excites the $HEM_{11\delta}$ mode, whereas a rectangular DRA gives the $TE_{01\delta}$.

2.5.4 Aperture Coupling

The very common and important coupling method to excite a dielectric resonator antenna is known as aperture coupling because it avoids spurious radiation from the feed line, as shown in Figure 2.31 [21].

Slots are cut into the ground plane and electromagnetic energy is provided below the ground plane where the feed network is presented after that a DR is placed. These arrangements are done to provide electromagnetic energy

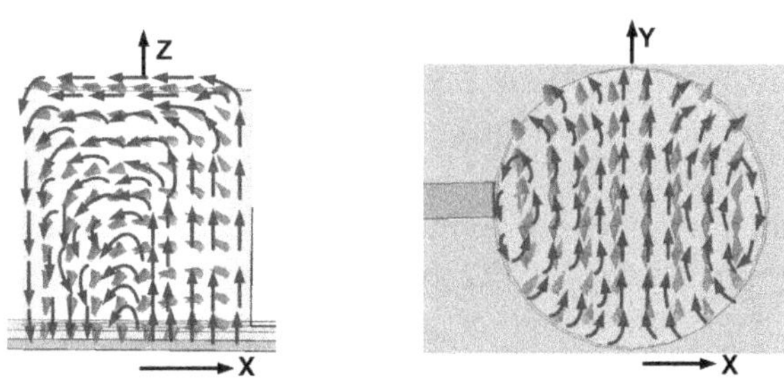

Figure 2.29 Distributions of electric and magnetic fields inside a cylindrical DRA to confirm the $HEM_{11\delta}$ mode: (a) E-field, side view in xz-plane, and (b) H-field, top view in xy-plane.

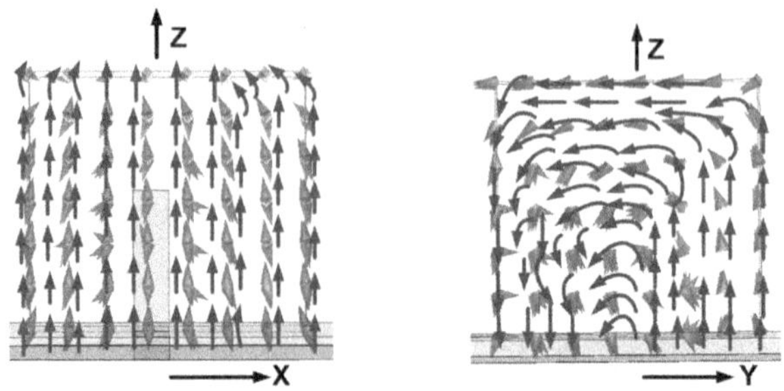

Figure 2.30 Distributions of electric fields inside a rectangular DRA to confirm the $TE_{01\delta}$ mode: (a) E-field, side view in xz-plane, and (b) E-field, side view in yz-plane.

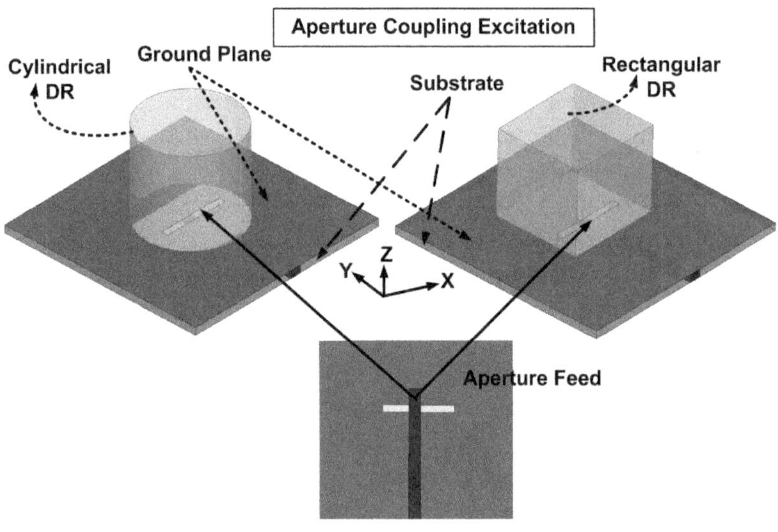

Figure 2.31 Excitation of a cylindrical and rectangular DRA by using aperture coupling: (a) cylindrical DRA, and (b) rectangular DRA.

to the DRA, as shown in Figure 2.32. To provide strong coupling, the slot is located in the region of strong magnetic fields and the slot area should be as small as possible to avoid excessive radiation underneath the ground plane. The aperture coupling provides horizontal magnetic dipole and it excites different modes in different shapes of DRA; for example, if a rectangular DRA is placed over a rectangular slot, it excites $TE_{11\delta}$ mode (Figure 2.33), while in a cylindrical DRA, it excites the $HEM_{11\delta}$ mode (Figure 2.34).

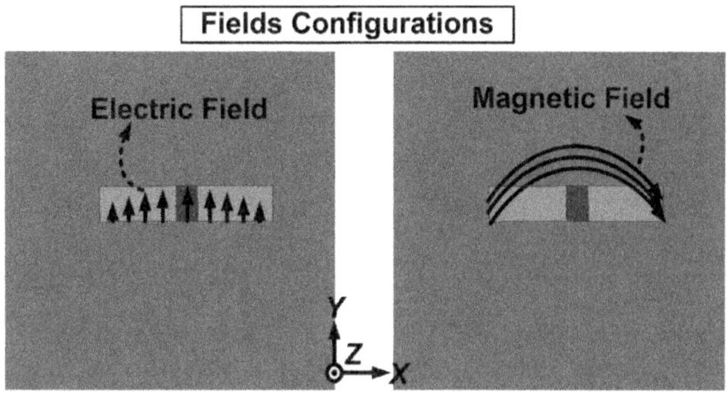

Figure 2.32 Electric and magnetic field profiles of aperture coupling technique: (a) electric fields, and (b) magnetic fields.

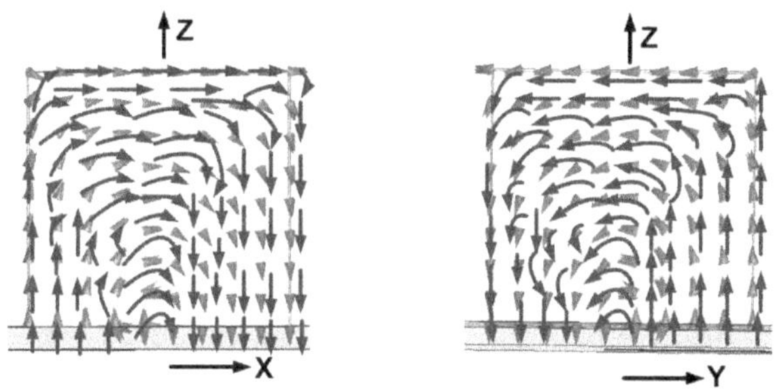

Figure 2.33 Distributions of electric fields inside a rectangular DRA to confirm the $TE_{11\delta}$ mode: (a) E-field, side view in xz-plane, and (b) E-field, side view in yz-plane.

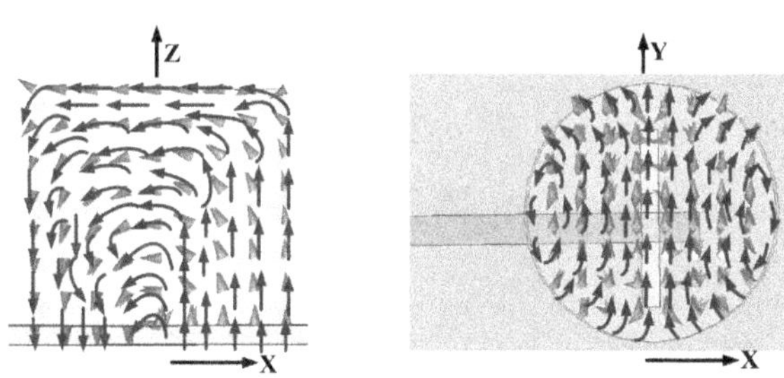

Figure 2.34 Distributions of electric and magnetic fields inside a cylindrical DRA to confirm the $HEM_{11\delta}$ mode: (a) E-field, side view in xz-plane, and (b) H-field, top view in xy-plane.

2.5.5 Coplanar Waveguide Feed Coupling

The coplanar waveguide (CPW) is one of the most suitable for microwave monolithic integrated circuits (MMICs) because it uses planar antenna geometry and is easily integrated into the MMICs. The main advantages of a CPW feed are that both feed line and ground plane are on the same plane and coupling takes place via slots, as shown in Figure 2.35 [22, 23]. Another advantage of a CPW feed is that it is a combination of the microstrip feed line and aperture couple where the radiation from the feed structure is negligible because it is excited in the odd mode of a coupled slot line. The modes of a DRA depend on the shape and location of the CPW feed line. Therefore, by changing the location of the CPW feed, the modes of a DRA can be changed, for example, in the case of a cylindrical DRA, for the CPW feed line in the center, $TM_{01\delta}$ mode is generated (Figure 2.36) while the off-centered feed excites $HEM_{11\delta}$ mode, as shown in Figure 2.37. From Figure 2.38, $TE_{11\delta}$ mode is excited in rectangular dielectric resonator antenna (RDRA) using this CPW feeding.

2.6 Circular Polarization Generation Approaches in DRAs

The process of circular polarization generation in DRAs has been discussed in Chapter 1. It was mentioned that two linearly polarized fields having equal magnitudes and a phase difference of 90° are required to excite CP radiation. These conditions can be obtained by various approaches. In the early years,

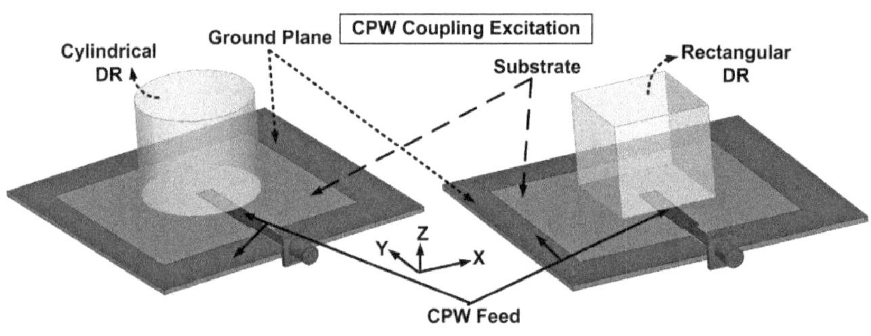

Figure 2.35 Excitation of RDRA by using CPW coupling: (a) 3D, and (b) top view.

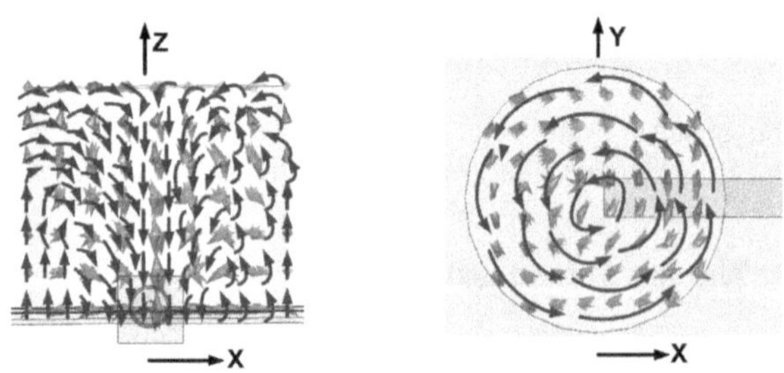

Figure 2.36 Distributions of electric and magnetic fields inside a cylindrical DRA to confirm the $TM_{01\delta}$ mode: (a) E-field, side view in xz-plane, and (b) H-field, top view in xy-plane.

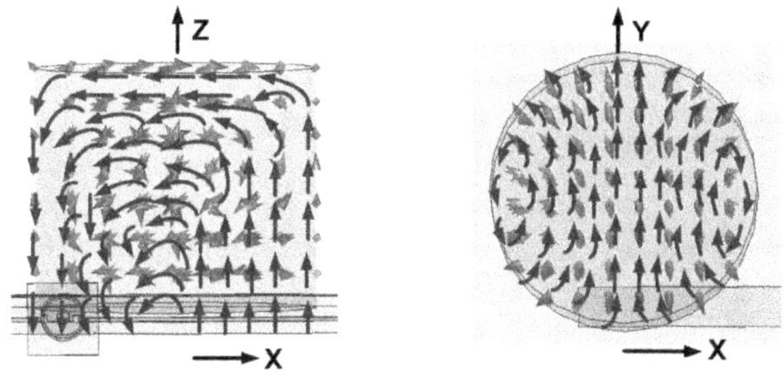

Figure 2.37 Distributions of electric and magnetic fields inside a cylindrical DRA done by HFSS to confirm the $HEM_{11\delta}$ mode: (a) E-field, side view in xz-plane, and (b) H-field, top view in xy-plane.

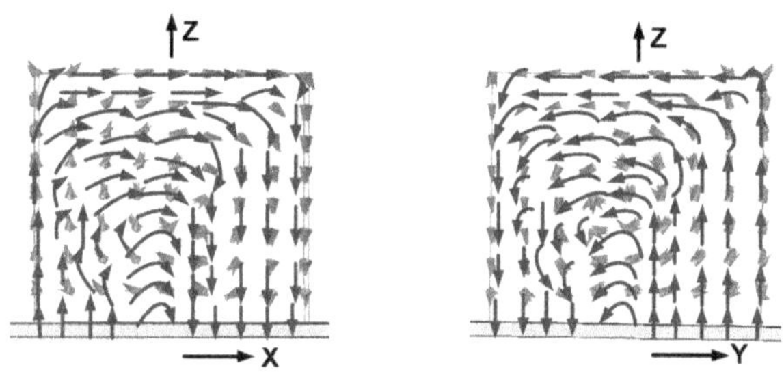

Figure 2.38 Distributions of electric fields inside a rectangular DRA to confirm the $TE_{11\delta}$ mode: (a) E-field, side view in xz-plane, and (b) E-field, side view in yz-plane.

one such approach of dual-point feed techniques was employed using power dividers such as a hybrid coupler [24], Wilkinson splitter [25], and T-junction splitter [26]. The hybrid coupler splits the incoming signal into two equal magnitude signals in phase quadrature, but this requires a larger area as shown in Figure 2.39. The Wilkinson and T-junction splitters also split the signal into two equal parts, but an external delay line is also required to get the correct phase difference. These two dividers require less area. An example of a T-junction splitter is shown in Figure 2.40. Even though it provided wide axial ratio bandwidth (ARBW), due to complex feed network complications researchers moved their attention to single-point feed networks. Another method adopted for getting wide axial ratio bandwidth was to arrange DRA elements in array form by using the sequential rotation technique [27].

In the current scenario, most of the research is developed by using single-feed techniques. By using single-feed techniques, a CP radiation can be obtained by exciting two quasi-degenerate orthogonal modes in the dielectric resonator in phase quadrature. Designing single-feed networks to generate CP is easy and feasible for practical use as compared to dual-feed point feeds at the cost of narrower axial ratio bandwidth. Thus, most of the research on circularly

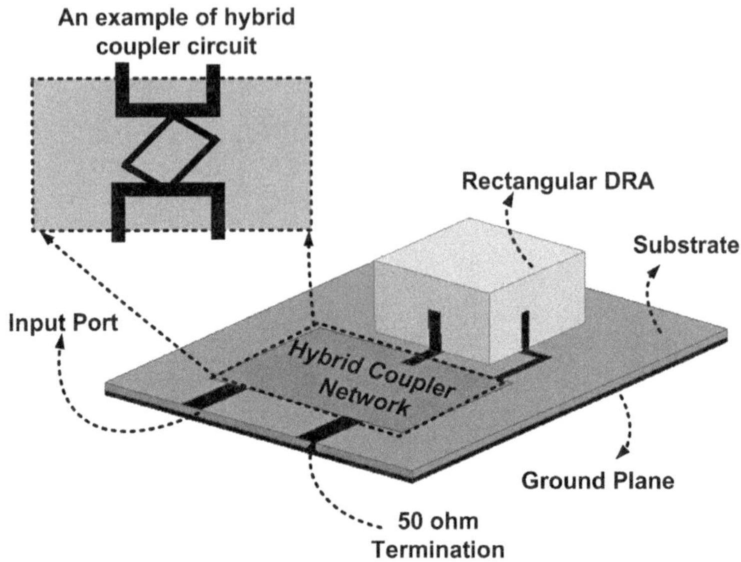

Figure 2.39 Example of a circularly polarized DRA using hybrid coupler. (Modified after [24].)

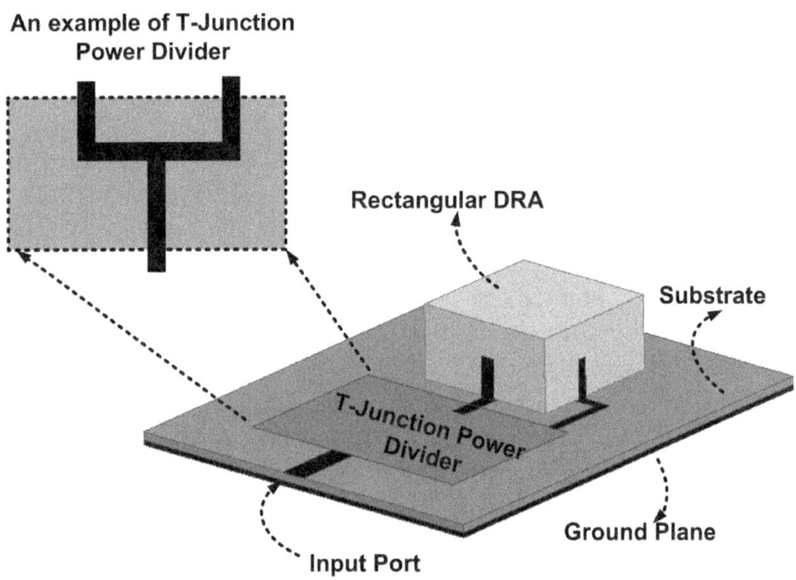

Figure 2.40 Example of a circularly polarized DRA using T-junction power divider. (Modified after [26].)

polarized DRAs is focused on developing antenna configurations employing single-feed networks to obtain CP radiation and improving their ARBW. The single-feed networks can be categorized based on different feeding mechanisms to DRAs, which are shown in Figure 2.41.

In this book, the subsequent chapters are devoted to each feeding technique separately, focusing on the procedure of CP analysis using a particular feeding technique and DRA modification in single-feed and dual-feed arrangements. A brief introduction for each technique is discussed below.

- *Microstrip coupling technique*: The microstrip feeding technique is one of the popular techniques used to excite the DRA for CP radiation. It can be shaped into various networks and also provide better impedance matching, which is easily acquired by adjusting the DRA position with respect to the length and width of the microstrip feed line, and it behaves like a horizontal magnetic dipole. This technique is presented in Chapter 3.
- *Conformal microstrip line technique*: This technique is a subset of the microstrip line technique in which conformal microstrip lines are attached to a DRA in various configurations to excite the desired conditions of CP. In this technique, the axial ratio is mostly dependent on the relative dimensions of the microstrip lines. This technique is presented in Chapter 4.
- *Coplanar waveguide technique*: By using the coplanar waveguide technique, only a few circularly polarized designs are available in the literature but they have several advantages as compared to microstrip lines, which will be discussed in Chapter 5. In this technique, ground plane and feed positions are in the same plane, which can easily be integrated into the MMICs.

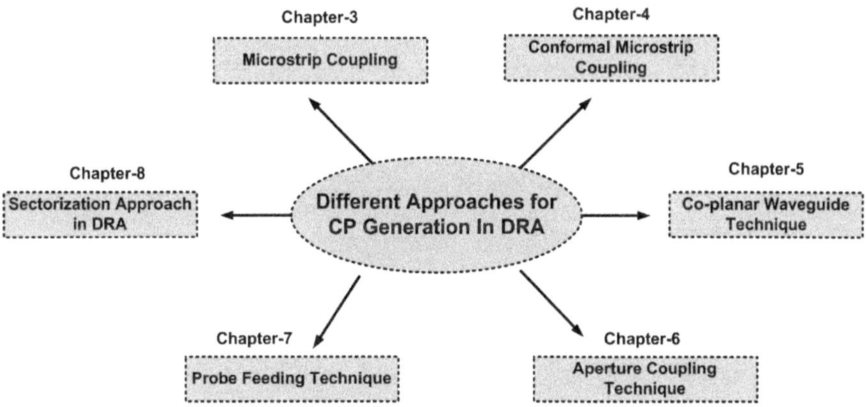

Figure 2.41 Different feeding techniques available for CP generation in DRAs.

- *Aperture coupling technique*: In Chapter 6, the most widely used aperture coupling technique is used to get CP radiation from DRAs. The aperture or slot coupling technique does not suffer from any spurious radiation from the feed line. This technique provides greater flexibility because desired slots can be cut into the ground plane on which the DRA is placed.
- *Probe-feeding technique*: In Chapter 7, the idea of using coaxial probe feeding is presented for the generation of circular polarization in different shapes of DRAs. Probe feeding provides direct coupling to 50Ω antenna systems without any requirement for matching networks. In this feeding, usually the dielectric resonator geometry is modified into various shapes to generate CP by using the concept of detuning the degenerate modes. Circularly polarized MIMO antennas, dual-band CP antennas, and array antennas follow the discussion. In this case, the axial ratio is mostly dependent on the geometrical dimensions of the DRA.
- *Sectorization technique*: In Chapter 8, the focus is on the sectorization approach in DRAs for CP generation. Some investigations have also been performed on different forms of sectored cylindrical and hemispherical dielectric resonator antennas formed using different sector angle (β) to improve the CP performance.

References

[1] Richtmyer, R. D., "Dielectric Resonators," *Journal of Applied Physics*, Vol. 10, 1939, pp. 391–398.

[2] Kajfez, D., A. W. Glisson, and J. James, "Computed Modal Field Distributions for Isolated Dielectric Resonators," *IEEE Transactions on Microwave Theory and Techniques*, Vol. 32, No. 12, 1984, pp. 1609–1616.

[3] Kajfez, D., and P. Guillon, *Dielectric Resonators*, Norwood, MA: Artech House, 1986.

[4] Sinha, D., and G. A. J. Amaratunga, "Electromagnetic Radiation under Explicit Symmetry Breaking," *Physics Review Letters*, Vol. 114, No. 147701,2015, pp. 1–7.

[5] Petosa, A., *Dielectric Resonator Antenna Handbook*, Norwood, MA: Artech House, 2007.

[6] https://www.ansys.com/en-in/.

[7] Glisson, A. W., D. Kajfez, and J. James, "Evaluation of Modes in Dielectric Resonators Using a Surface Integral Equation Formulation," *IEEE Transactions on Microwave Theory and Techniques*, Vol. 31, No. 12, 1983, pp. 1023–1029.

[8] Van Bladel, J., "On the Resonances of a Dielectric Resonator of Very High Permittivity," *IEEE Transaction on Microwave Theory and Techniques*, Vol. 23, 1975, pp. 199–208.

[9] Van Bladel, J., "The Excitation of Dielectric Resonators of Very High Permittivity," *IEEE Transaction on Microwave Theory and Techniques*, Vol. 23, 1975, pp. 208–215.

[10] Mongia, R. K., "Theoretical and Experimental Resonant Frequencies of Rectangular Dielectric Resonators," *IEEE Proceedings-H*, Vol. 139, No. 1, 1992, pp. 98–104.

[11] McAllister, M. W., and S. A. Long, "Resonant Hemispherical Dielectric Antenna", *IET Electronics Letters*, Vol. 20, No. 16, 1984, pp. 657–659.

[12] Long, S. A., M. W. McAllister, and L. C. Shen, "The Resonant Cylindrical Dielectric Cavity Antenna," *IEEE Antennas and Wireless Propagation Letters*, Vol. 31, No. 3, 1983, pp. 406–412.

[13] McAllister, M. W., S. A. Long, and G. L. Conway, "Rectangular Dielectric Resonator Antenna," *IET Electronics Letters*, Vol. 19, No. 6, 1983, pp. 218–219.

[14] Luk, K. M., and K. W. Leung, "Dielectric Resonator Antenna," Baldock, U.K.: Research Studies Press, 2003.

[15] Chauthaiwale, P., R. K. Chaudhary, and K. V. Srivastava, "Circularly Polarized Bowtie-shaped Dielectric Resonator Antenna Excited with Asymmetric Cross Slot," *Microwave and Optical Technology Letters*, Vol. 57, 2015, pp. 1732–1727

[16] Mongia, R. K., A. Ittipiboon, "Theoretical and Experimental Investigations on Rectangular Dielectric Resonator Antennas," *IEEE Transactions on Antennas and Propagation*, Vol. 45, No. 9, 1997, pp. 1348–1356.

[17] Mongia, R. K., and P. Bhartia, "Dielectric Resonator Antennas-A Review and General Design Relations for Resonant Frequency and Bandwidth," *International Journal of Microwave and Millimeter-Wave Computer-Aided Engineering*, Vol. 4, No. 3, 1994, pp. 230–247.

[18] McAllister, M. W., S. A. Long, and G. L. Conway, "Rectangular Dielectric Resonator Antenna," *IET Electronics Letters*, Vol. 19, No. 6, 1983, pp. 218–219.

[19] Petosa, A., and A. Iittipiboon, "Dielectric Resonator Antennas: A Historical Review and the Current State of the Art," *IEEE Antennas and Propagation Magazine*, Vol. 52, No. 5, 2010, pp. 91-47.

[20] Leung, K. W., "Conformal Strip Excitation of Dielectric Resonator Antenna," *IEEE Transactions on Antennas and Propagation*, Vol. 48, No. 6, 2000, pp. 961–967.

[21] Huang, C. Y., J. Y. Wu, and K. W. Wong, "Cross Slot Coupled Microstrip Antenna and Dielectric Resonator Antenna for Circular Polarization," *IEEE Transactions on Antennas and Propagation*, Vol. 47, No. 4, 1999, pp. 605–609.

[22] Kranenburg, R. A., S. A. Long, and J. T. Williams, "Coplanar Waveguide Excitation of Dielectric Resonator Antennas," *IEEE Transactions on Antennas and Propagation*, Vol. 39, No. 1, 1991, pp. 119–122.

[23] Chen, Y. C., S. L. Yao, and K. C. Chen, "Dual Band Hybrid CPW Fed Planar Monopole/Dielectric Resonator Antenna," *IEEE 9th Malaysia International Conference* on, December 15–17, 2009.

[24] Leung, K. W., W. C. Wong, K. M. Luk, and E. K. N. Yung, "Circular-Polarised Dielectric Resonator Antenna Excited by Dual Conformal Strips," *Electronics Letters*, Vol. 36, No. 6, 2000, pp. 484–486.

[25] Kumari, R., and R. K. Gangwar, "Circularly Polarized Cylindrical Dielectric Resonator Antenna Excited by Dual Conformal Strips Along with Modified Wilkinson Power Divider for High Gain Application," *Microwave and Optical Technology Letters*, Vol. 59, 2017, pp. 908–913.

[26] Wong, W. C., and K. W. Leung, "Circularly Polarized Dielectric Resonator Antenna Excited by Dual Conformal Strips of Unequal Lengths," *Microwave and Optical Technology Letters*, Vol. 29, No. 5, 2001, pp. 348–350.

[27] Yang, S. S., et al., "Study on Sequential Feeding Networks for Subarrays of Circularly Polarized Elliptical Dielectric Resonator Antenna," *IEEE Transactions on Antennas and Propagation*, Vol. 55, No. 2, 2007, pp. 321–333.

3

Circular Polarization Analysis in Dielectric Resonator Antennas Using the Microstrip Coupling Technique

This chapter presents the microstrip coupling technique and its usefulness in different DRA designs. This coupling technique has potential advantages for the generation of circular polarization by using different configurations of DRAs, such as single, multiple, modified, multiple-input and multiple-output, array, and filtering structures. The step-by-step design guidelines are also included in this chapter at the end of every approach for better understanding of the microstrip coupling technique.

3.1 Introduction to the Microstrip Coupling Technique

This chapter introduces various microstrip coupling approaches that are potentially useful for wireless communication applications. A resonator is a crucial element of microwave circuits. The main requirements of microwave resonators are a low loss or high-quality factor, and temperature stability. The methodology is here to focus on microstrip coupling for different DRA design aspects that are applicable for single, multi, modified, multiple-input, and multiple-output array configurations and filtering applications. This coupling offers several advantages and is widely used to avoid unwanted higher-order modes, which is one of the unique advantages of this coupling mechanism. It also provides better impedance matching, which is easily acquired by adjusting the DRA position with respect to the length and width of the microstrip feed line, as shown in Figure 3.1. The microstrip coupling technique is divided into two categories [1]: (a) direct coupling, and (b) side-coupling (indirect coupling).

Both direct and side-coupling are used for different configuration of DRAs to get circularly polarized radiations, as depicted in Figures 3.2 and 3.3, respectively.

In the case of direct coupling, the dielectric resonator is placed directly above the microstrip line network, as shown in Figure 3.2. The amount of coupling is adjusted by the portion of the microstrip line length present beneath the dielectric resonator, which is signified by symbol "D" in Figure 3.2 for a direct coupling approach.

In the case of side coupling, the dielectric resonator and the microstrip line is not in contact where the coupling strength between them is tuned by the distance "D" as shown in Figure 3.3.

Moreover, in both cases, the strength of coupling is also dependent on the dielectric constant of the DRA, and for higher values of dielectric constant (>20), the strongest coupling can be achieved [1]. Therefore, both microstrip coupling and dielectric constant are related to each other for generating various application purposes like wideband, circularly polarized, multiple-input and

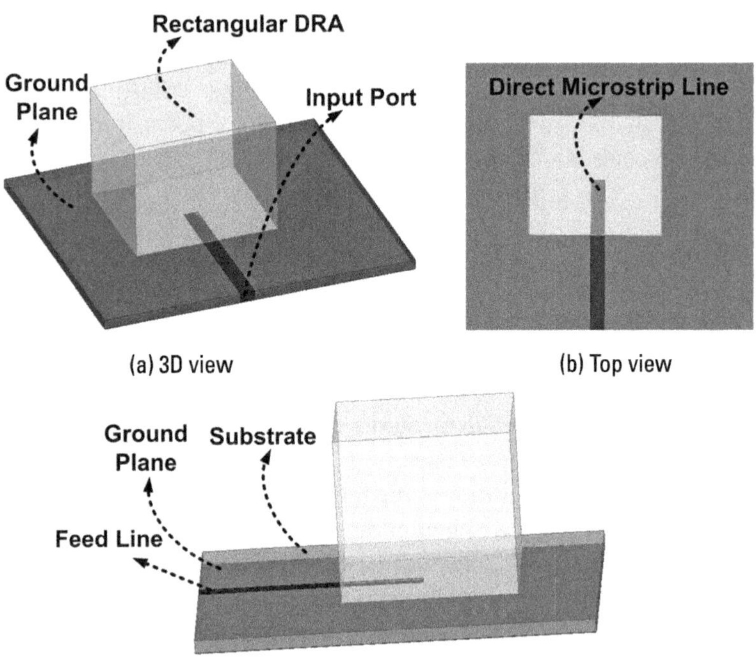

Figure 3.1 General arrangement of microstrip coupling technique in dielectric resonator antenna.

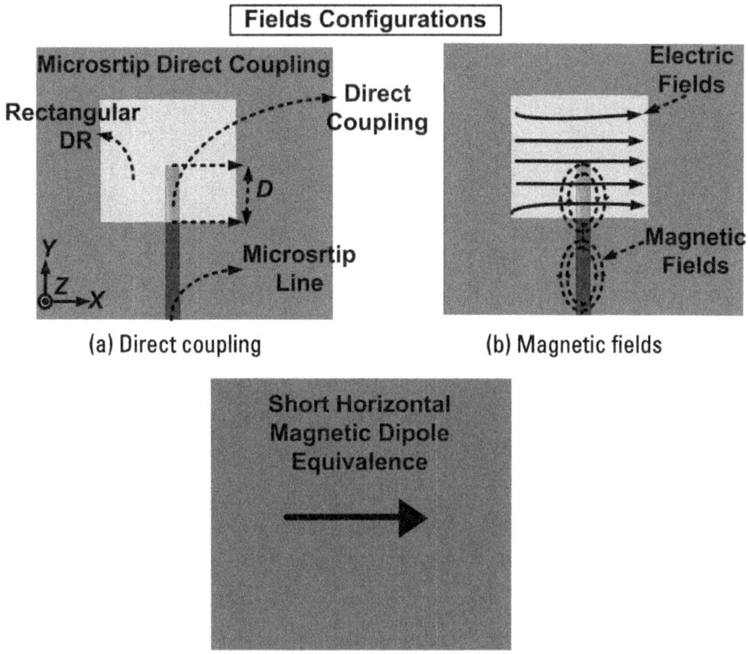

(a) Direct coupling

(b) Magnetic fields

(c) Horizontal magnetics

Figure 3.2 Direct microstrip line coupling to the dielectric resonator antennas, fields, and radiation model: (a) direct coupling, (b) magnetic fields, and (c) horizontal magnetic.

multiple-output (MIMO), and other operations [2]. Microstrip coupling feeding is one of the easiest and most versatile methods that has been employed in different DRA structures. By versatile, it means that there is flexibility in choosing the shape of the microstrip line network to generate CP. This section summarizes the best and most widely used excitations mechanisms while providing many references so that designers can choose the most and appropriate configurations. For example: (1) it excites the $TE_{01\delta}$/ $TE_{11\delta}$ modes and other higher modes in a rectangular DRA if the microstrip feed line is placed along the side of the DRA, while (2) $HEM_{11\delta}$/ $HEM_{21\delta}$ and other higher modes can be excited in a cylindrical DRA if the microstrip feed line is placed in the middle of a cylindrical DRA, which behaves like a horizontal magnetic dipole [1, 2].

The generation of CP in a DRA has been mentioned in Chapter 1, where mainly two conditions are important for the generation of circular polarization: (1) excitation of orthogonal modes with equal magnitude, and (2) phase shift between orthogonal modes should be ±90° [3].

In the literature there are mainly two approaches used to achieve wideband CP.

The first approach is to excite degenerate modes (two modes having the same resonance frequency) in a DRA and obtaining odd multiples of 90° phase

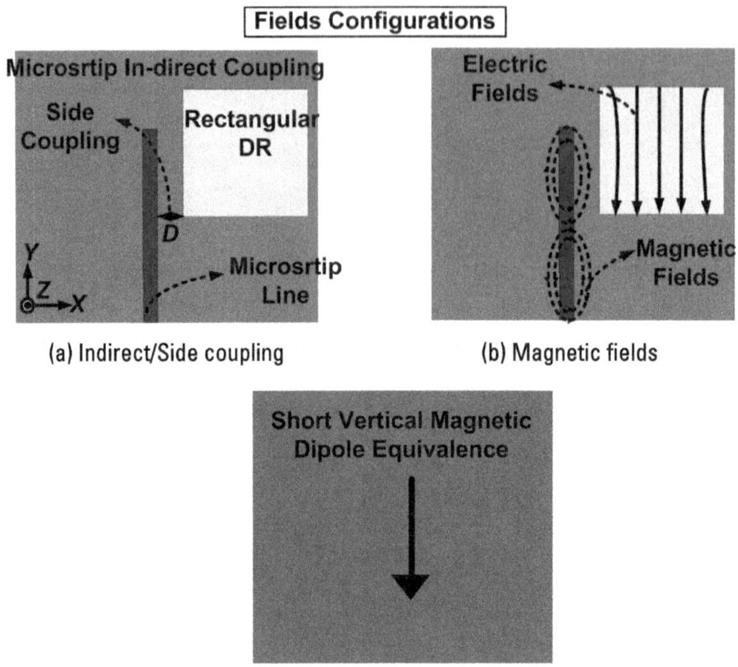

(a) Indirect/Side coupling

(b) Magnetic fields

(c) Vertical magnetic

Figure 3.3 Side microstrip line coupling to the dielectric resonator antennas, fields, and radiation model: (a) side coupling, (b) magnetic fields, and (c) vertical magnetic.

difference between them. Such approaches have been used in [4]. The second approach is to excite two modes that are orthogonal in nature (at two different frequencies), and odd multiples of 90° phase difference between them is obtained to realize CP. Such approaches have been used in [5].

3.2 Design Example Using Microstrip Coupling for CP Generation

In this section, a design example with methodology is presented for the generation of circular polarization in a DRA using microstrip coupling techniques as depicted in Figures 3.4 and 3.5, respectively.

3.2.1 Antenna Configuration

The proposed circular polarization antenna design includes a dielectric resonator, a ground plane, and a 50-Ω matching microstrip line, as shown in Figure 3.5. The dimensions of the rectangular dielectric resonator are D_L = 20 mm × D_W= 25 mm × D_H= 22 mm, which represent the length, width, and height, respectively, and its relative permittivity (ε_{DRA}) is 9.8. A circular bend type metal

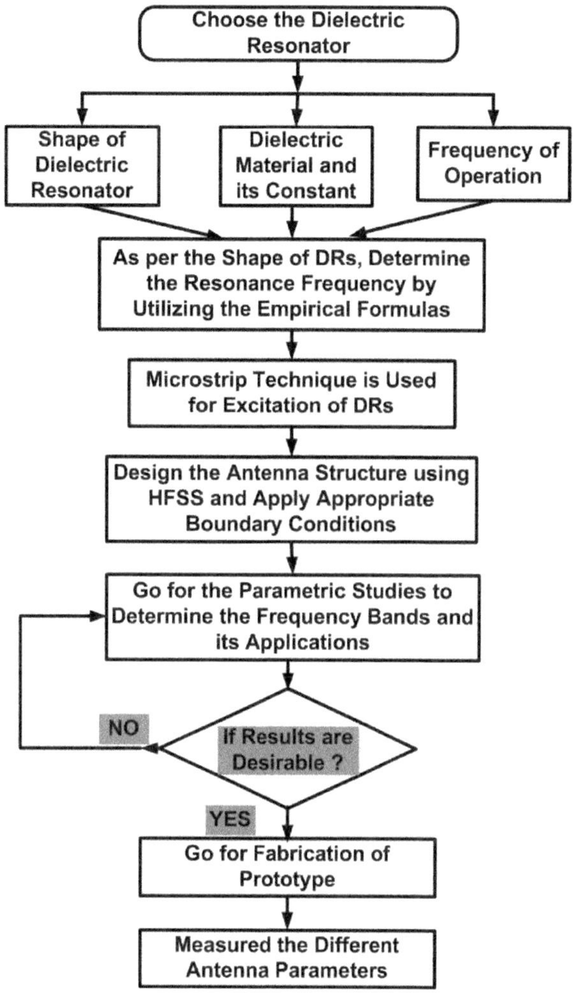

Figure 3.4 Design steps and methodology of microstrip coupling techniques.

strip of width (CF_W = 2 mm) is attached to one side of the feed line and connected to the main microstrip feed having length (M_{FL} + MF_{L1} = 23.5 mm), width (M_{FW} = 2.5 mm), and height (MF_H = 0.035 mm), respectively, which is placed on the FR4 glass epoxy substrate. The arc-shaped microstrip feed line having outer (r_1 = 12.5 mm) as well as the inner radius (r_2 = 11 mm) and partial ground plane (G_{PG} = 17 mm) are located on the top and bottom layers of an FR4 substrate with relative permittivity (ε_{Sub}) is 4.4 and its length (S_L = 40 mm), width (S_W = 50 mm), and height (S_H = 1.6 mm), respectively.

$$MF_L + MF_{L1} = 12 + 11.5 = 23.5 mm \tag{3.1}$$

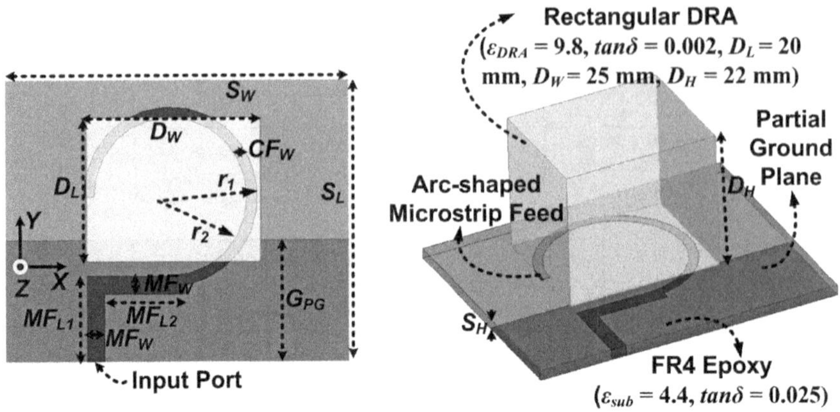

(a) (b)

Figure 3.5 Proposed example for a rectangular DRA fed by arc-shaped microstrip feedline: (a) top view, and (b) 3D view. (Modified after [6].)

3.2.2 Operating Mechanism

To examine the capabilities of the proposed design, the performance of a rectangular DRA is simulated by using ANSYS HFSS for all microstrip feeding configurations given in Figure 3.6. Initially, the rectangular DRA is excited by the quarter-circular (90°) microstrip feed line in the first configuration (Sub-Ant. 1), then by the quarter-circular microstrip feed line is extended to the semicircular (180°) microstrip feed line in the second configuration (Sub-Ant. 2), by adding another quarter-circular in the semicircular microstrip feed line it is extended to the three quarter-circular (270°) microstrip feed line in the third configuration (Sub-Ant. 3- Proposed), and by complete circular (360°) microstrip feed line in the fourth configuration (Sub-Ant. 4), as shown in Figure 3.6.

Other dimensions of the four antennas stay the same to have a better comparison between the antenna performances. Figure 3.7 shows the input impedance bandwidths (IBWs) and ARs for all four antenna configurations.

Case-1 (Antenna-1): The Sub-Ant. 1 is designed by taking the initial dimensions and permittivity of the ceramic material of the rectangular DR, which has been discussed in Chapter 2. Here, an example of circular polarization is discussed so the formula of DWM is rarely applicable for the resonance frequency of a DRA. In this antenna construction, the main focus has been given to the circular polarization instead of the linearly polarized antenna. By taking this into consideration, the dimensions of the DRA, its permittivity (dielectric constant), ground plane dimensions, and microstrip line geometry has been chosen.

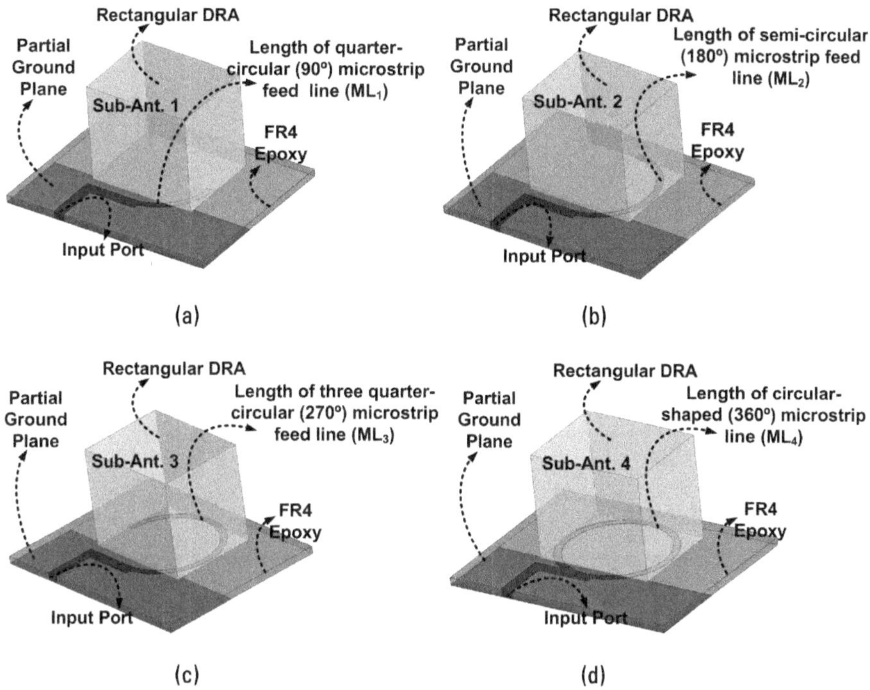

Figure 3.6 The rectangular DRA stage evolutions and its feeding configurations. (a) Antenna 1 (Sub-Ant. 1): quarter-circular. (b) Antenna 2 (Sub-Ant. 2): semicircular. (c) Antenna 3 (Sub-Ant. 3-Proposed): three quarter-circular. (d) Antenna 4 (Sub-Ant. 4): complete circular microstrip feed line.

Antenna configuration-1 has been designed by placing the quarter-circular (90°) microstrip feed line at the corner of the DRA surface called Sub-Antenna-1 and shown in Figure 3.6(a).

The optimized feed parameters are $MF_L = 23.5$ mm and $CF_W = 2$ mm. The width of the feeding strip is 2.5 mm. The present antenna configuration is linearly polarized having very low impedance matching and is also not suitable for generation of degenerate mode pair in these configurations. The corresponding input impedance and ARBW for this antenna configuration are shown in Figures 3.7(a) and 3.7(b), respectively. In this case, the length of the quarter-circular microstrip line is $(M_{L1}) = 11$ mm, and free space wavelength is $(\lambda_0) = 3$ GHz. By utilizing (3.1), the phase difference between the two linearly polarized waves is $\approx 0.22\pi = \pi \times 11/50$, which does not achieve the condition of circular polarization in the Sub-Ant. 1. Therefore, this condition is not suitable for getting the axial ratio and impedance bandwidth.

$$\textit{Phase difference} = \frac{2\pi}{\lambda} \times \textit{Path difference} \tag{3.2}$$

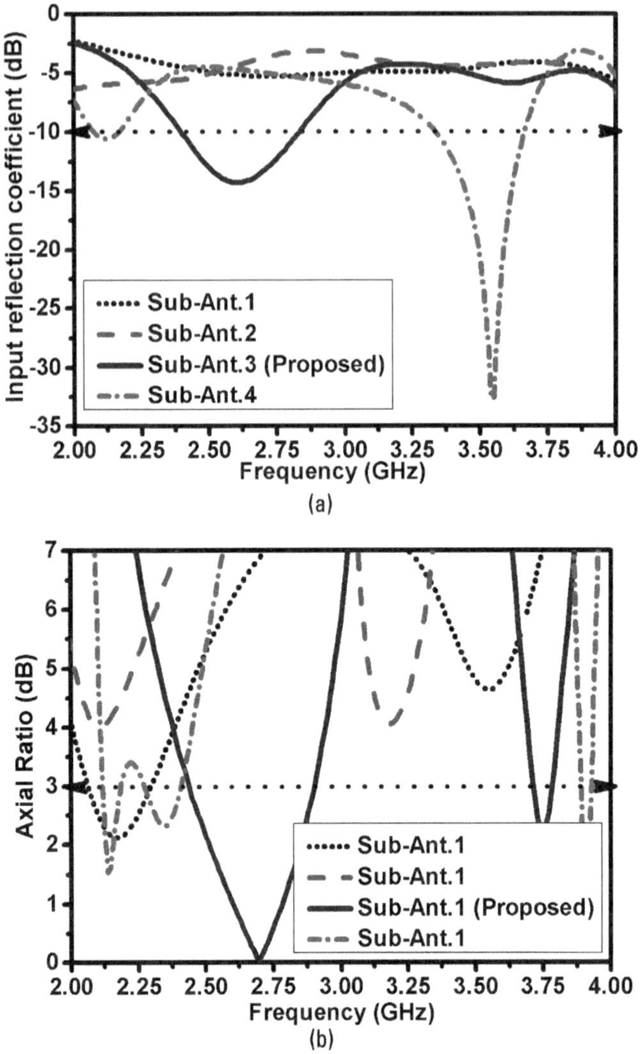

Figure 3.7 Simulated (a) input reflection coefficient, and (b) axial ratio plot for different stages for all subantennas.

Case-2 (Antenna-2): The Sub-Ant. 2 is a modification of Sub-Antenna-1. This design has been obtained by changing the feeding position and extended to the semicircular (180°) microstrip feed line in the second configuration. All other antenna parameters remain constant, as shown in Figure 3.6 (b). The orthogonal degenerate mode pair has not been excited, as can be realized from Figure 3.7(b) while Sub-Ant. 2 also displays a poor IBW, which is depicted in Figure 3.7(a). By using the semicircular position, the matching impedance is due less due to the fact that the embedded rectangular DRA could not be excited by using semi-circular microstrip line.

In this antenna step, the length of the quarter-circular microstrip line ($M_{l,}2$) = 22 mm, and free space wavelength (λ_0) = 3 GHz. By utilizing (3.1), the phase difference between the two linearly polarized waves is $\approx 0.44\pi = \pi \times 11/25$, which again does not achieve the condition of circular polarization in the Sub-Ant. 2. Hence, this condition is also not suitable for getting the axial ratio and impedance bandwidth.

Case-3 (Antenna-3): The Sub-Ant. 3 (Proposed antenna) has been designed by further modifying antenna configuration-2 to obtain the circular polarization radiation, as shown in Figure 3.6. The modifications involve the addition of quarter-circular (90°) to the semicircular (180°), which extended to the three quarter-circular (270°) microstrip feed line. Using this position of three quarter-circular microstrip feed line, the orthogonal degenerate mode pair has been excited. This is responsible for circular polarization wave generation. Further, the degenerate $TE^x_{\delta 11}$ and $TE^y_{1\delta 1}$ mode has been excited at 2.63 GHz and shown in Figure 3.8. Using this excitation position of the feed through DRA in the presented antenna, the input impedance bandwidth ($|S_{11}| \leq$ 10dB) of 16.85% (2.39–2.83 GHz) along with axial ratio bandwidth ($|AR| \leq$ 3dB) of 17.66% (2.43–2.90 GHz) over the same frequency range have been achieved as shown in Figures 3.7(a) and 3.7(b), respectively. A successful overlapped CP bandwidth $|S_{11}| \leq$ 10dB and ($|AR| \leq$ 3dB) of 15.38% (2.43–2.83 GHz) has been achieved by the presented antenna. In the whole CP bandwidth, a useful average CP gain of 2.69 dBic has been reached as demonstrated in Figure 3.12(a). Here, the length of the quarter-circular microstrip line is (M_{L3}) = 33 mm, and free space wavelength is (λ_0) = 3 GHz. By utilizing (4.1), the phase difference between the two linearly polarized waves is $\approx 0.66\pi = \pi \times 33/50$ which fulfills the condition of circular polarization in the Sub-Ant. 3. Therefore, this condition is more suitable for getting the axial ratio and better impedance bandwidth.

Case-4 (Antenna-4): The Sub-Ant. 4 has been constructed by adding a quarter-circular (90°) microstrip line in the existing three quarter-circular (Sub-Ant. 3) microstrip feed line to make it a circular-shaped (360°) microstrip line as illustrated in Figure 3.6(d). The size of the circular-shaped microstrip line has been optimized by iterative design procedure, which has been opted by the parametric study. The configuration medications have been made for further improvement of impedance-matching bandwidth and axial ratio bandwidth, although the Sub-Ant. 4 has a dual-band character with much less 3-dB axial ratio axial at the lower band, as depicted in Figure 3.7. The impedance bandwidths ($|S_{11}| \leq$ 10dB) at the lower and upper resonances are at 4.71% (2.07–2.17 GHz) and 9.14% (3.34–3.66 GHz) along with lower resonances axial ratio bandwidth ($|AR| \leq$ 3dB) of 2.79% (2.12–2.18 GHz), respectively. At this antenna step, the length of the quarter-circular microstrip line is (M_{L4})

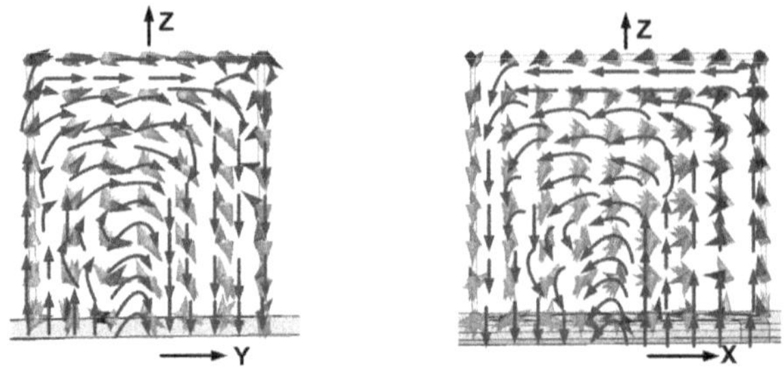

Figure 3.8 Simulated electric fields of Sub-Ant. 3-Presented antenna (a) $TEx_{\delta 11}$ and (b) $TE^{y}_{1\delta 1}$ at 2.65 GHz.

= 44 mm, and free space wavelength is (λ_0) = 3 GHz. By utilizing (3.2), the phase difference between the two linearly polarized waves is $\approx 0.88\pi = \pi \times 22/25$, which does not completely fulfill the condition of circular polarization in the Sub-Ant. 4. However, it is suitable to generate a very small axial ratio bandwidth in the lower band frequency. As observed, Sub-Ant. 3 (Proposed antenna) performs better than all other Sub-Ants. in terms of the –10 dB impedance and 3-dB axial ratio bandwidths because of the formation of orthogonal modes and effective coupling of the electromagnetic energy (electric source) in the rectangular DRA.

3.2.3 Electric-Field Distributions in the Rectangular DRA

The generation of degenerate orthogonal mode can be clearly understood by observing the field distribution, as shown in Figure 3.8. With this feeding configuration, the fundamental orthogonal degenerate mode of $TE^{x}_{\delta 11}$ along with $TE^{y}_{1\delta 1}$ are generated, which are responsible for the generation of the CP wave at a frequency of 2.65 GHz.

The vector *E*-field distribution on top of the rectangular DRA with different phases is demonstrated in Figure 3.9 for the circular polarization mechanism in the rectangular DRA. As observed in Figure 3.9, by changing the phases from 0 to 270 degrees, the electric fields turn in a clockwise direction at 2.65 GHz. As a result, a left-hand circular polarization (LHCP) radiation is achieved.

3.2.4 The Role of Partial Ground Plane

To further understand the design rules of the proposed antenna, the key parameter (i.e., ground variation) is investigated in this section to make it clear how they influence the antenna performance. Figure 3.10(a) depicts the effect of the ground plane on the input reflection coefficient. It is observed that as ground plane increases, the matching of the impedance bandwidth decreases (full ground plane) the higher frequency because of the maximum generation of

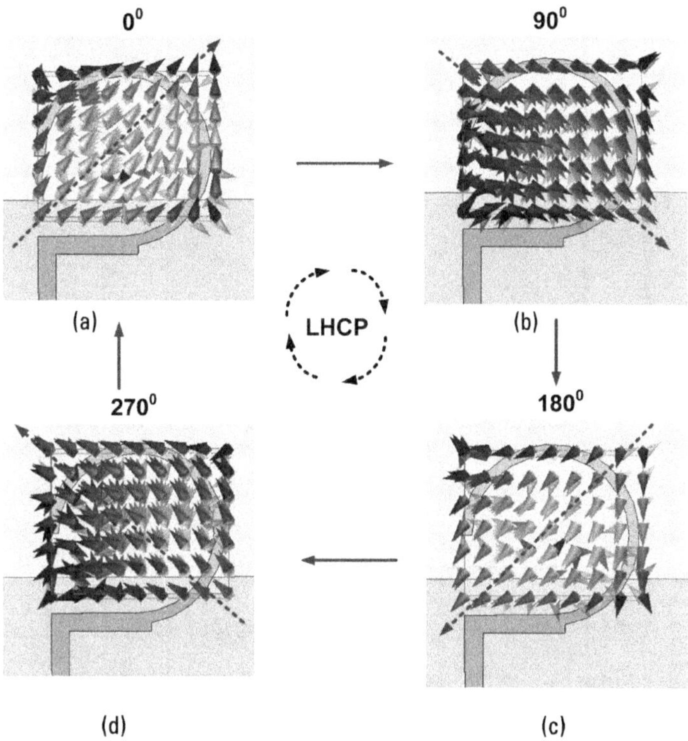

Figure 3.9 Simulated E-field distribution at the top of the rectangular DR observed along the +Z-direction with different phases, (a) 0°, (b) 90°, (c) 180°, and (d) 270°, at 2.65 GHz.

capacitance (store of the energy). It can be seen that as ground plane increases, the degenerate orthogonal mode has been achieved at a ground plane equal to 17 mm, and remaining other conditions, it does not fulfill the conditions of the circular polarization, as shown in Figure 3.10(b).

3.2.5 Analysis of Simulated Results

A simulated photograph of the proposed antenna is shown in Figure 3.5. The simulated input reflection coefficient, as well as axial ratios of the rectangular DRA, is shown in Figure 3.11.

The –10-dB simulated impedance bandwidths ($|S_{11}| \leq 10$dB) and 3-dB axial ratio bandwidth ($|AR| \leq 3$dB) in boresight direction are given as 16.85% (2.39–2.83 GHz) and 17.66% (2.43–2.90 GHz), as represented in Figures 3.11(a) and 3.11(b), respectively. The presented antenna also achieved an overlapped CP bandwidth ($|S_{11}| \leq 10$dB) and ($|AR| \leq 3$dB) of 15.38% (2.43–2.83 GHz).

The simulated LHCP and right-hand circular polarization (RHCP) radiation patterns at 2.65 GHz are shown in Figure 3.12. In the broadside radiation

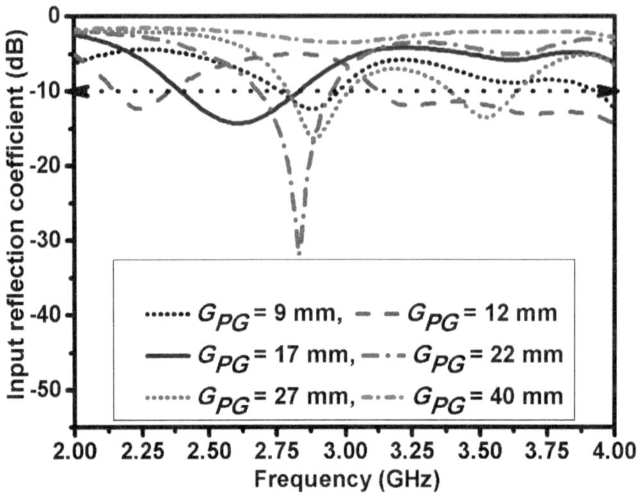

(a)

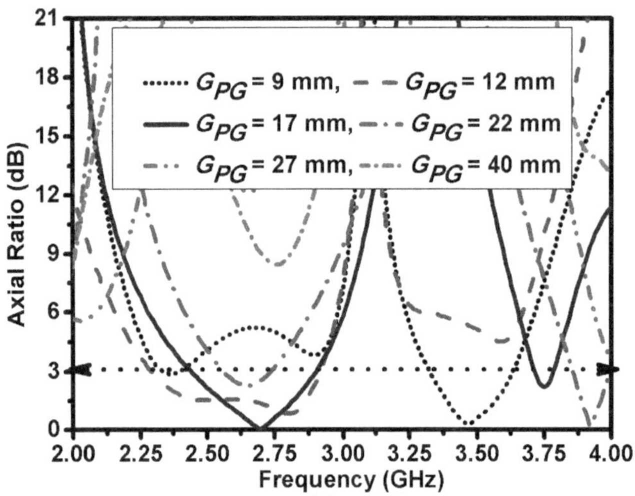

(b)

Figure 3.10 Simulated results with different ground plane, (a) input reflection coefficient, and (b) axial ratio plot.

pattern (*xz*-plane), the LHCP field is stronger than the RHCP counterparts by more than 49.84 dB, and also shows the stable CP radiation patterns throughout the working frequencies, as shown in Figure 3.12(a). It is also observed in Figure 3.12(b) that the LHCP field is stronger than the RHCP in both the broadside and end fire directions. The simulated gains and radiation efficiency of CP antenna is depicted in Figure 3.13. The simulated boresight average gain is 2.10 dBic, as shown in Figure 3.13(a), while the average boresight radiation efficiency is 85.75% in the CP working band, as shown in Figure 3.13(b).

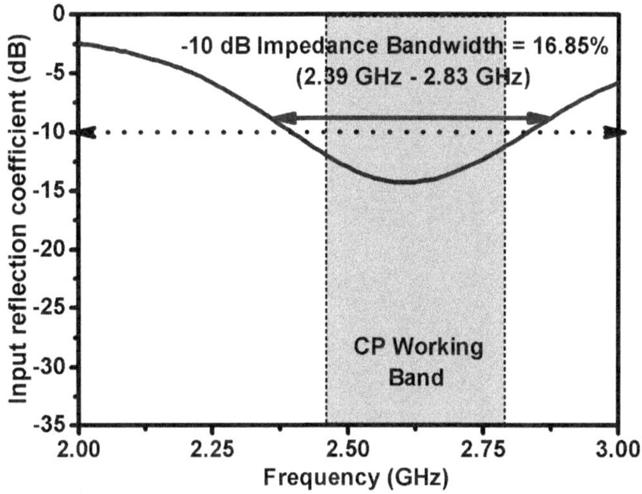

(a)

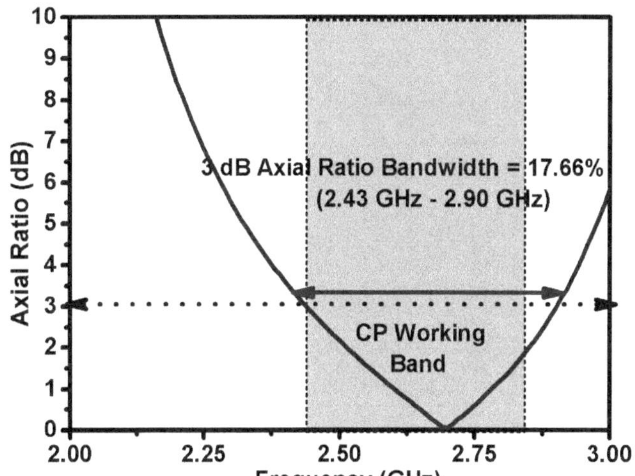

(b)

Figure 3.11 Simulated and measured reflection coefficients and ARs of the proposed antenna: (a) input reflection coefficient, and (b) axial ratio plot.

3.3 CP Generation in Conventional DRA Configurations

In this section, a rigorous analysis of the microstrip line fed single-element DR antenna is discussed in the context of circular polarization. It can be further divided into three types: single-point-fed convention single DRA, single-point-fed along with shorted/open-ended stubs in a single DRA, and single-point-fed along with parasitic metallic plates/strip in a single DRA.

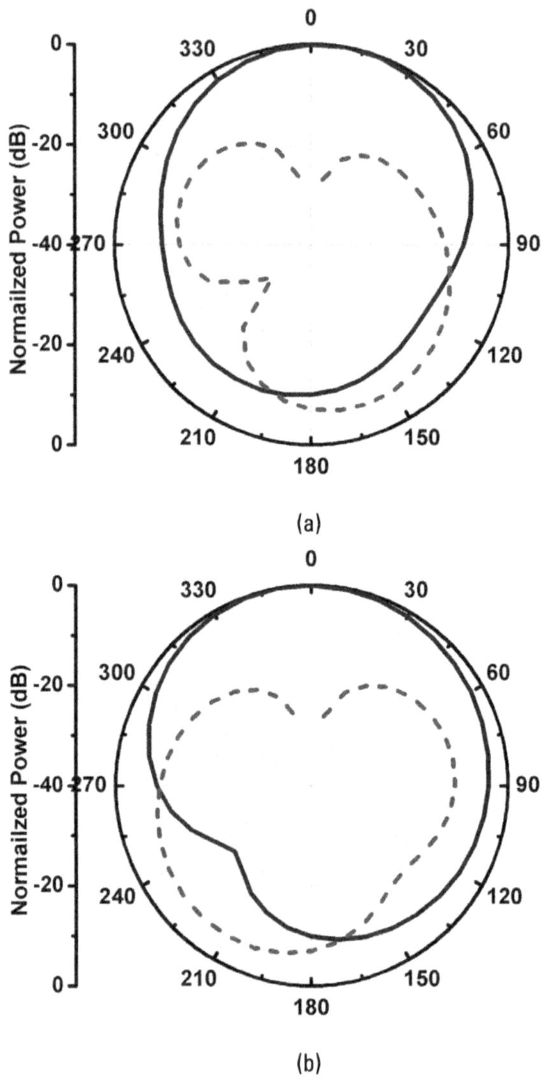

Figure 3.12 Simulated LHCP and RHCP radiation patterns of the prototype at 2.65 GHz.

3.3.1 Single-Point-Fed

In order to produce CP waves, orthogonal electric fields with a 90° phase difference are employed between the two linearly polarized waves. For example, Figure 3.14 shows the construction of a CP-DRA in which the circularly polarized generation is achieved by using a cubic shaped dielectric resonator excited by a long modified microstrip feed line that is identified as a question-mark-shaped microstrip feed [6]. The question-mark-shaped microstrip feed and partial ground plane help the generation of degenerate orthogonal $TE^{x}_{\delta 11}$

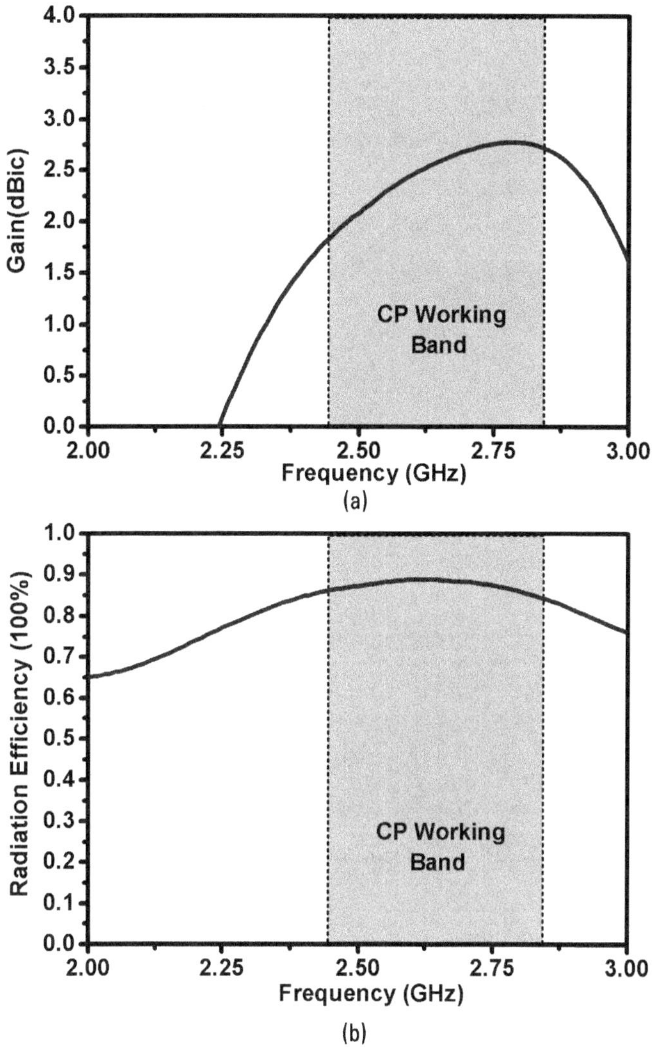

Figure 3.13 Simulated gain and radiation efficiency of the proposed antenna: (a) gain and (b) radiation efficiency plot in boresight direction.

and $TE^y_{1\delta 1}$ modes in the cubic DRA to realize circular polarization. This type of antenna configuration gives a wide range of ARBW of 20.62% in broadside direction [6].

Another example is shown in Figure 3.15 [7]. This design uses a meandered-line inductor microstrip feeding configuration in showing a simple structure and produces large ARBW. In this antenna structure, both meandered-line inductor feed and partial ground plane are utilized for achieving circular polarization, as shown in Figure 3.15.

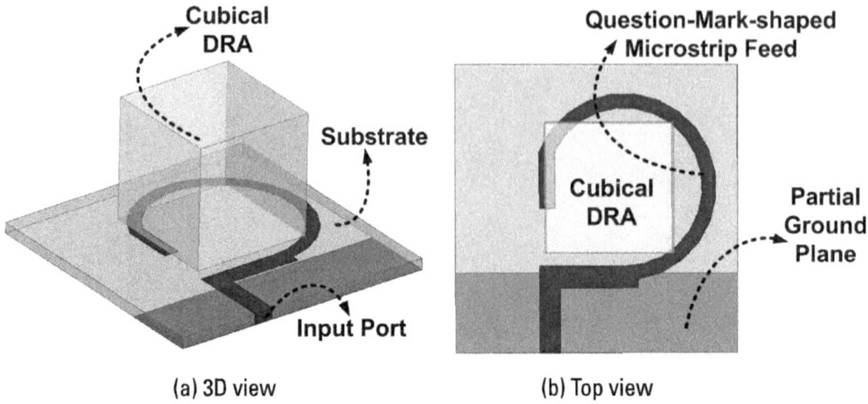

Figure 3.14 A question-mark-shaped microstrip coupled cubic DRA for wideband circular polarization [6].

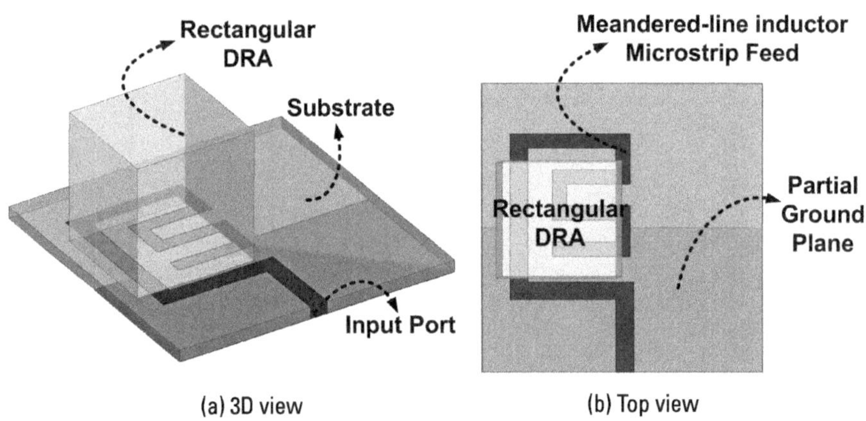

Figure 3.15 A meandered-line inductor microstrip coupled rectangular DRA for wideband circular polarization [7].

The other advantage of this design is that by controlling the length of the meandered-line inductor, the authors have proposed three different CP DRA for various wireless applications, which are industrial, scientific and medical (ISM) 2400, Worldwide Interoperability for Microwave Access (WiMAX), and wireless local area network (WLAN) band. The degenerate orthogonal mode (i.e., $TE^x_{\delta 11}$ and $TE^y_{1\delta 1}$) are responsible for the generation of circular polarization in the structure and offers an ARBW of 27.95% in broadside direction.

A hybrid DRA is used for generation of circularly polarized fields in the antenna configuration. The hybrid antenna is a combination of an asymmetrical square ring-shaped microstrip line and a rectangular DRA [9]. Similar types of antenna results could also be found in the combination as represented in

Figure 3.16. Generation of orthogonal mode is possible because of the symmetricity in the square ring of microstrip line and partial ground plane, and is also responsible for creating opposite sense characteristics in CP radiation. The fundamental TE_{111} mode is responsible for the creation of CP waves with dual-sense characteristics having an ARBW of 11.61% and 5.84% in $\theta = 0^0$, $\varphi = 0^0$, (i.e., broadside direction).

3.3.2 Single-Point-Fed with Shorted/Open-Ended Stubs

To realize wide ARBW in CP antennas without an external power divider network, microstrip feeding structures along with shorted or open-ended stubs are utilized to excite the orthogonal modes in DRAs [10–14]. In [10], the Y-shaped microstrip line is designed for achieving the CP fields in the compact antenna structure, whose radius is only 4.5 mm. Similar results may be obtained by using the combination as shown in Figure 3.17. The Y-shaped microstrip line has been constructed by simply connecting two quadrate shorted stubs, and the CP field in the rectangular DRA is generated by adjusting the lengths and widths of these two stubs. Therefore, this antenna structure provided ARBW of 1.18%, which covers the frequency range from 4.187 to 4.237 GHz having a minimum axial ratio of 0.8 dB at 4.21 GHz.

In another example in [11], a multiple-point stub has been used in the microstrip line along with an asymmetrical ground plane for the formation of a degenerate orthogonal mode in the rectangular DRA, as shown in Figure 3.18 [11]. The wider ARBW in rectangular DRAs can be realized using a new radiating stub microstrip feed with an asymmetrical ground plane. The combination of three radiating stubs in the microstrip feed line and the asymmetrical ground plane is responsible for the generation of two different modes (i.e., $TE_{11\delta}$ and

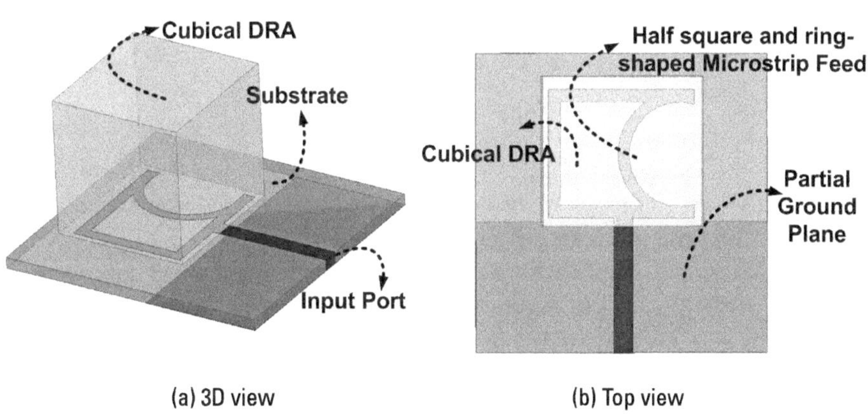

Figure 3.16 An asymmetrical half-square and ring-shaped microstrip coupled cubical DRA for wideband circular polarization. (Modified after [9].)

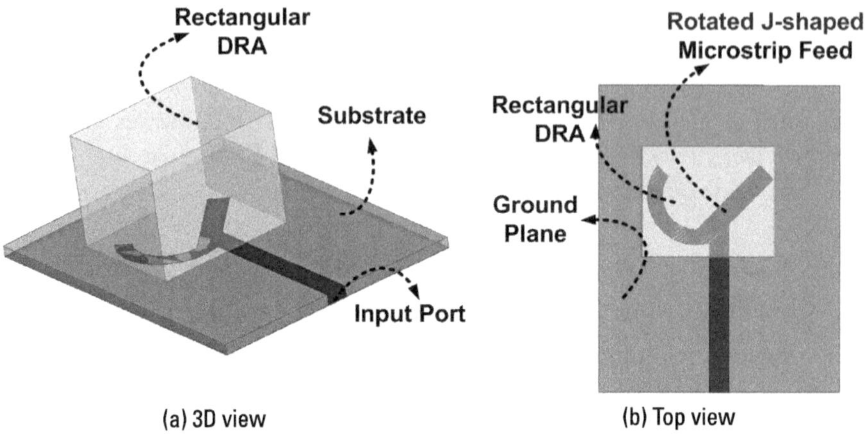

Figure 3.17 A rotated J-shaped microstrip coupled rectangular DRA for small band circular polarization. (Modified after [10].)

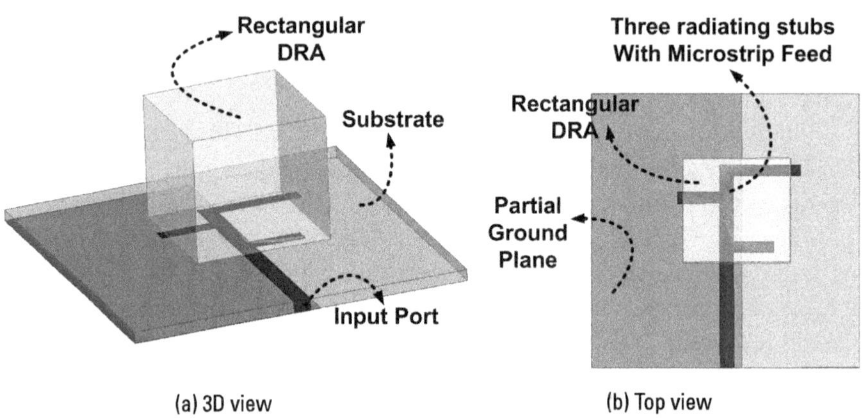

Figure 3.18 Three radiating stubs with microstrip coupled rectangular DRA for small band circular polarization [11].

$TE_{12\delta}$) in the rectangular DRA. The $TE_{11\delta}$ ($TE^x_{11\delta}$ and $TE^y_{11\delta}$) and $TE_{12\delta}$ ($TE^x_{12\delta}$ and $TE^y_{12\delta}$) modes are responsible for enhancing the impedance bandwidth by using the mode-matching concepts as well as the $TE_{11\delta}$ ($TE^x_{11\delta}$ and $TE^y_{11\delta}$) mode are responsible for achieving the degenerate orthogonal modes in the presented antenna configuration for circular polarization. Such antenna arrangements give a wider ARBW of 23.32% in the broadside direction.

In [12], the generation of the orthogonal degenerate mode is executed by a flag-shaped microstrip line network that is applied in a rectangular DRA. In this, the proposed design utilizes TE^x_{111} and TE^y_{111} orthogonal mode in the rectangular DRA for the generation of circularly polarized fields.

By truncating three notches from the flag-shaped microstrip line and applying them to excite the rectangular DRA, they drastically enhance the axial ratio bandwidth from 16.45% to 58.36% [12]. The truncated notches are simply the stubs that also create L-shape microstrip feed in this antenna structure, and along with partial ground plane, play an important role in achieving a broadband axial ratio bandwidth of 47.27% $\theta = 0^0$, $\varphi = 0^0$ (i.e., broadside direction). Analogous outcomes could be obtained by using the antenna structure, as illustrated in Figure 3.19.

3.3.3 Single-Point-Fed along with Parasitic Metallic Plates/Strip

As compared to the techniques presented in previous examples, another approach is also most applicable for generation of CP fields by using parasitic metallic plates/strips on the surface of a DR to reduce the excess electromagnetic fields as reported in [15–18]. The authors in [15] loaded a rectangular metallic patch with a suitable aspect ratio for obtaining the circular polarization in a low-profile square DRA. Improved outcomes may be achieved by using Figure 3.20. The degenerate orthogonal modes in the presented antenna scheme can easily be excited with various sizes of the rectangular loading patch on top of the square DRA. This design configuration is suitable for low-profile square DR antenna with a high relative permittivity of 79 for obtaining the CP fields. By simply selecting the suitable aspect ratio of loaded rectangular patch on a square DRA, the fundamental mode TE_{111} has been found, and it could be split into two orthogonal near-degenerate resonant modes for achieving the CP fields. The obtained 3-dB ARBW is about 25 MHz or 1.1% in broadside direction. In [17], circularly polarized broadband is obtained by using microstrip line feed and is placed at the right edge of a rectangular DRA; therefore, it excites or-

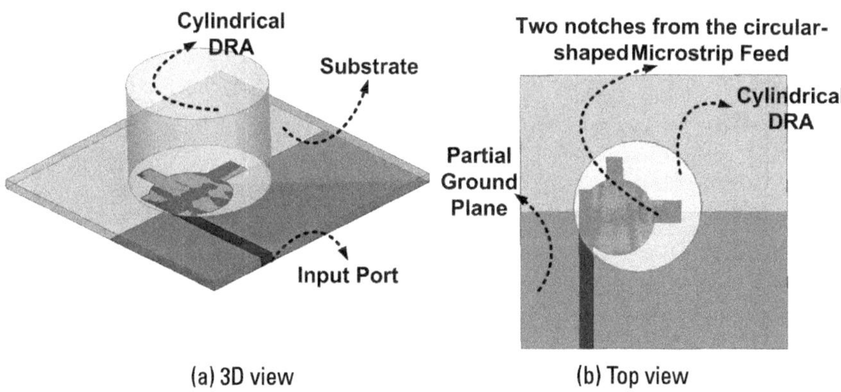

Figure 3.19 Two notches from the circular-shaped microstrip coupled cylindrical DRA for small band CP. (Modified after [12].)

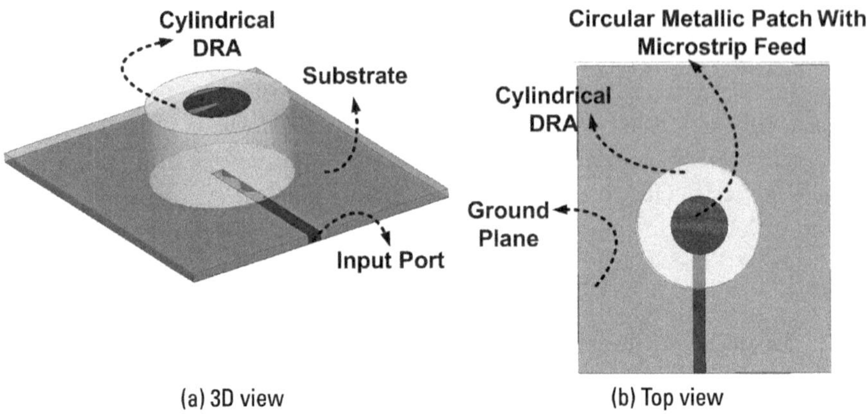

Figure 3.20 A circularly polarized low-profile cylindrical DRA with a loading circular patch and microstrip coupled for small band circular polarization. (Modified after [15].)

thogonal modes. The glueless concept has also been used to achieve mechanical stability and enhancement of the ARBW without affecting the radiation characteristics. This antenna configuration provides the 3-dB ARBW of 38.85% in broadside direction and covering 6.08 to 9.01 GHz. For achieving mechanical stability and enhancing the axial bandwidth, the glueless method along with two metallic patches and a pin are placed on both sides of the rectangular DRA. The pin is inserted inside the rectangular DRA and shorted it to metallic patch on the substrate. Figure 3.21 is also used to achieve similar results.

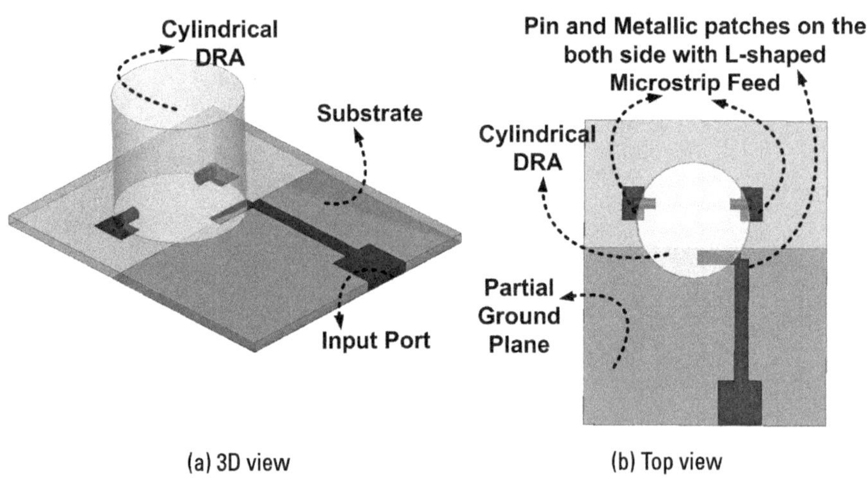

Figure 3.21 The glueless method along with two metallic patches and a pin inserted on both sides of a cylindrical DRA for broadband circular polarization. (Modified after [17].)

Table 3.1
A Brief Comparison of Various CP Designs Using the Single-Point-Fed Microstrip Line Technique in Conventional-Shaped DRAs

Ref	DRA Shape, Number of DRs (ε_r)	Microstrip Shape	IMBW (%)	ARBW (%)	f_c (GHz)	Modes Excited
[6]	Cubic, 1 (9.8)	Question-mark-shaped	35.35	20.62	3.14	$TE^x_{\delta 11}$ and $TE^y_{1\delta 1}$
[7]	Rectangular, 1 (9.8)	Meandered-line inductor	20.67	27.95	2.67	$TE^x_{\delta 11}$ and $TE^y_{1\delta 1}$
[8]	Rectangular, 1 (37)	Simple	8.1	1.8	5.18	$TE^x_{\delta 11}$ and $TE^y_{1\delta 1}$
[9]	Rectangular, 1 (9.8)	Asymmetrical square ring-shaped	55.18	11.61 and 5.84	4.53	TE_{111}
[10]	Cylindrical, 1 (38)	Y-shaped	6.8	1.18	4.21	Not mentioned
[11]	Rectangular, 1 (9.8)	Three radiating stubs	44.78	23.32	3.75	$TE_{11\delta}$ and $TE_{12\delta}$
[12]	Rectangular, 1 (9.8)	Truncating three notches from the flag shape	66.66	47.27	2.5	TE^x_{111} and TE^y_{111}
[13]	Square, 1 (9.8)	Notch is truncated from the rectangular patch	75.86	43.75	3.7	TE^x_{111} and TE^y_{111}
[14]	Rectangular, 1 (4.4)	Four-line feed using a circular loop type feed	11.76	3.89	2.5	$TE_{11\delta}$
[15]	Rectangular, 1 (79)	Loading of a rectangular patch	Not mentioned	1.1	2.22	TE_{111}
[16]	Cylindrical, 1 (20.8)	Conducting strip loaded on DRA top surface	16.6	3.14	2.89	TM_{110}
[17]	Rectangular, 1 (9.8)	Pin and metallic patch on the substrate and inserted pin of the DR are shorted to each other	63.92	38.85	8.51	Not mentioned
[18]	Cylindrical, 1 (9.8)	Modified J-shaped ground plane	41.44	29.91	3.74	$HEM_{11\delta}$

IMBW = fractional –10 dB impedance bandwidth, ARBW = fractional 3-dB axial ratio bandwidth, ε_r = dielectric constant of DRA, f_c = center frequency with regard to axial ratio plot.

A brief summary of all designs reported is provided in Table 3.1. Here, the conventional shape of a DRA has been used, which is excited by microstrip feeding techniques along with stubs or metallic parasitic strip. The antenna design step can be consolidated by seeing each configuration. Moreover, it is mainly based on four different steps for CP generation using microstrip feeding techniques:

1. The dimensions of the DRA have been chosen based on the resonance frequency by using a different formula for different convention shape of the dielectric resonator, which is given in Chapter 2;
2. For the generation of pure circular polarization, the path difference must be $\lambda/2$ or a 90° phase shift between two different linearly polarized signals, as given in (3.1);
3. Generally, microstrip feed will only provide a phase shift of 45° or a path difference of $\lambda/8$ distance, so it requires an additional 45° phase shift or $\lambda/8$ path difference to fulfill conditions of circular polarization, which is obtained by arranging the position of DRAs or directly using the stubs or metallic parasitic strips;
4. Therefore, generation of circularly polarized is totally based on the 90° phase shift or $\lambda/4$ path difference between the two signals, which can be obtained by arranging the different combinations like using stubs, parasitic element, and partial ground plane in the conventional shape of DRAs.

3.4 CP Generation in Nonconventional DRA Configurations

The techniques described in the previous examples use only the conventional shape of a single DRA for obtaining the CP fields in the respective antenna configurations. In this section, the nonconventional (modified shape) of a single DRA has been utilized for the generation of circular polarization in the respective antenna design systems. It is worth mentioning here that shape modification adds to the complexity of the fabrication and verification of electromagnetic modes. However, it offers some degree of freedom to the designer to design a circularly polarized antenna in several ways by using the three-dimensional geometry of the DRA. Further, this approach can also be divided into two categories for better understanding: single-point-fed with modified single DRA and dual-fed with modified single DRA.

3.4.1 Single-Point-Fed

The conventional to nonconventional shape of the dielectric resonator has been used for the generation of orthogonal modes in the DRA with the help of microstrip feeding techniques as reported in [19, 20]. In [19], the presented antenna design employs a hybrid concept for getting wideband circular polarization applications, and it is a combination of printed monopole antenna loaded with the dielectric resonator. The dielectric-loaded monopole antenna is used to provide enhancement of axial ratio bandwidth and impedance bandwidth along with good gain and high radiation efficiency compared to alone printed monopole antenna structure. Comparable antenna results could be also analyzed as shown in Figure 3.22.

Initially, the circular cylinder shape of the dielectric resonator is chosen to achieve circular polarization, after that a half-circular cylindrical shape of the dielectric resonator is used for size reduction, which is suitable for mobile/portable communication applications. Here, two different degenerate orthogonal $TE_{01\delta}$ and $HEM_{11\delta}$ modes have been generated, which are responsible for the generation of wide axial ratio bandwidth. Loading the dielectric resonator with a printed monopole antenna has achieved 3-dB axial ratio bandwidth of 35% covering the frequency band from 5.4 to 7.65 GHz.

The parasitically coupled to a printed rectangular loop-based wideband circularly polarized ring rectangular DRA was proposed in [20]. It has been constructed by using a thin dielectric substrate loaded with a ring rectangular DRA and excited a with narrow microstrip line.

A partial ground plane along with conducting loops and lines connected to the ground plane are placed on the other side of a thin dielectric substrate. Basically, the presented antenna configuration is linearly polarized, and the CP fields conducting loops are printed on the other side of the ring rectangular DRA. Hence, it produced the degenerate orthogonal modes with the required quadratic phase shift for the generation of circular polarization. The numerous loops are wrapped around the ring rectangular DRA that act as an open conducting ring that is parasitically coupled through a fringing capacitor. A second strip with an F-shaped microstrip that has been connected to the first strip and partial ground plane in order to increase the bandwidth, parallel outcomes may occur by using the antenna geometry as shown in Figure 3.23. Therefore, it forms two additional loops, which are very much useful for broadening the CP bandwidth as well as improving antenna gain. The final design has supported

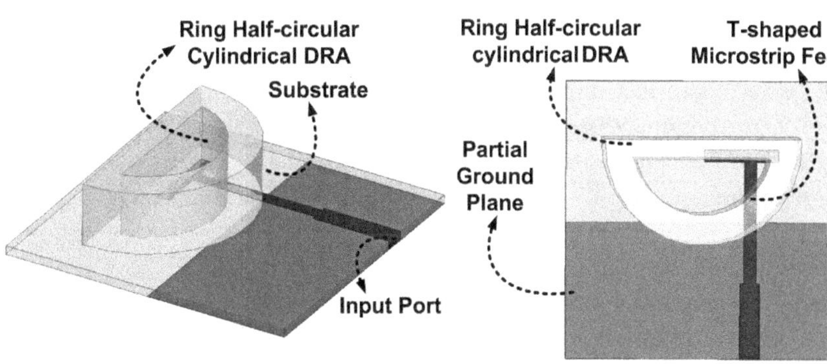

(a) 3D view (b) Top view

Figure 3.22 A hybrid monopole ring half-circular cylindrical DRA for wide circular polarization. (Modified after [19].)

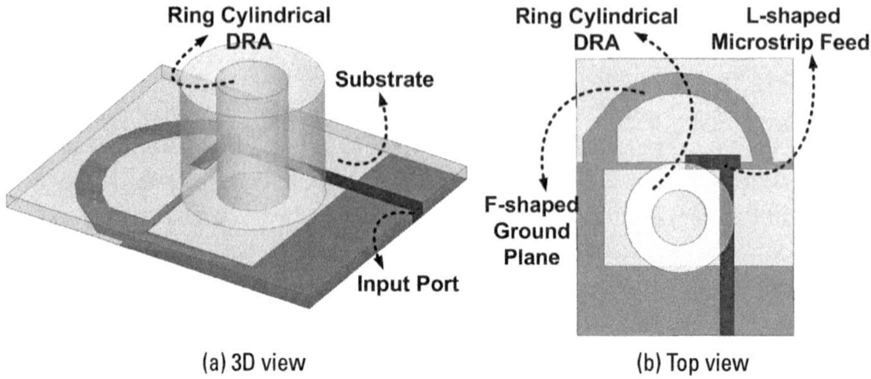

Figure 3.23 Parasitically coupled with F-shaped strip connected to a ground plane ring cylindrical DRA for broadband circular polarization. (Modified after [20].)

many wireless systems, such as Wi-MAX, WLAN, and Wi-Fi, and achieved 3-dB axial ratio bandwidth with about 51% from 3.4 to 5.75 GHz.

3.4.2 Dual-Point-Fed

In the previous sections, the antenna structures are based on single-point-fed microstrip lines with modified DRAs for the generation of orthogonal modes. However, there is another approach in which a dual-point feed can also be utilized for achieving degenerate orthogonal modes in DRAs. The dual-point feed is a combination of the same feeding techniques such as microstrip and microstrip, a combination of different feeding techniques such as microstrip and coaxial probe, or combinations of other feeding techniques. A dual-band antenna with two different radiation patterns has been achieved by using a single ring cylindrical DRA. Here, the authors have also discussed the concept of exciting two different modes independently in a single ring cylindrical DRA in [21]. The antenna design procedure is done in two steps: the first one is based on circular polarization and focuses on building the feeding network to excite the orthogonal degenerate $HEM_{11\delta}$ mode, while the second one handles the excitation technique to be used for the $TM_{01\delta}$ mode (linearly polarized) and improves the coupling by properly choosing the probe dimensions and position of feed. Figure 3.24 depicts similar types of structures that are suitable to achieve similar antenna results. The degenerate orthogonal $HEM_{11\delta}$ mode has been generated because of L-shaped microstrip lines of optimized width, which gives the quadrature-phase difference between the two signals and are printed on a circular substrate. For better coupling and desired radiating mode, the L-shaped microstrip lines are placed underneath the ring cylindrical DRA. The presented dual-band antenna configuration gives a 3-dB axial-ratio bandwidth of 3.2% in broadside direction (i.e., in principle axis).

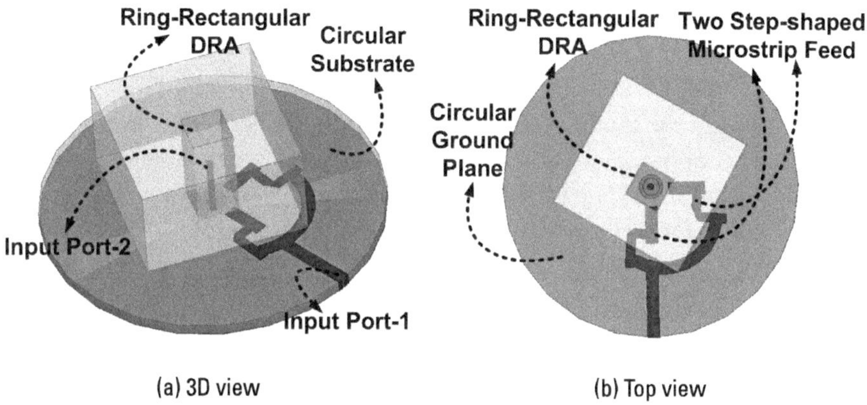

Figure 3.24 Two step-shaped microstrips coupled with a circular ground plane ring rectangular DRA for small band circular polarization. (Modified after [21].)

All the designs based on the modified shape DRAs are summarized in Table 3.2, and are excited by the microstrip feeding technique. It is easy to get the same amplitude and phase shift of 90° between the two signals by using the modified shapes of DRAs because the feed and sometimes the ground plane help to provide the CP condition of the antennas. By seeing the antenna design based on the modified shape and ground plane, it can be divided into four different steps for obtaining the CP fields, which are

1. The dimensions of the modified DRA have been chosen based on the optimizing process and use of the frequency range.

Table 3.2
A Brief Comparison of Various CP Designs Using Single-/Dual-Point-Fed Microstrip Techniques with Modified Shaped Single DRA

Ref	DRA Shape, Number of DRs (ε_r)	Microstrip Shape	IMBW (%)	ARBW (%)	f_c (GHz)	Modes Excited
[19]	Half-circular–cylindrical, 1 (10.2)	Hybrid printed monopole antenna	89.57	35	8.15	$TE_{01\delta}$ and $HEM_{11\delta}$
[20]	Ring rectangular, 1 (10.2)	Parasitically coupled with F-shaped strip connected to ground plane	54	51	4.67	Not mentioned
[21]	Ring cylindrical, 1 (38)	Two L-shaped and circular ground plane	7.34	3.2	3.13	$HEM_{11\delta}$

IMBW = fractional −10-dB impedance bandwidth, ARBW = fractional 3-dB axial ratio bandwidth, ε_r = dielectric constant of DRA, f_c = center frequency with regard to axial ratio plot.

2. For the generation of pure circular polarization, the path difference must be $\lambda/2$ or a 90° phase shift between two different linearly polarized signals, as given in (4.1).
3. Mostly the modified ground plane and microstrip feed line will provide a phase shift of 45° or path difference of $\lambda/8$ distance, and the remaining 45° phase shift or $\lambda/8$ path difference is given by the modified shape of the DRA.
4. Therefore, the combination of modified shape and microstrip feed along with ground plane gives a 90° phase shift or $\lambda/4$ path difference between the two signals. Hence, it gives the circularly polarized fields the respective antenna configurations, as reported in Table 3.2.

3.5 CP Generation in Multiple DRA Configurations

This section extends the previous single DRA works to multiple DRAs for circularly polarized design and its application in various wireless systems. Referring to the single DRA geometry, here the main idea is to reduce the equivalent permittivity of the dielectric material. The microstrip direct feed line is used to assist the integration of DRAs and help the realization of circular polarization through the feeding networks, although multiple DRAs can also be separated into two classes for better clarity: single-point-fed multiple DRAs and dual-point-fed multiple DRAs.

3.5.1 Single-Point-Fed

A hybrid multiband Z-shaped cylindrical DRA is proposed in [22] with a partial ground plane and excited with dual C-shaped microstrip feed lines. The dual C-shaped patch also acts as a radiator and supports excitation of $TE_{01\delta}$ mode in Z-shaped cylindrical DRA. The Z-shaped DRA is made of two cylindrical DRA, which comes from the offset between an upper and lower layer of a cylindrical DRA and is responsible for the reduction in quality factor. A similar antenna structure could be used for getting the same results as shown in Figure 3.25. This design of cylindrical DRAs have also produced $\pi/2$ phase shift between the field lines and the path difference of field lines between upper and lower cylindrical DRAs is $\lambda/4$, which generates circular polarization. The combination of dual C-shaped feed lines and Z-shaped cylindrical DRAs provide five different frequency bands and are capable of generating circular polarization in one of the bands. Here, $TE_{01\delta}$ mode is responsible for the generation of degenerate orthogonal mode and shows the 3-dB axial ratio bandwidth of

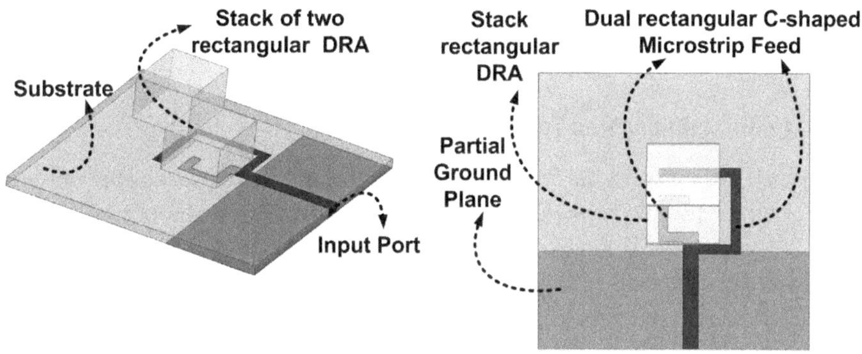

(a) 3D view (b) Top view

Figure 3.25 Dual rectangular C-shaped patch microstrip feed excited stack rectangular DRA for small band circular polarization. (Modified after [22].)

6.17% for the fifth band, which starts from 7.85–8.35 GHz in the direction of $\theta = 90°$ and $\varphi = 0°$.

In [23], the authors have proposed a hybrid dual-cylindrical DRA for the sub-6-GHz band, which is excited by an S-shaped microstrip feed for achieving wide circular polarization band.

Here, degenerate orthogonal $HE_{11\delta}$ mode is excited by utilizing the S-shaped microstrip feed line along with the dual-cylindrical DRA of the presented antenna, as shown in Figure 3.26. The wideband circular polarization has been constructed by using two annular shapes feeding networks loaded with two cylindrical DRAs. Because of the hybrid antenna nature, both the annular and single S-shaped microstrip feed line along with both the cylindrical DRAs are jointly radiating structures. Therefore, the hybrid dual-cylindrical DRAs

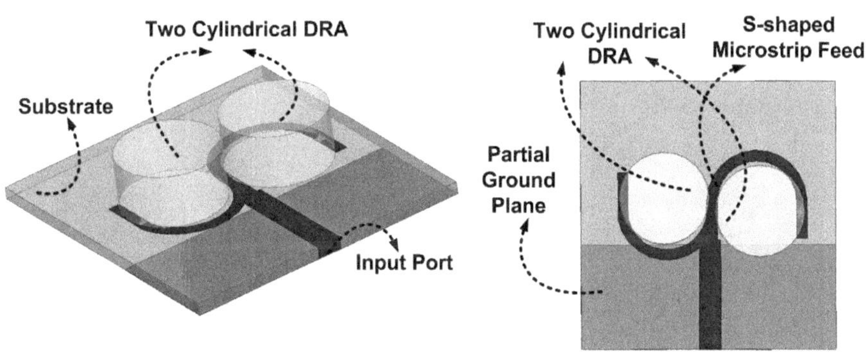

(a) 3D view (b) Top view

Figure 3.26 Dual cylindrical DRA excited with S-shaped microstrip feeding for circular polarization [23].

give the 3-dB axial ratio bandwidth of 15.2% covering the frequency range of 3.52 to 4.1 GHz in the broadside direction.

3.5.2 Dual-/Multiple-Point-Fed

Since the earth is spherical in shape, a circularly polarized spaceborne antenna has been very useful for the coverage of earth applications as reported in [24]. This is because the distance traveled from the satellite to various points on the earth are different and follow different path distances to cover various points, as well as being affected by atmospheric attenuation. To maintain uniformity effective signal strength on the surface of the earth, the radiation pattern of the antenna must be formed according to the earth's shape. To fulfill this gap a specially designed cavity-backed patch antenna loaded with two layers of a cylinder DRA has been presented. Similar results can be also achieved by using Figure 3.27.

The dielectric cylinder has a concave structure at the top; therefore, the refraction of the electromagnetic wave can be controlled and used for the shaping of the radiation pattern. The present antenna structure is excited by the sequentially feeding network with four probes and is composed of three Wilkinson power dividers, which provided the CP radiation pattern. It shows the 3-dB axial ratio bandwidth of 5.94%, which cover from 4.9 to 5.2 GHz in the elevation angle of ±50°.

All the design based on the multiple DRA configurations have been summarized in Table 3.3, excited by a different combination of the microstrip feed line. Using the modified microstrip feed line along with multiple DRAs for getting the circular polarization of reported structure is shown in Table 3.3. The configuration of antenna design is based on the multiple DRAs and modified

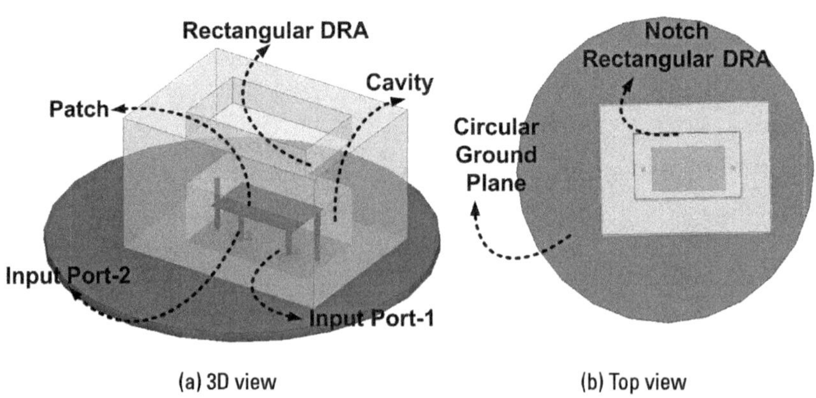

Figure 3.27 Cavity back notched rectangular DRA excited with dual feed for circular polarization. (Modified after [24].)

Table 3.3
A Brief Comparison of Various CP Designs Using Single-/Dual-Point-Fed Microstrip Techniques with Multiple DRAs

Ref	DRA Shape, Number of DRs (ε_r)	Microstrip Shape	IMBW (%)	ARBW (%)	f_c (GHz)	Modes Excited
[22]	Z-shaped/ two cylindrical, 2 (9.8)	Dual C-shaped patch microstrip feed	16.56	6.17	7.85	$TE_{01\delta}$
[23]	Cylindrical, 2 (9.8)	S-shaped microstrip feed	93.06	15.2	4.04	$HE_{11\delta}$
[24]	Dielectric cylinder with inverted frustum cone-shaped concave structure, 2 (2.1)	Sequentially feeding network with four probes and composed of three Wilkinson power dividers	37.15	46.0	5.06	Not mentioned

IMBW = fractional –10-dB impedance bandwidth, ARBW = fractional 3-dB axial ratio bandwidth, ε_r = dielectric constant of DRA, f_c = center frequency with regard to axial ratio plot.

microstrip feed along with the ground plane. Therefore, it can be divided into four different steps for obtaining the CP radiation patterns:

1. The dimensions of multiple DRAs have been chosen based on the optimizing process and uses of the frequencies.
2. For the generation of pure circular polarization, the path difference must be $\lambda/2$ or have a 90° phase shift between two different linearly polarized signals, as given in (3.1).
3. Mostly, the microstrip feed line along with ground plane provides a phase shift of 45° or a path difference of $\lambda/8$ distance, and another 45° phase shift or a $\lambda/8$ path difference is taken by the multiple DRAs.
4. Therefore, the combination of the multiple DRAs and modified microstrip feed with ground plane provides a 90° phase shift or a $\lambda/4$ path difference between the two signals. Hence, it shows the circularly polarized waves in the respective design antenna.

3.6 Microstrip Coupling in MIMO-DRA Configurations for CP

Due to high data rate demands, large-signal capacity, link reliability, and stability without increasing the channel bandwidth, MIMO is becoming the most attractive technology currently in wireless communication systems. The promising configuration of MIMO technology is that it can use transmitter and receiver multiple uncorrelated signals with equal power levels and be able to reduce the effects of cochannel as well as adjacent channel interferences. Initially, MIMO antennas are employed using microstrip technology, but at higher frequencies the conductor losses are very high and the radiation efficiency also decreases. Due to these reasons the dielectric resonant antenna is more preferred, which utilize a dielectric material as a radiator. Therefore, the diversity performances of the MIMO antennas can be realized by following parameters: envelope correlation coefficient (ECC), diversity gain (DG), mean effective gain (MEG), total active reflection coefficient (TARC), and channel capacity loss (CCL), respectively [25–27].

The circularly polarized based MIMO DRA has been reported in [28, 29] and excited by a microstrip feed line. A cylindrical DRA along with hybrid MIMO configuration having high isolation has been presented [28] for triple-band applications. The presented design is based on hybrid antenna structure and capabilities for improving the impedance matching as well as enhancing the axial ratio bandwidth. This MIMO DRA consists of two symmetric folded microstrip feeding structures on top of the substrate at two different ends to excite the two cylindrical DRAs. The two inverted L-shaped strips are connected on the ground plane, which is the other side of the microstrip feed line for enhancement of the isolation between two ports. Similar results can be also achieved by using Figure 3.28.

The metallic plate is placed in the ground plane between the two cylindrical DRAs, which act as an electromagnetic reflector for enhancing the isolation between two antennas. The generation of circular polarization depends on the open stub-folded microstrip lines, which generate a $\lambda/4$ path distance or a $\pi/2$ phase shift between the electric field lines. The presented MIMO configuration gives the axial ratio bandwidth 3.55% in the frequency range of 5.55–5.75 GHz. The design MIMO DRA shows good performances in the envelop correlation coefficient, mean effective gain, and effective diversity gain, respectively.

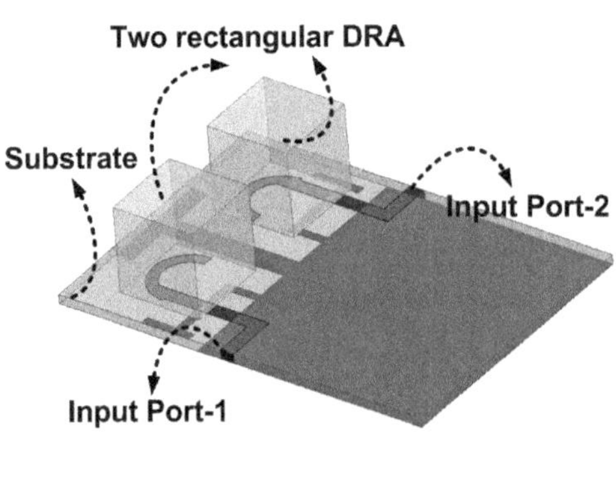

(a) 3D view

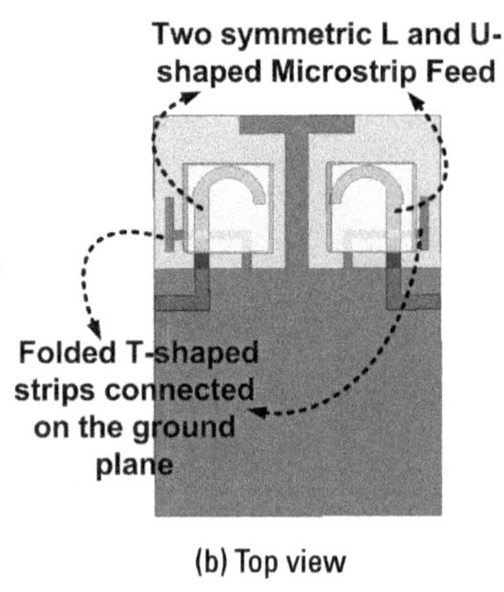

(b) Top view

Figure 3.28 Symmetric L- and U-shaped microstrip feeds with a folded T-shaped excited cylindrical MIMO DRA for small band circular polarization. (Modified after [28].)

In [29], two cylindrical DRAs have been introduced for dual-polarized triple-band hybrid MIMO antenna application. Mainly, the MIMO DRAs comprise a pair of two symmetrical modified Y-shaped microstrip feeding lines, two cylindrical DRAs, and two metallic strips. Similar antenna results could also be generated as presented in Figure 3.29.

The two metallic strips and a slot on the ground plane are used to improve the isolations between two ports, which are made of some branches with the same widths and lengths. The generation of circular polarization totally depends on the complete antenna structure, which is capable of exciting orthogonal modes. The $HEM_{11\delta}$ modes are simply the orthogonal degenerate modes for this antenna structure and are responsible for obtaining CP fields. A slot and two metallic strips are introduced on the ground plane to get better diversity performance and also improve the isolations between the two ports. The presented antenna is based on a two-port microstrip feeding system, and metallic strips act as electromagnetic reflectors to achieve high isolation. Hence, the MIMO DRA shows that the 3-dB axial ratio bandwidths of the upper-frequency band is 4.18%. It also shows the better MIMO performance in terms of diversity parameters like ECC (less than 0.04), MEG (less than 3 dB), DG (near about 10 dB), and CCL(less than 0.4 b/s/Hz). The presented MIMO cylindrical DRA with two identical antenna elements are used for WLAN and WiMAX applications. A summary based on the circularly polarized MIMO DRAs is given in Table 3.4, which are excited by a different arrangement of the microstrip feed line. By using the altered microstrip feed line along with DRAs, they are capable of producing a degenerate orthogonal mode in the antenna design structure.

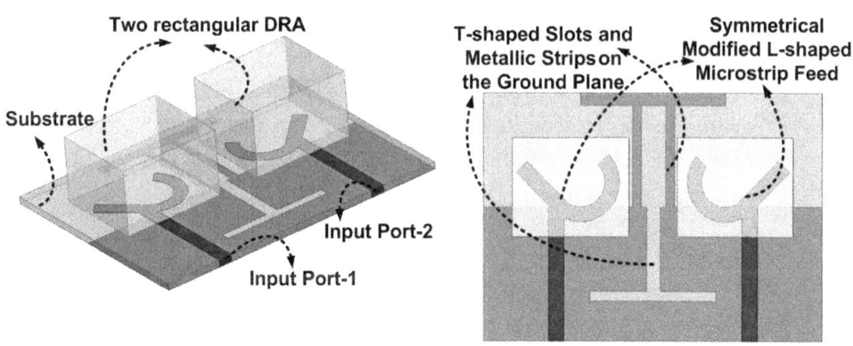

(a) 3D view (b) Top view

Figure 3.29 L-shaped microstrip feed with T-shaped slot and metallic strips introduced on the ground plane to excited cylindrical MIMO DRAs for small band circular polarization. (Modified after [29].)

Table 3.4
A Brief Comparison of MIMO CP Designs Using the Microstrip Technique with DRAs

Ref	DRA Shape, Number of DRs (ε_r)	Microstrip Shape	IMBW (%)	ARBW (%)	f_c (GHz)	Modes Excited
[28]	Two cylindrical, 2 (9.8)	Two symmetric folded microstrip feeding along with two inverted L-shaped strips have connected on the ground plane	35.69	3.55	5.62	$HE_{11\delta}$
[29]	Two cylindrical, 2 (9.8)	Symmetrical modified Y-shaped microstrip feeding line along with a slot and two metallic strips are introduced on the ground plane	14	4.18	5.70	$HEM_{11\delta}$

IMBW = fractional −10-dB impedance bandwidth, ARBW = fractional 3-dB axial ratio bandwidth, ε_r = dielectric constant of DRA, f_c = center frequency with regard to axial ratio plot.

The process of antenna design is based on using multiple DRAs and different combinations of microstrip feed lines along with the ground plane. Thus, it can be separated into four different parts for obtaining the CP waves, which are

1. The dimensions of different shapes of DRAs are chosen based on the optimizing process and uses of the frequency bands.
2. For the generation of exact circular polarization wave, the path difference should be a $\lambda/2$ or 90° phase shift between two linearly polarized signals, as given in (3.1).
3. In MIMO configuration, the different combinations of microstrip feed and ground plane provide a path difference of $\lambda/8$, and different DRAs obtain another path difference of $\lambda/8$.
4. Hence, the final combination of the MIMO DRAs and combination of microstrip feed with ground plane gives a $\lambda/4$ path difference between the two signals. Therefore, it shows circularly polarized fields in the respective MIMO DRAs.

3.7 Microstrip Coupling in Array Configurations for CP

The rapid growth of modern wireless communication technologies, such as 5G and millimeter-wave applications, increases the demand for the circularly polarized antenna array, which offers the advantages of high gain, low profile, and immunity to multipath fading, and therefore is well suited for mobile communication, remote control, satellite, and radar systems. In the array antenna, the feed network plays an important role and determines the overall performance of the antenna. Hence, it can obtain the radiation pattern of the antenna with the desired specifications and polarization [30, 31]. The available standard formulas can be found from the radiation pattern and array factor of an array of N-identical elements. The broadband circularly polarized array DRA in [32] consists of 4 and 16 elements (i.e., 4 × 4 configurations). This is the minimum number of elements required to achieve the desired axial ratio bandwidth, beamwidth, and desired sidelobe levels. The array arrangement of the DRA pair-unit is used for the construction of the subarray; a similar kind of structure is shown in Figure 3.30. Therefore, each element of this pair-unit has orthogonal feed points with uniform distribution of power levels and are responsible for the generation of circular polarization. The constructed subarray shows approximately five times the axial ratio bandwidth compared to a single circularly polarized DRA element. After that, four subarrays are constructed with a 4 × 4 element, which is also excited by a power divider mounted on the rear panel of the array. Therefore, it shows approximately five times larger ARBW compared to an ordinary array. The presented planar array consisting of a DRA pair-unit shows the ARBW of 15% in boresight direction.

In [33], a high-gain circularly polarized rectangular DRA array has been proposed for microwave image sensing, which is excited by a folded microstrip line. In this array structure, two microstrip lines have been inserted below the rectangular DRA for exciting the degenerate orthogonal modes that are responsible for the CP fields. A power divider is also incorporated with the quarter-wavelength transformer to feed the DRA. The microstrip line has been inserted

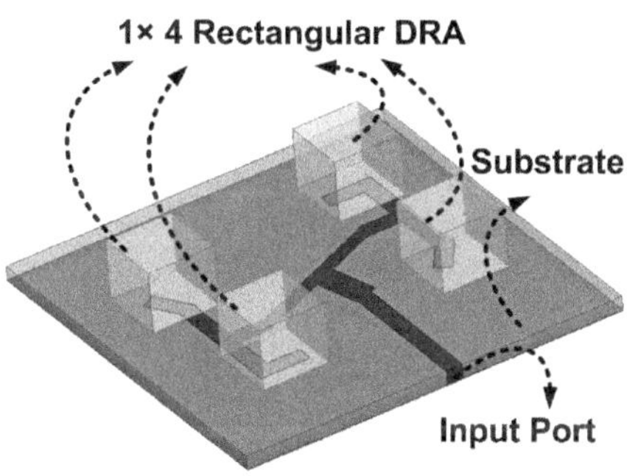

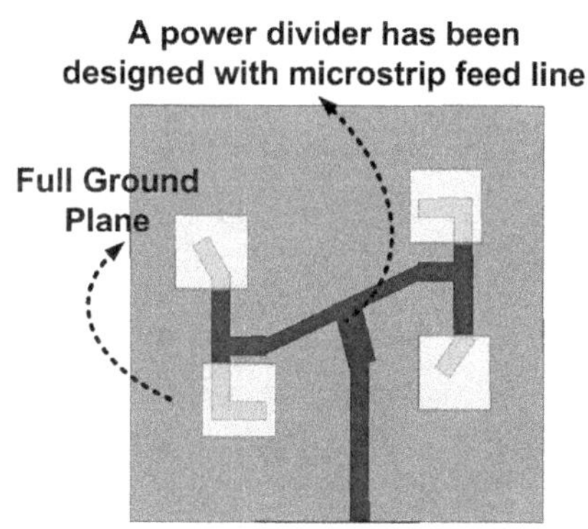

Figure 3.30 A power divider has been designed with the quarter-wavelength transformer to excite a modified rectangular DRA array for wideband circular polarization. (Modified after [32].)

horizontally below the rectangular DRA, which add an extra path length of $\lambda + \lambda/4$ ($360° + 90°$ phase shift) compared to a vertically inserted microstrip line below the rectangular DRA. The $\lambda/4$ length is not sufficient to feed the horizontal DRA for this configuration and hence, $\lambda + \lambda/4$ length has been suitable for this design to feed the horizontal DRA elements. Circular polarization is achieved in this configuration by exciting two orthogonal TE_{111} modes with a 90° phase shift within the rectangular DRA. After that, a 1 × 4 DRA array has been designed by using the same dimensions of all four DRA elements and a $\lambda/4$ transformer is used to have equal power distribution to all output ports. The same antenna results can be analyzed by using Figure 3.31. The distance between the DRA elements is kept fixed so that mutual coupling between the two DRA elements is very small (<–22 dB) over the whole operating band. The circularly polarized DRA array gives the 3-dB axial ratio bandwidth of 10% (6.2–6.85 GHz) in the broadside direction ($\theta = 0°$, $\varphi = 0°$).

Table 3.5 provides a complete summary of circularly polarized array DRAs, which are excited by different power dividers connected through the microstrip feed line. In these designs, the power divider with a quarter-wavelength transformer is used to get orthogonal degenerate modes in the rectangular DRA for the generation of CP fields. The array antenna has been designed by the combination of the microstrip feed line, ground plane, and the rectangular DRA. Hence, the design process for getting CP is divided into four steps:

1. The dimensions of different shapes of DRAs have been chosen based on the optimizing process and selection of the frequency bands.
2. For the generation of exact CP wave, the path difference should be a $\lambda/2$ or 90° phase shift between two linearly polarized signals, as given in (3.1).

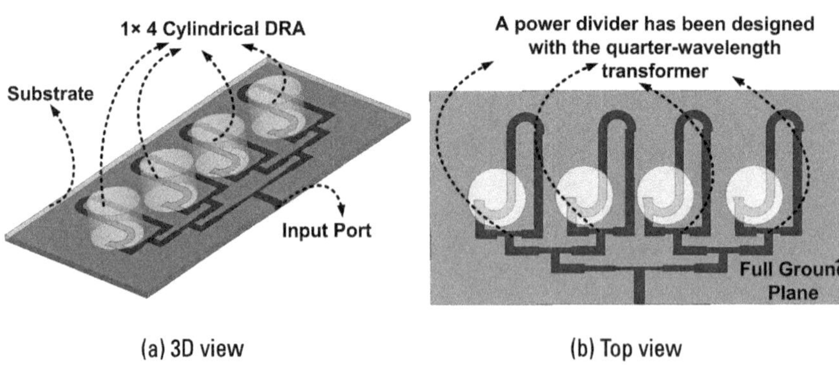

(a) 3D view (b) Top view

Figure 3.31 A power divider has been designed with the quarter-wavelength transformer to excite a cylindrical DRA array for wideband circular polarization. (Modified after [33].)

Table 3.5
A Brief Comparison of Array CP Designs Using Microstrip Techniques with DRAs

Ref	DRA Shape, Number of DRs (ε_r)	Microstrip Shape	IMBW (%)	ARBW (%)	f_c (GHz)	Modes Excited
[32]	Hexagonal, 1×4, 4×4 (9.4)	Power divider is mounted on the microstrip feeding	Not mentioned	15	Not mentioned	Not mentioned
[33]	Rectangular, 1×4 (20)	A power divider has been designed with the quarter-wavelength transformer to feed the DRA	17	10	6.77	TE_{111}

IMBW = fractional –10-dB impedance bandwidth, ARBW = fractional 3-dB axial ratio bandwidth, ε_r = dielectric constant of DRA, f_c = center frequency with regard to axial ratio plot.

3. In an array configuration, a power divider has been designed with the quarter-wavelength transformer to feed the DRAs for obtaining CP fields.
4. Therefore, the final array DRA design depends on the power divider along with if the ground plane gives a $\lambda/4$ path difference between the two signals. Therefore, it gives circularly polarized fields in the respective array DRAs.

3.8 Microstrip Coupling in Filtering DRA Configurations

Currently, the filtering antenna that accommodates the filter and antenna together is in rapid development in modern wireless communication. This is because of the increasing demand to enhance the performance of the RF front end in a system and help the miniaturization as well as the size reduction of the system. According to their design methods, the filtering antennas can be separated into three different categories. The first approach is the simplest one and utilizes the antenna feedline to achieve a filtering response. The second approach utilizes the coupling matrix theory to design filtering antenna by using a traditional filter synthesis method [34, 35]. Here, the antenna works both as a radiator and resonator of a bandpass filter (BPF). In the latter approach, the parasitic elements of the patch antenna have been introduced, such as shorting pins, slots, and stacked patches. Due to the parasitic elements, the antenna currents could be made of equal magnitude but out of phase at different frequencies, which cancels the radiated fields. Hence, radiation nulls are obtained. The third method can also be suitable for the DRA to give the filtering DRA. A wideband circularly polarized DRA with filtering response has been reported in [36]. A filtering quadrature coupler has been used to feed a hollow DRA for the realization of filtering and harmonic suppression functions as well as it gives circular polarization. The filtering CP antenna is presented, comprising the hollow rectangular DRA and underlying patch quadrature coupler, which are arranged within the hollow region of the DRA for obtaining compactness; similar results can be analyzed as shown in Figure 3.32. The filtering quadrature coupler is based on a snowflake-shaped patch where operating modes are flexibly controlled by the loading slots. For getting excellent bandpass filtering, wide bandwidth, and wide harmonics suppression characteristics, quarter wavelength coupled line sections have been implemented to the coupler. The hollow DRA has the following advantages when designing filtering antenna. First, the hollow region decreases the Q-factor and effective dielectric constant and results enhance the impedance bandwidth. Second, the hollow region is also provided with the space to locate the underlying coupler. Therefore, this design has been combined with the quadrature filtering coupler for the realization of wideband filtering circularly polarized DRAs. The bandpass filtering circularly polarized DRA gives the wide 3-dB axial ratio bandwidth of 27.8% in $\theta = 0°$ $and \varphi = 0°$ plane. A brief summary of filtering circularly polarized DRAs using modified microstrip technique is shown in Table 3.6. In the DRAs, the filtering antenna is a new concept along with the generation of circularly polarized radiation in the structure. Very few research articles have been published about the circularly polarized technique, which also is not really part of the microstrip feeding technique.

There are three ways to implement the filtering antenna, which were discussed earlier, in which the third method is most suitable for the DRAs. Here, the underlying patch quadrature coupler microstrip technique is used for

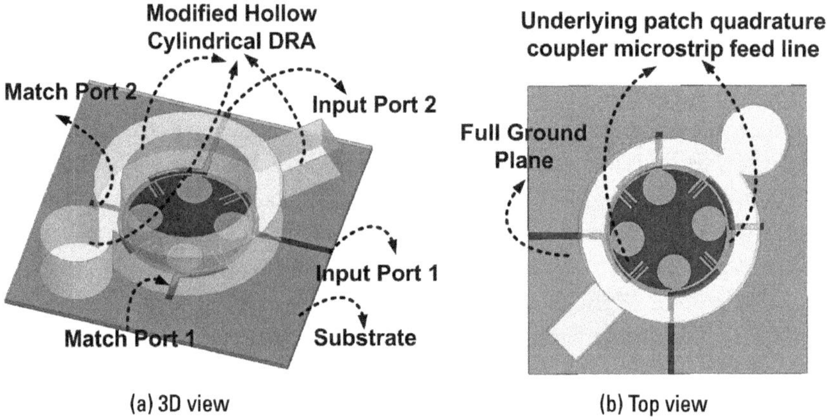

(a) 3D view (b) Top view

Figure 3.32 An underlying patch quadrature coupler to excite modified hollow cylindrical DRA filter for wideband circular polarization. (Modified after [36].)

Table 3.6
The Filtering CP Designs Using the Microstrip Technique with DRAs

Ref	DRA Shape, Number of DRs (ε_r)	Microstrip Shape	IMBW (%)	ARBW (%)	f_c (GHz)	Modes Excited
[36]	Modified hollow rectangular, 1 (9.7)	Underlying patch quadrature coupler microstrip feeding	27.8	27.8	1.85	Not mentioned

IMBW = fractional −10-dB impedance bandwidth, ARBW = fractional 3-dB axial ratio bandwidth, ε_r = dielectric constant of DRA, f_c = center frequency with regard to axial ratio plot.

obtaining the degenerate orthogonal modes in a modified hollow rectangular DRA for getting the circularly polarized field. The filtering circularly polarized DRAs can be designed by the arrangement of microstrip excitation techniques such as power divider or quadrature coupler, or the modified ground plane along with the modified shape of the DRAs. Thus, the implementation procedure is also separated into four different parts for achieving the filtering CP DRAs:

1. The dimensions of different shapes of DRAs have been chosen based on the optimizing process and uses of the frequency bands.
2. For the generation of pure circular polarization radiation, the path difference must be a $\lambda/2$ or 90° phase shift between two linearly polarized signals, as given in (3.1).
3. In filtering arrangements, a quadrature coupler or power divider has been used for obtaining the quarter-wavelength for getting CP fields.
4. Hence, the final filtering DRAs scheme depends on the quadrature coupler or power divider along with the ground plane to get the $\lambda/4$ path difference. Therefore, it generates circularly polarized fields in the respective filtering DRAs.

3.9 Summary

In this chapter, the various approaches for designing circularly polarized DRAs were studied and analyzed based on the microstrip feeding technique. Different shapes of DRA were chosen based on the performance of the respective designs and achieving circular polariz ation radiation by using different combinations of microstrip feeding techniques (e.g., simple divider, stubs, parasitic metallic patch or strip, power divider, and quadrature coupler). Hence, researchers were able to design various combinations of circularly polarized DRAs for single-

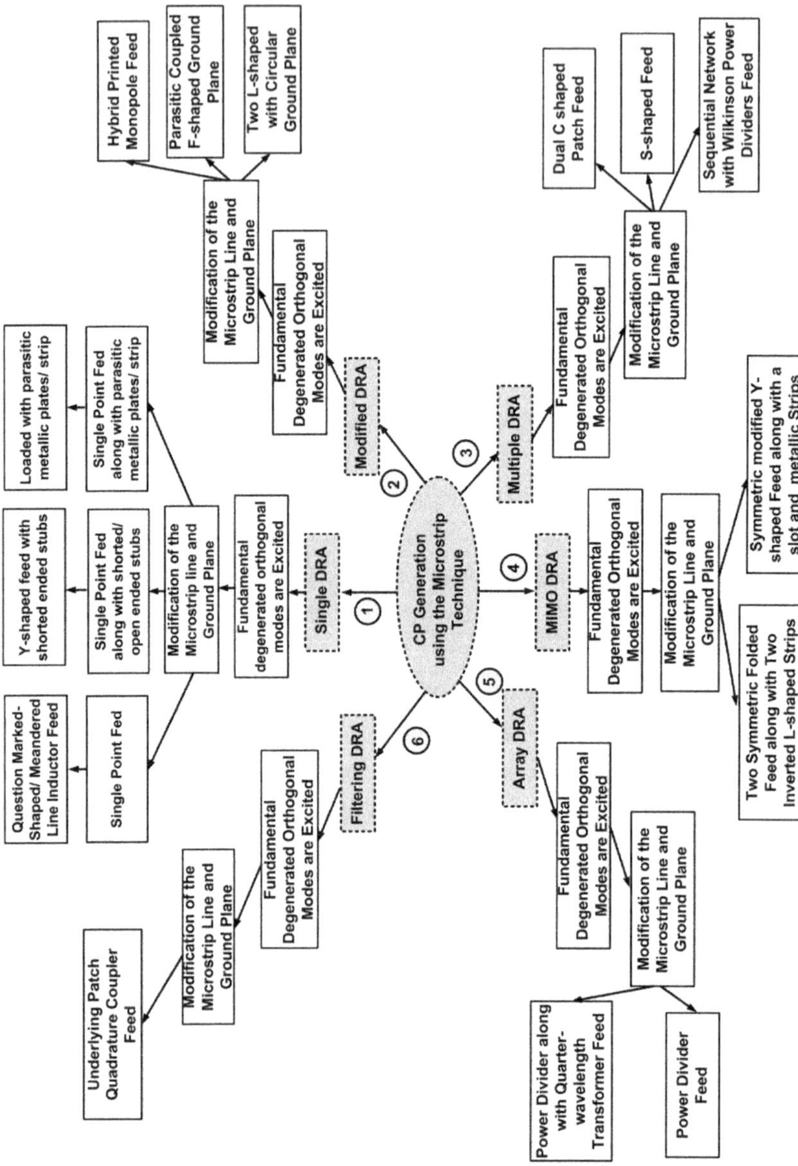

Figure 3.33 Complete outline of the microstrip coupling technique used for generation of CP in DRAs.

band, dual- and triple-band, MIMO, array, and filtering applications. Initially, the design methodology of the circularly polarized DRA was included in this chapter for simplification and better overall understanding. This chapter began with the single DRA modification and goes to multiple DRA configurations, which iare excited by the microstrip feeding line. To generate the circular polarization in these DRAs, the stubs (shorted or open-ended) and parasitic metallic (plates or strips) have been utilized along with the microstrip feeding line. Likewise, modified shaped and multiple DRAs along with single or double microstrip feeding lines were used for getting circular polarization. Then these microstrip techniques were expanded and applied in MIMO, array, and filtering DRA structures for obtaining circular polarization fields, which were detailed in separate sections. A complete outline of the microstrip coupling technique used in CP DRAs is shown in Figure 3.33. Each section of this chapter included a table for better understanding and quick review of the generation of circular polarization and its effect on antenna parameters.

References

[1] Petosa, A., *Dielectric Resonator Antenna Handbook*, Norwood, MA: Artech House, 2007.

[2] Luk, K M., and K. W. Leung, *Dielectric Resonator Antenna*, Baldock, U.K.: Research Studies Press, 2003.

[3] Balanis, C. A., *Antenna Theory: analysis and Design*, Hoboken, NJ: John Wiley & Sons, 2005.

[4] Patel, P., B. Mukherjee, and J. Mukherjee, "Wideband Circularly Polarized Rectangular Dielectric Resonator Antennas Using Square-Shaped Slots," *IEEE Transactions on Antennas and Propagation*, Vol. 15, 2016, pp. 1309–1312.

[5] Fakhte, S., H. Oraizi, R. Karimian, and R. Fakhte, "A New Wideband Circularly Polarized Stair-Shaped Dielectric Resonator Antenna," *IEEE Transactions on Antennas and Propagation*, Vol. 63, No. 4, 2015, pp. 1828–1832.

[6] Kumar, R., and R. K. Chaudhary, "A Wideband Circularly Polarized Cubic Dielectric Resonator Antenna Excited with Modified Microstrip Feed," *IEEE Antennas and Wireless Propagation Letters*, Vol. 15, 2016, pp. 1285–1288.

[7] Kumar, R., and R. K. Chaudhary, "Wideband Circularly Polarized Dielectric Resonator Antenna Coupled with Meandered-Line Inductor for ISM/WLAN Applications," *International Journal of Microwave and Millimeter-Wave Computer-Aided Engineering*, Vol. 27, No. 7, 2017, pp. 1–7.

[8] Zhang, B., J. Zhang, C, Liu, and Zhi Wu, "Investigation of Microstrip-Fed Low-Profile Broadband Circularly Polarized Rectangular Dielectric Resonator Antennas," *Microwave and Optical Technology Letters*, Vol. 59, No. 11, 2017, pp. 2767–2772.

[9] Sahu, N. K., A. Sharma, and R. K. Gangwar, "Design and Analysis of Wideband Composite Antenna with Dual-sense Circular Polarization Characteristics," *Microwave and Optical Technology Letters*, Vol. 60, No. 8, 2018, pp. 2048–2054.

[10] Li, L. X., et al., "Circularly Polarized Ceramics Dielectric Resonator Antenna Excited by Y-Shaped Microstrip," *Microwave and Optical Technology Letters,* Vol. 51, No. 10, 2009, pp. 2416–2418.

[11] Kumar, R., R. Chowdhury, and R. K. Chaudhary, "A New Dual-Mode Wideband Circularly Polarized Rectangular Dielectric Resonator Antenna Coupled with Stubs and Asymmetric Ground Plane for WiMAX Applications," *International Journal of RF and Microwave Computer-Aided Engineering,* Vol. 27, No. 9, 2017, p. 21150.

[12] Kumari, R., and R. K. Gangwar, "Circularly Polarized Rectangular Dielectric Resonator Antenna Fed by a Flag Shape Microstrip Line for Wideband Applications," *Microwave and Optical Technology Letters,* Vol. 60, No. 10, 2018, pp. 2577–2584.

[13] Kumari, R., and R. K. Gangwar, "Wideband Circularly Polarized Square Dielectric Resonator Antenna for WLAN/WiMAX Applications," *International Journal of RF and Microwave Computer-Aided Engineering,* Vol. 29, No. 2, 2019, p. 21492.

[14] Singhwal, S. S., B. K. Kanaujia, A. Singh, and Jugul Kishor, "Novel Circularly Polarized Dielectric Resonator Antenna for Microwave Image Sensing Application," *Microwave and Optical Technology Letters,* Vol. 61, No. 7, 2019, pp. 1821–1827.

[15] Hsiao, F. R., T. W. Chiou, and K. L. Wong, "Circularly Polarized Low-Profile Square Dielectric Resonator Antenna with a Loading Patch," *Microwave and Optical Technology Letters,* Vol. 31, No. 3, 2001, pp. 157–159.

[16] Kumar, A. V. P., V. Hamsakutty, J. Yohannan, and K. T. Mathew, "A Strip Loaded Dielectric Resonator Antenna for Circular Polarisation," *Microwave and Optical Technology Letters,* Vol. 48, No. 7, 2005, pp. 1354–1356.

[17] Kumar, A. V. P., V. Hamsakutty, J. Yohannan, and K. T. Mathew, "Broadband Circularly Polarized Edge Feed Rectangular Dielectric Resonator Antenna Using Effective Glueless Technique," *Microwave and Optical Technology Letters,* Vol. 62, No. 10, 2020, pp. 3333–3341.

[18] Kumar, R., and R. K. Chaudhary, "A New Bidirectional Wideband Circularly Polarized Cylindrical Dielectric Resonator Antenna using Modified J-shaped Ground Plane for WiMAX/LTE Applications," *Radioengineering,* Vol. 28, No. 2, 2019, pp. 391–398.

[19] Haraz, O. M., A. R. Sebakand, and T. A. Denidni, "A Hybrid Printed Monopole Antenna Loaded with Dielectric Resonator for Wideband and Circular Polarization Applications," *International Journal of RF and Microwave Computer-Aided Engineering,* Vol. 22, No. 5, 2012, pp. 588–593.

[20] Khalily, M, M. K. A. Rahim, and A. A. Kishk, "Planar Wideband Circularly Polarized Antenna Design with Rectangular Ring Dielectric Resonator and Parasitic Printed Loops," *IEEE Antennas and Wireless Propagation Letters,* Vol. 11, 2012, pp. 905–908.

[21] Hady, L. K., A. A. Kishk, and D. Kajfez, "Dual-Band Compact DRA with Circular and Monopole-Like Linear Polarizations as a Concept for GPS and WLAN Applications," *IEEE Transactions on Antennas and Propagation,* Vol. 57, No. 9, 2009, pp. 2591–2597.

[22] Sharma, A., and R. K. Gangwar, "Circularly Polarised Hybrid Z-shaped Cylindrical Dielectric Resonator Antenna for Multiband Applications," *IET Microwaves, Antennas & Propagation,* Vol. 10, No. 12, 2016, pp. 1259 –1267.

[23] Sani, M. M., R. Chowdhury, and R. K. Chaudhary, "Hybrid Wideband Dual-cylindrical Circularly Polarized Dielectric Resonator Antennas Excited with S-shaped Microstrip Feed for Sub-6 GHz Frequency Range," *International Journal of RF and Microwave Computer-Aided Engineering,* Vol. 30, No. 2, 2020, pp. 01–11.

[24] Ren, X., S. Liao, and Q. Xue, "A Circularly Polarized Spaceborne Antenna with Shaped Beam for Earth Coverage Applications," *IEEE Transactions on Antennas and Propagation,* Vol. 67, No. 4, 2019, pp. 2235–2241.

[25] Zhou, Q., H. Dai, "Joint Antenna Selection and Link Adaptation for MIMO Systems," *IEEE Trans. Veh. Technol.*, Vol. 55, 2006, pp. 243–255.

[26] Reddy, T. S., R. Kumar, and R. K. Chaudhary, "Isolation Enhancement and Radar Cross Section Reduction of MIMO Antenna with Frequency Selective Surface," *IEEE Transactions on Antennas and Propagation,* Vol. 66, No. 3, 2018, pp. 1595–1600.

[27] Zou, L., D. Abbott, and C. Fumeaux, "Omnidirectional Cylindrical Dielectric Resonator Antenna with Dual Polarization," *IEEE Antenn Wirel. Propag. Lett.,* Vol. 11, 2012, pp. 515–518.

[28] Sharma, A., G. Das, and R. K. Gangwar, "Dual Polarized Triple Band Hybrid MIMO Cylindrical Dielectric Resonator Antenna for LTE2500/WLAN/WiMAX Applications," *International Journal of RF and Microwave Computer-Aided Engineering,* Vol. 26, No. 9, 2016, pp. 763–772.

[29] Sahu, N. K., G. Das, and R. K. Gangwar, "Dual Polarized Triple-band Dielectric Resonator Based Hybrid MIMO Antenna for WLAN/WiMAX Applications," *Microwave and Optical Technology Letters,* Vol. 60, No. 4, 2018, pp. 1033–1041.

[30] Birand, M. T., and R. V. Gelsthorpe, "Experimental Millimetric Array Using Dielectric Radiators Fed by Means of Dielectric Waveguides," *Electron. Lett.,* Vol. 17, 1981, pp. 633–635.

[31] Asem, S., A.-Zoubi, A. A. Kishk, and A. W. Glisson, "A Linear Rectangular Dielectric Resonator Antenna Array Fed by Dielectric Image Guide with Low Cross Polarization," *IEEE Transactions on Antennas and Propagation,* Vol. 58, No. 3, 2010, pp. 697–705.

[32] Haneishi, M., and H. Takazawa, "Broadband Circularly Polarised Planar Array Composed of A Pair Of Dielectric Resonator Antennas," *Electronics Letters,* Vol. 21, No. 10, 1985, pp. 437–438.

[33] Rana, B., and S. K. Parui, "Microstrip Line Fed Wideband Circularly-Polarized Dielectric Resonator Antenna Array for Microwave Image Sensing," *IEEE Sensors Letters,* Vol. 1, No. 3, , 2017.

[34] Choudhary, D. K., and R. K. Chaudhary, "Compact Lowpass and Dual-band Bandpass Filter with Controllable Transmission Zero/Center Frequencies/Passband Bandwidth," *IEEE Transactions on Circuits and Systems II: Express Briefs,* Vol. 67, No.6, 2020, pp. 1044–1048.

[35] Chuang, C. T., and S. J. Chung, "A Compact Printed Filtering Antenna Using a Ground-Intruded Coupled Line Resonator," *IEEE Trans. Antennas Propag.*, Vol. 59, No. 10, 2011, pp. 3630–3637.

[36] Xiang, B. J., S. Y. Zheng, Y. M. Pan, and Y. X. Li, "Wideband Circularly Polarized Dielectric Resonator Antenna with Bandpass Filtering and Wide Harmonics Suppression Response," *IEEE Transactions on Antennas and Propagation*, Vol. 65, no. 4, 2017, pp. 2096–2101.

4

Circular Polarization Analysis in Dielectric Resonator Antenna Using Conformal Coupling Technique

In this chapter, the details of the conformal coupling technique are analyzed for the generation of circularly polarized radiations in different antenna combinations along with different shapes of DRAs. It is the most suitable feeding technique and allows for various arrangement of DRAs such as single, modified, multiple, multiple-input and multiple-output, array, and filtering to excite the degenerate orthogonal modes, and hence it shows the circularly polarized fields. Here, other approaches, such as wearable, reconfigurable, and use of parasitic and modified shapes, are included and are discussed for helping to obtain the orthogonal modes. A brief design example of antenna procedure is also incorporated to provide better understanding of the conformal coupling technique.

4.1 Introduction to the Conformal Coupling Technique

The concept of the conformal coupling technique is being introduced because of its great advantage over the coaxial probe and microstrip coupling techniques for generation of orthogonal modes. In the literature, the conformal strip technique is the better option compared to previously discussed feeding techniques. This is used to easily excite the orthogonal modes in any shape of DRA to obtain the circularly polarized fields [1, 2]. It has replaced the disadvantage of coaxial probe techniques, such as holes in the ground plane and substrate, to excite the DRA or sometimes holes in the dielectric resonator itself to excite it for the generation of desired modes. It is very difficult to have a hole that perfectly fits the size of the coaxial probe because an air gap exists between probe and DRA and causes the measured results to deviate from theoretical analysis. So the conformal strip is a better solution for avoiding all these problems. In the case of the microstrip feeding technique, there is a need for a high dielectric constant ($\varepsilon_r > 20$) for achieving strong coupling into the DRA but this adversely affects the impedance bandwidth. This creates a huge problem for wideband applications because a low dielectric constant is required for obtaining wideband operations. There is another problem because if the dielectric constant of the DRA is lowered, it significantly reduces the amount of coupling [3]. The conformal strip completely removes these drawbacks that happen due to the coaxial probe and microstrip techniques. The major advantage of the conformal strip technique is that it allows tight coupling through the surface of DRA. Hence, it is more suitable and efficient in electromagnetic energy coupling than the other two techniques (i.e., coaxial probe and microstrip [1]).

There are two ways to use the conformal strip: associated with a microstrip feed line as shown in Figure 4.1 or associated with a coaxial probe feed as shown in Figure 4.2.

The distributions of electric and magnetic fields are the same as a coaxial probe, and can be also identified in Figures 4.3 and 4.4, which are associated with the microstrip feed line as well as coaxial probe coupling. In the case of circularly polarized generation of various shapes of DRAs, it can easily excite the orthogonal modes and also allow the complete electric current to flow on the surface of DRAs, as shown in Figure 4.1.

For example, if a conformal strip is placed on the side of the DRA, it has excited the $TM_{01\delta}$ or $TM_{11\delta}$ and other higher modes in rectangular or cylindrical DRAs, while if a cylindrical DRA is used, the conformal strip excites the $HEM_{11\delta}$ or $HEM_{21\delta}$ and other higher modes whereas a rectangular DRA allows the $TE_{01\delta}$ or $TE_{11\delta}$ and other higher modes [3].

The conditions of circular polarization in DRAs are stated in Chapter 1. Mainly two situations are important for achieving circular polarization: excitation of orthogonal modes with equal magnitude and the displayed phase shift between them is ±90° [3].

In terms of the conformal coupling technique, many research articles have been published to generate circular polarizations in DRAs and are divided into different configurations.

A literature survey reveals that there are fundamentally two methods for generation of circular polarization:

1. To excite the degenerate mode in a DRA and those two modes obtained with the same resonance frequencies having the same ampli-

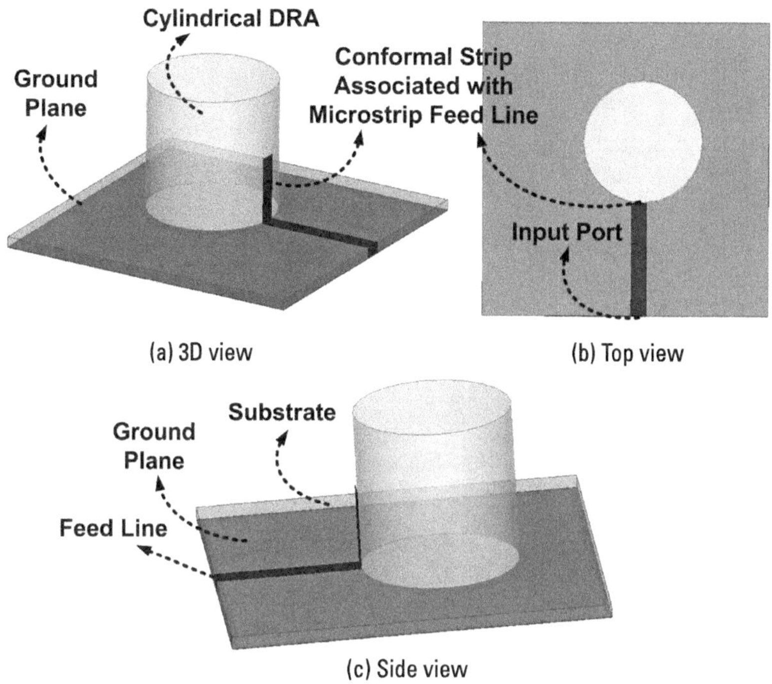

Figure 4.1 General arrangement of the conformal coupling technique associtaed with microstrip line and loaded with dielectric resonator antenna: (a) 3D view, (b) top view, and (c) side view.

tude with an odd multiple of 90° phase difference between them. Such a method was utilized in [4].

2. To esxcite the degenerate mode in a DRA and those two modes obtained at two different frequencies having the same amplitude with an odd multiple of 90° phase difference between them. Such a method was utilized in [5].

4.2 Design Methodology of Conformal Coupling Technique CP-DRAs

Figure 4.5 shows a flow chart of the conformal coupling technique, which is used for the generation of orthogonal modes in the various shapes of dielectric resonators for several purposes (such as antenna, arrays antenna, filter, multiple-input and multiple-output, wearable, etc.).

Based on the antenna designed example, here DRAs have been investigated to demonstrate the generation of the circularly polarized method. So, a

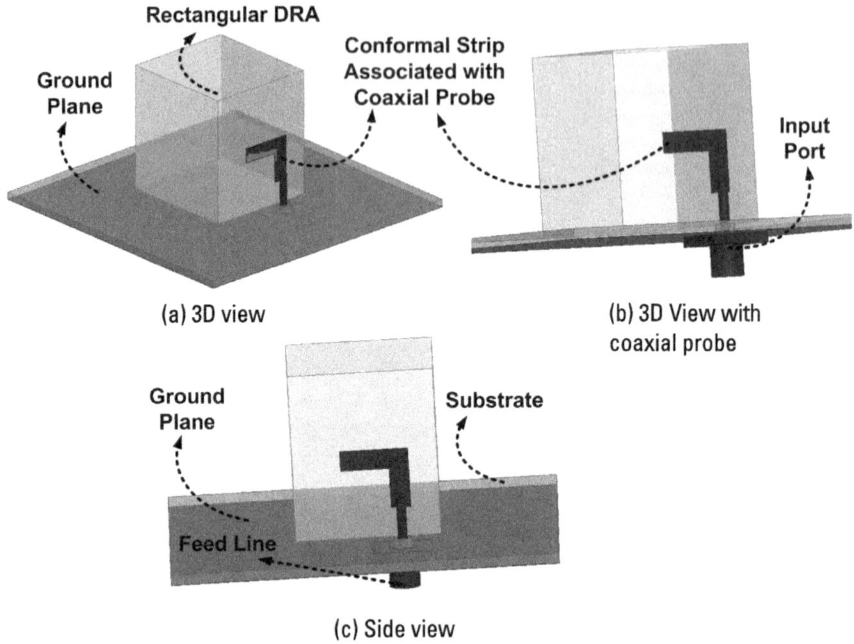

Figure 4.2 General arrangement of the conformal coupling technique associtaed with coaxial probe and loaded with dielectric resonator antenna: (a) 3D view, (b) top view, and (c) side view.

design step and its procedure by using the conformal coupling technique have been presented in Figure 4.5.

4.2.1 Antenna Element Configuration

Figure 4.6 shows the proposed example antenna configuration that is based on the cylindrical DRA structure. The cylindrical DR (ε_{DRA} = tanδ = 0.002) should be seated on a thick rectangular FR4 class epoxy substrate (εsub = 4.4 tanδ = 0.025) with backside ground plane. The proposed antenna is centrally fed by a dual vertical conformal strip associated with an L-shaped microstrip line. Here, two metallic strips are attached to two corners of the cylindrical DRA to realize the circular polarization in the principal plane. The height of two metallic strips (VC_{S1} = 9 mm, and VC_{S2} = 19 mm) are different to create the offset that offers phase shift between the field lines, and hence generation of circular polarization in the cylindrical DRA. The radius and height of the cylindrical DRA is chosen as D_r = 12 mm and D_H = 22 mm, whereas the length, width, and height of FR4 substrate is taken as S_L = 55 mm × S_W = 55 mm = 55 mm × S_H = 1.6 mm, as shown in Figure 4.6, respectively. The main microstrip feed line having a length of (MF_{L1} + MF_{L2} + MF_{L3} = 22 + 6 + 8 = 36 mm), width (MF_W = 2 mm) and a height of (MF_H = 0.035 mm) is connected with

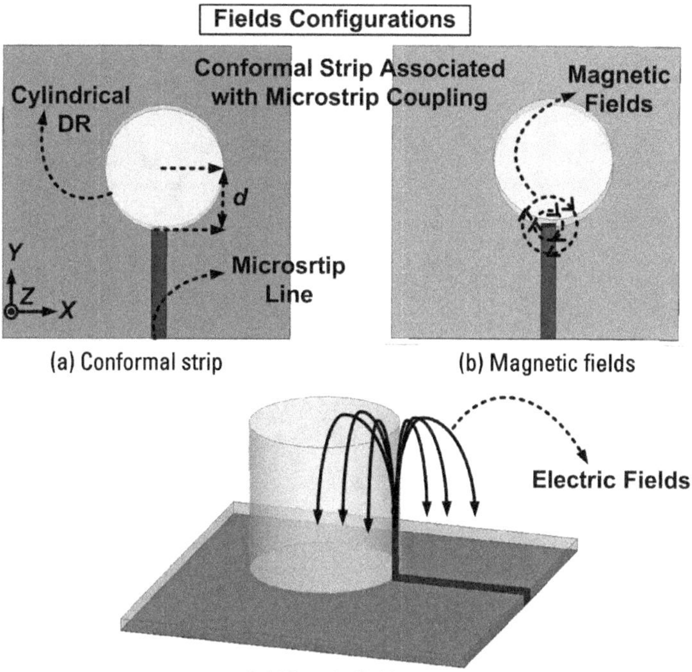

(a) Conformal strip (b) Magnetic fields

(c) Electric fields

Figure 4.3 Conformal strip associated with microstrip line coupling to the dielectric resonator antennas, fields, and radiation model: (a) conformal strip, (b) magnetic fields, and (c) electric fields.

two vertical conformal strips to feed the cylindrical DRA and the opposite side fully ground plane is placed. The size of the ground plane is the same as the size of the FR4 substrate.

The dimension of the cylindrical DRA is calculated by empirical formulas for the given resonant frequencies of the fundamental mode, which were given in Chapter 2.

4.2.2 Operating Mechanism

In order to illustrate the proposed design procedure, a cylindrical DRA is considered as a preliminary shape and the effect of a different conformal shape is observed by utilizing the ANSYS HFSS of all antenna configurations. As shown in Figure 4.7, a cylindrical DRA is excited by the microstrip feed line as shown in the first configuration known as Sub-Ant. 1, the simple microstrip feed line is extended with a bend (i.e., L-shaped microstrip feed line in the second configuration, Sub-Ant. 2) by adding a vertical strip at the end of an L-shaped microstrip feed line, which is called conformal technique in the third configuration (Sub-Ant. 3), in a similar way by adding a vertical strip at the ori-

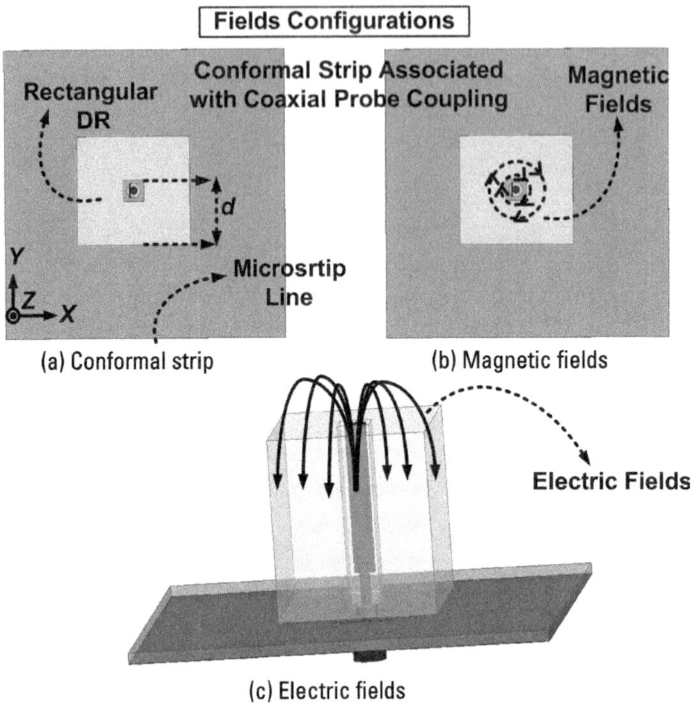

Figure 4.4 Conformal strip associated with coaxial probe coupling to the dielectric resonator antennas, fields, and radiation model: (a) conformal strip, (b) magnetic fields, and (c) electric fields.

gin of an L-shaped microstrip feed line that is also called conformal technique in the fourth configuration (Sub-Ant. 4), and by adding two vertical strips at the origin as well as at the end of an L-shaped microstrip feed line called final conformal technique in the fifth configuration (Sub-Ant. 5-proposed). All other dimensions of the subantennas are stably preserved to provide a better comparison among the all subantenna performances. The IBWs and ARs for all five subantennas are given in Figure 4.8.

Case 1: Sub-Ant. 1 is constructed by using the preliminary dimensions and the alumina material for a cylindrical DR, which was included in Chapter 2. The DWM is used for calculating the initial resonance frequency of the cylindrical DRA but as per the generation of circular polarization the formulas are less applicable. The design antenna configuration focuses on the circular polarization instead of the linearly polarized antenna. By considering the circular polarization, we have chosen the dimensions, permittivity (dielectric constant), and shape of all the three main components (i.e., ground plane, feed line, and dielectric resonator).

A basic cylindrical DRA is used here for constructing the configuration of Sub-Ant. 1 and excited with a microstrip feed line at the center of the DRA

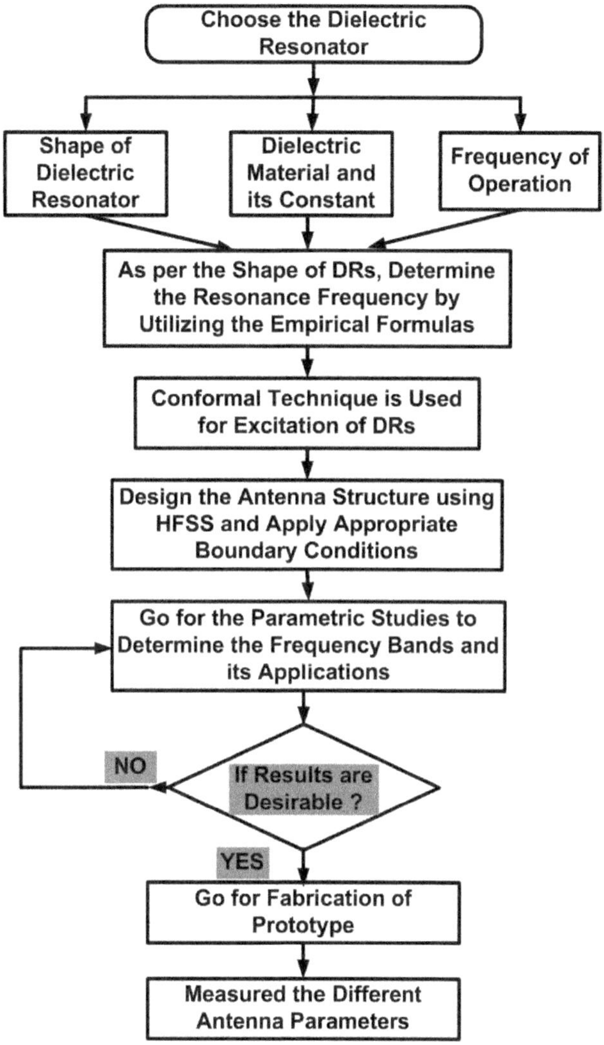

Figure 4.5 Design steps and methodology of the conformal coupling technique.

as presented in Figure 4.7(a). The optimized feed parameters (i.e., length and width) are $MF_{L1} = 22$ mm and $MF_W = 2$ mm, respectively. The present antenna configuration is linearly polarized with less impedance matching and is unable to generate a degenerate orthogonal mode. Therefore, input impedance bandwidth and axial ratio bandwidth for Sub-Ant. 1 has not been achieved, as shown in Figures 4.8(a) and 4.8(b), respectively.

Calculation of orthogonal mode:

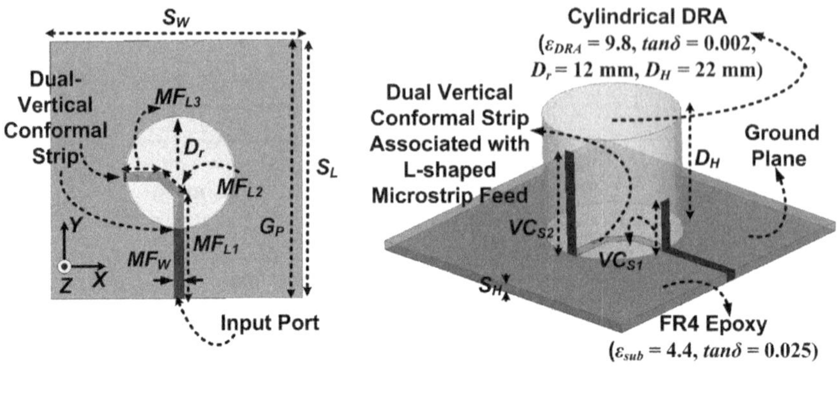

(a) (b)

Figure 4.6 Proposed example for a cylindrical DRA fed by two vertical strips connected at the origin and end point of an L-shaped microstrip feed line: (a) top view, and (b) 3D view. (Modified after [11].)

$$Phase\ difference = \frac{2\pi}{\lambda} \times Path\ difference \tag{4.1}$$

Here, the free space frequency (f_0) is considered as 3 GHz and the length of microstrip line (MF_{L1}) in the given antenna structure is 22 mm. By utilizing (4.1), the phase difference between the two linearly polarized fields is $0.44\pi\left(=2\pi\times\frac{22}{100}\right)$, which is insufficient to fulfill the condition of circular polarization of the Sub-Ant. 1. Hence, this condition is not suitable for generating the axial ratio bandwidth.

Case 2: The Sub-Ant. 2 is designed by the extension of a microstrip feed line, called the Sub-Ant. 1, and all the remaining antenna components are the same as presented in Figure 4.7(b). Sub-Ant. 2 has been obtained by changing the current path and is obtained from the addition of the length of the microstrip feed line. Hence, it becomes an L-shaped microstrip feed line, which covers the maximum current path to excite the DRA. The goal of this antenna arrangement also failed because the generated electric field is not sufficient enough to excite the cylindrical DRA. Therefore, it is unable to form the orthogonal degenerate mode and also gives poor matching of IBW, as shown in Figure 4.8.

Calculation of the orthogonal mode is shown in (4.1).

Here, the length of an L-shaped microstrip feed line is ($MF_{L1} + MF_{L2} + MF_{L3} = 22 + 6 + 8$) = 36 mm, and the width is (MF_W) of 2 mm with a free space frequency of (f_0) = 3 GHz. From (4.1), the phase difference between the two linearly polarized waves is $0.72\pi\left(=2\pi\times\frac{36}{100}\right)$, which does not fall under

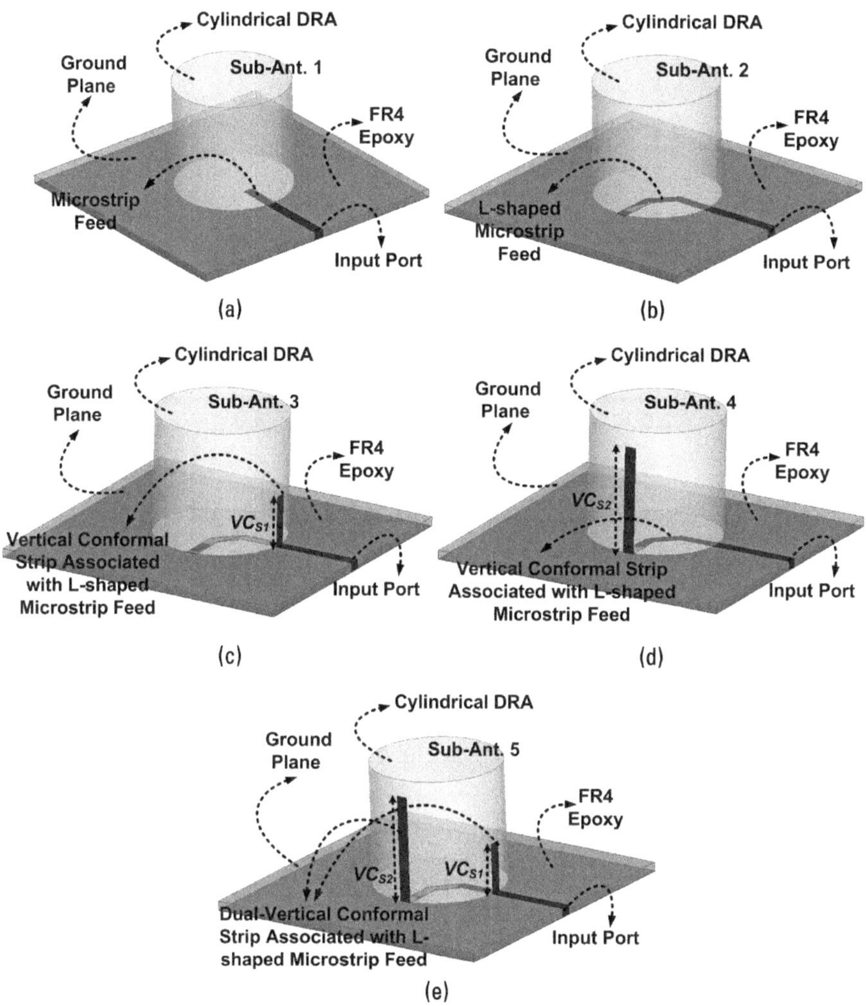

Figure 4.7 The cylindrical DRA stage evolutions and its conformal feeding configurations. (a) Antenna 1 (Sub-Ant. 1), simple feed. (b) Antenna 2 (Sub-Ant. 2), L-shaped. (c) Antenna 3 (Sub-Ant. 3-Proposed), vertical strip at the end of the L-shaped feed line. (d) Antenna 4 (Sub-Ant. 4), vertical strip at the origin of the L-shaped feed line. (e) Antenna 5 (Sub-Ant. 5-Proposed), two vertical strips at the origin as well as end of the L-shaped feed line.

the condition of generation of circular polarization. Hence, Sub-Ant. 2 does not show the axial ratio bandwidth, as shown in Figure 4.8(b).

Case 3: The Sub-Ant. 3 has been assembled by adding a vertical stub ($VC_{S1} = 9$ mm) with the existing L-shaped (Sub-Ant. 2) microstrip feed line at the start point of the face of cylindrical DRA, as shown in Figure 4.7(c). The open stub helps to maximize electromagnetic energy inside the Sub-Ant. 3 configuration and hence it shows better impedance matching as depicted in Figure

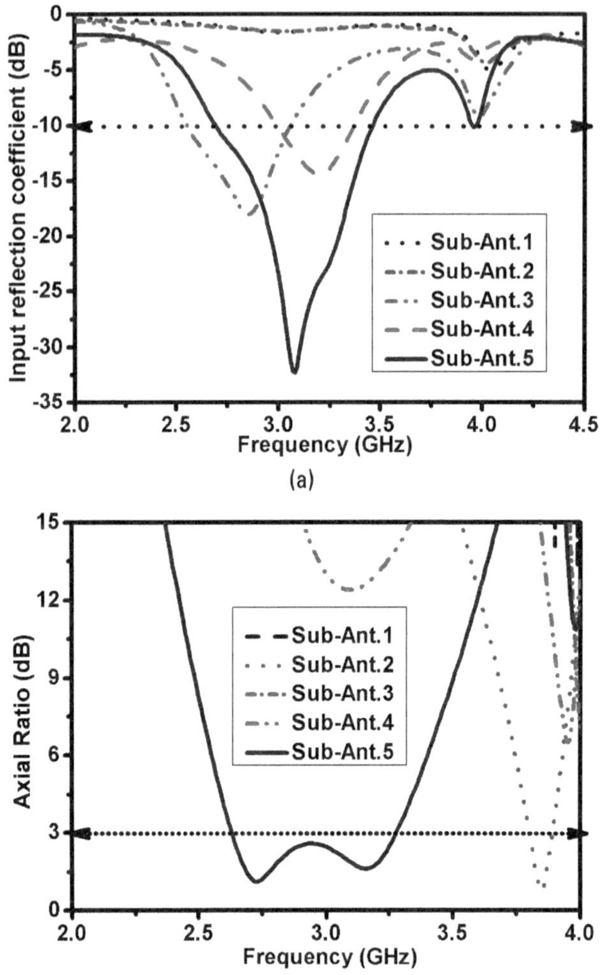

Figure 4.8 Different stage development of all Sub-Ant. design example simulated: (a) input reflection coefficient, and (b) axial ratio versus frequency plots.

4.8(a). In the case of obtaining axial ratio bandwidth, the feeding mechanism of Sub-Ant. 3 is not suitable to provide the condition for the generation of orthogonal modes, as illustrated in Figure 4.8(b).

Here, the combination of length of an L-shaped microstrip line along with vertical stub ($MF_L1 + MF_{L2} + MF_{L3} + VC_{S1} = 22 + 6 + 8 + 9 = 45$ mm) is calculated as 45 mm and *f* free space frequency (f_0) is taken by 3 GHz. By utilizing (4.1), the phase difference between the two linearly polarized fields is approximately $0.90\pi\left(=2\pi\times\frac{45}{100}\right)$, which also does not fulfill the condition of generation of circular polarization in the Sub-Ant. 3.

Case 4: In order to further enhance the performance of the antenna, the same vertical stub has been placed at the end of the L-shaped microstrip feed line, which is touching the left side of the face of the cylindrical DRA and is called the Sub-Ant. 4, as shown in Figure 4.7(d). In this configuration, the height of the vertical stub (VC_{S2} = 19 mm) is greater than the previous design, which has the advantage of lowering the axial ratio bandwidth but at the same time also deteriorates the impedance bandwidth and matching. The result is a deficient impedance bandwidth and matching due to the placement of the vertical stub at the end and hence it is inadequate to provide proper electromagnetic coupling to the cylindrical DRA, as illustrated in Figure 4.8(a). The mechanism of the placement of vertical stub at the Sub-Ant. 4 helps to lower the axial ratio bandwidth and also lay out the frequency range at almost same range of impedance bandwidth, as depicted in Figure 4.8(b).

Next, the combined length of an L-shaped microstrip line along with vertical stub ($MF_{L1} + MF_{L2} + MF_{L3} + VC_{S1} = 22 + 6 + 8 + 9 = 45$ mm) is calculated as 55 mm and free space frequency (f_0) is taken at 3 GHz with a free space wavelength (λ_0). By utilizing (4.1), the phase difference between the two linearly polarized waves is $\approx 1.1\pi = 2\pi \times \frac{55}{100}$, which is also incomplete to obtain the condition of circular polarization in the Sub-Ant. 4.

Case 5: Finally, Sub-Ant. 5 (Proposed Antenna) has been constructed by the combination of Sub-Ant. 3 and Sub-Ant. 4, as shown in Figure 4.7(e). It is based on the concept of Sub-Ant. 3 and Sub-Ant. 4 as well as proved by the calculation of the phase difference between the two linearly polarized fields. Further, the degenerate orthogonal $HEM^{x}_{\delta 11}$ and $HEM^{y}_{1\delta 1}$ mode has been obtained at 3.01 GHz, which is responsible for generation of axial ratio bandwidth, as shown in Figure 4.9. The presented results show that by employing the dual vertical stripes on the face of the DRA for obtaining wider input impedance ($|S_{11}| \leq 10$ dB) and axial ratio bandwidths ($|AR| \leq 3$ dB) of 25.08% (2.69 GHz

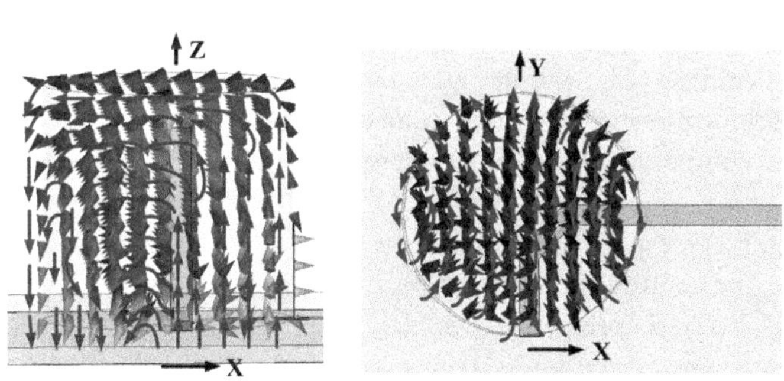

Figure 4.9 The simulated electric and magnetic fields inside the Sub-Ant. 5 (i.e., Presented Antenna has been plotted to confirm the HEM11δ mode): (a) E–field, side view in xz-plane, and (b) H–field, top view in xy-plane at 3.01 GHz.

– 3.46 GHz) and 22.03% (2.63–3.28 GHz), both almost achieve the same frequency range as shown in Figure 4.8(a) and (b), respectively. The overlapped CP bandwidth of 19.79% (2.69–3.28 GHz) has been achieved by the designed antenna configuration and it also gives a useful average CP gain of 6.47 dBic.

Next, the final length of dual vertical conformal strips along with an L-shaped microstrip line feeding ($MF_{L1} + MF_{L2} + MF_{L3} + VC_{S1} + VC_{S2} = 22 + 6 + 8 + 19 + 9 = 64$ mm) is 64 mm and free space frequency is (f_0) = 3 GHz. By utilizing (4.1), the phase difference between the two linearly polarized waves is $\approx 1.28\pi = 2\pi \times \frac{64}{100} \approx (2) \times 0.64\pi$, which fulfills the condition of circular polarization in the Sub-Ant. 5. Therefore, Sub-Ant. 5 offers the axial ratio bandwidth in the proposed conformal feeding structure.

It is observed from all the Sub-Ant. configurations that performance of Sub-Ant. 5 (Proposed Antenna) is far better because of the coupling of EM energy (electric source) in the cylindrical DRA. At the same time, this configuration also formed the orthogonal mode due to the help of dual vertical strips associated along with the L-shaped microstrip feed line.

4.2.3 E-field Distributions of the Cylindrical DRA

To further verify the CP mechanism of Sub-Ant.5, the simulated *E*-field distributions of the cylindrical DRA is plotted at 3.01 GHz. It is found in Figure 4.9 that orthogonal modes $HEM^{x}_{\delta 11}$ and $HEM^{y}_{1\delta 1}$ are the main sources of getting the circular polarization with reference to the coordinate axes.

The simulated electric field distributions at 3.01 GHz of the presented dual vertical strips associated along with the L-shaped microstrip feed line excited with the cylindrical DRA is illustrated in Figure 4.10 at four different phase angles. It can be seen that the E-field is rotated clockwise from 0° to 360°. Hence, it verifies the RHCP radiation of the proposed antenna.

4.2.4 Key Parameter-Feeding Network

To achieve circular polarization performance the conformal strip feeding technology is applied in the present antenna configuration. The conformal feeding network is composed of two different strips and added at the two different position of an L-shaped microstrip feed line with different lengths for getting a phase shift. Here, the variation of length of two different strips has been studied to find out the condition of circular polarization, as shown in Figure 4.11.

It is observed from Figure 4.11(a) that variation in the length of the strips slightly affects the impedance matching and impedance bandwidth of the designed antenna. But the generation of the degenerate orthogonal mode completely depends on the length of the strips, as depicted in Figure 4.11(b). Hence, to satisfy the circularly polarization condition the height of both the

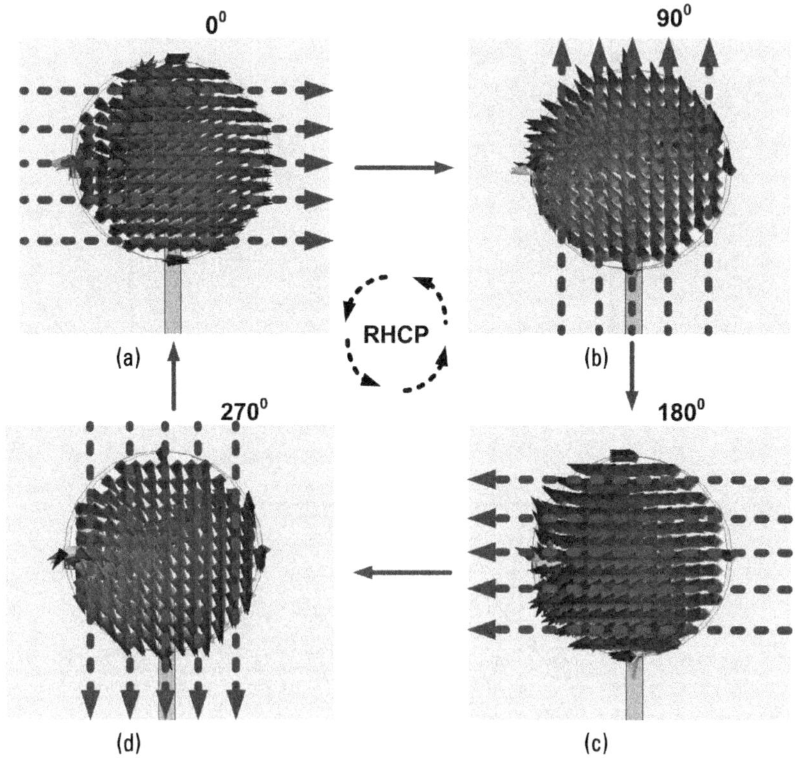

Figure 4.10 Simulated E-field distribution at the top of the cylindrical DRA observed along the Z-direction with different phases, (a) 0°, (b) 90°, (c) 180°, and (a) 270° at 3.01 GHz.

strips must be 9 mm (VC_{S1}) and 19 mm (VC_{S2}), respectively, and other remaining conditions are not useful for this feeding network.

4.2.5 Simulated Near- and Far-Field Results Analysis

A dual vertical conformal strip of the proposed antenna is shown in Figure 4.6. The simulated input reflection coefficient and axial ratio bandwidth at the broadside direction has been illustrated in Figure 4.12.

As can be seen in Figure 4.12(a), the simulated -10-dB impedance bandwidth ($|S_{11}| \leq 10$ dB) of the dual conformal strip cylindrical DRA is 25.08% (2.69–3.64 GHz). In addition to that, the simulated 3-dB axial ratio bandwidth ($|AR| \leq 3$ dB) in boresight direction has been observed as 22.03% (2.63–3.28 GHz), as represented in Figure 4.12(b). It can also be observed that the results of impedance-and axial ratio-bandwidths are matched reasonably well and a wide overlapped CP bandwidth of 19.79% (2.69–3.28 GHz) has been realized.

The simulated RHCP and LHCP patterns at 3.01 GHz have been given in Figure 4.13 over the E-plane (elevation plane) and H-plane (azimuth plane).

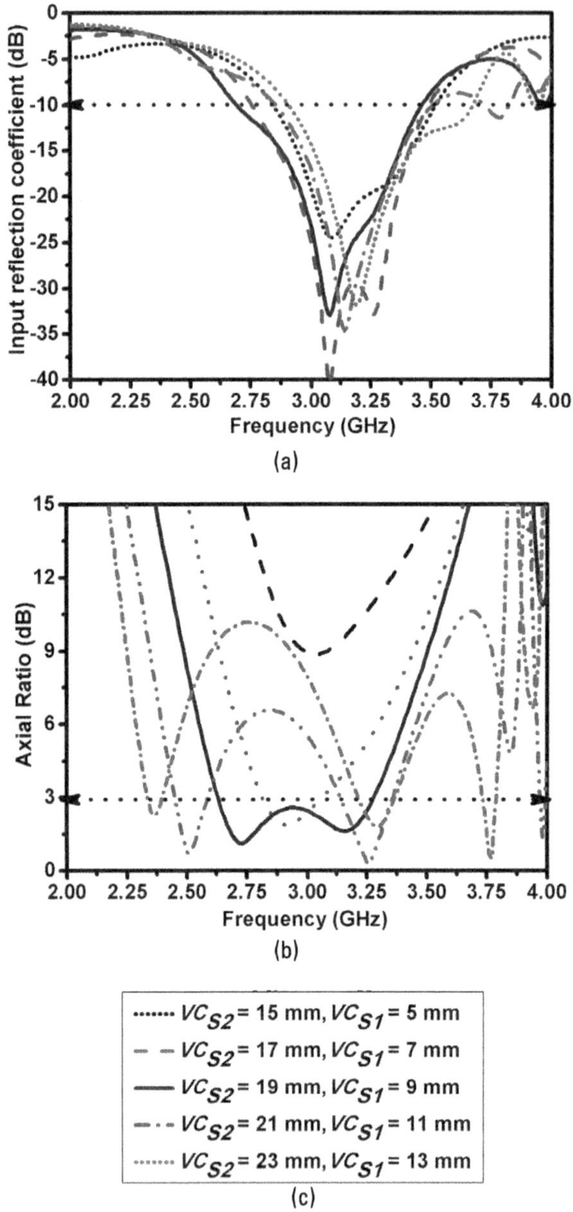

Figure 4.11 Simulated result of dual vertical strips with different lengths: (a) input reflection coefficient, (b) axial ratio plot, and (c) scale.

It can be seen in Figure 4.13 that the proposed antenna indeed has broadside RHCP radiation patterns at these CP frequency ranges.

The overall CP gain of the dual vertical conformal strips cylindrical DRA is higher than 6 dBic over the entire band from 2.69 to 3.28 GHz, as shown

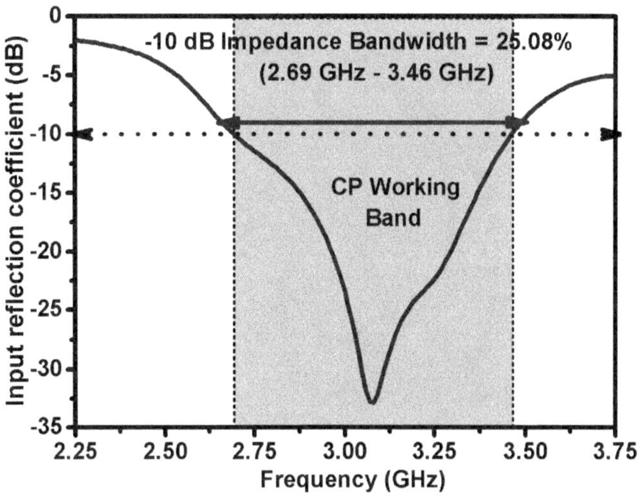

(a)

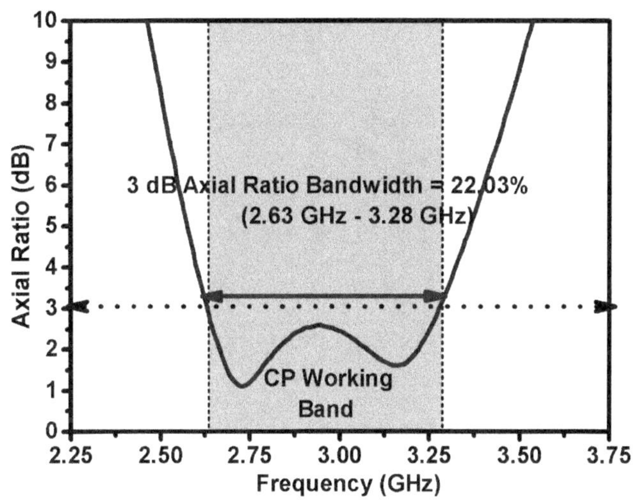

(b)

Figure 4.12 Simulated reflection coefficients and ARs of the presented antenna: (a) input reflection coefficient, and (b) axial ratio plot versus frequency.

in Figure 4.14(b). The radiation efficiency is also higher than 95% across the entire CP frequency band of interest, as depicted in Figure 4.14(b).

4.3 Conformal Coupling in a Conventional DRA

The only conventional shape of a DRA excited with conformal feeding technique is studied in this section for the generation of circular polarization. This section is further divided into three different sections based on the addition of

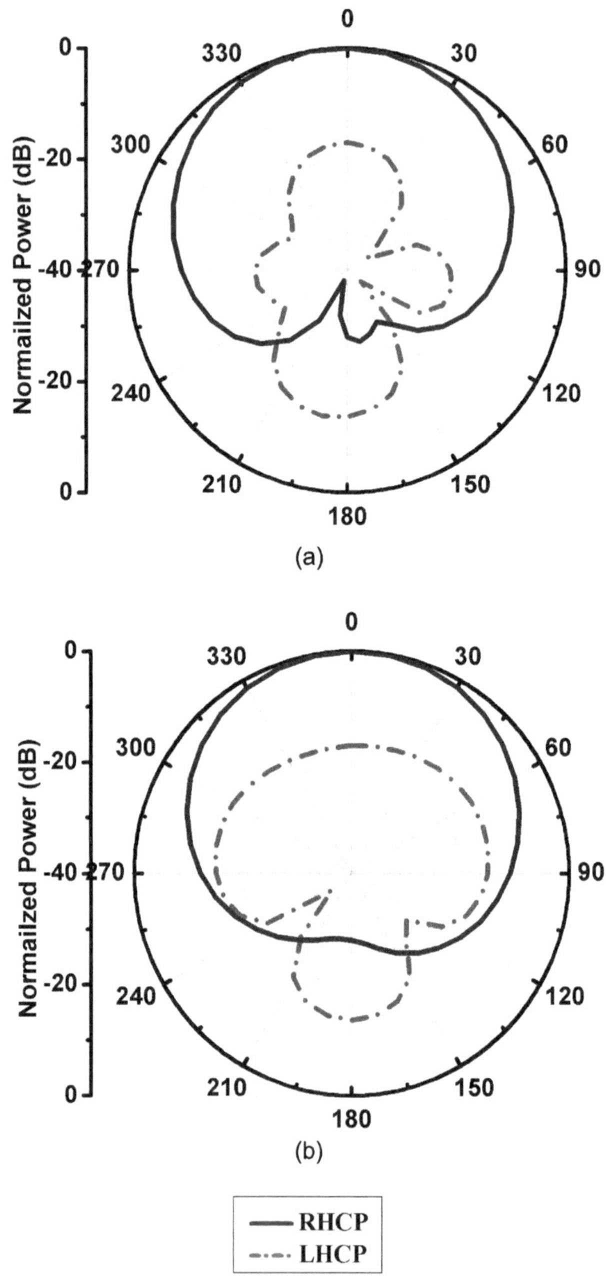

Figure 4.13 Simulated RHCP and LHCP radiation patterns of the prototype antenna at 3.01 GHz.

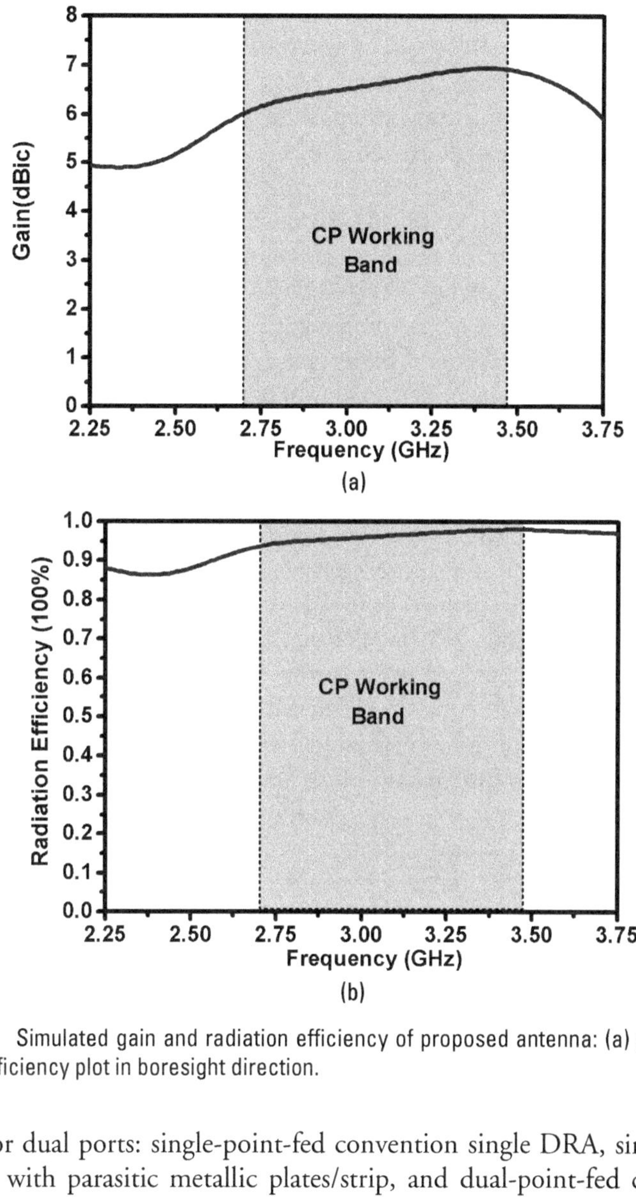

Figure 4.14 Simulated gain and radiation efficiency of proposed antenna: (a) gain, and (b) radiation efficiency plot in boresight direction.

parasitic or dual ports: single-point-fed convention single DRA, single-point-fed along with parasitic metallic plates/strip, and dual-point-fed convention DRA.

4.3.1 Single-Point Conformal-Fed Convention Single DRA

In [6], Wong et al. first investigated the conformal strip technique and it is used in a cylindrical DRA for obtaining the circular polarization field. In this presented antenna, the two unequal lengths of vertical conformal strips are utilized in two different places of a cylindrical DR. These strips are able to produce two linearly field components having the same amplitude and a 90° phase difference between them, hence it shows the 3-dB axial ratio bandwidth is 3.5%. The excitation of two near degenerate orthogonal modes in a cylindrical DRA directly depends on embedding two pairs of unequal slots in the finite ground plane of the feed substrate is excited and the conformal strip has been reported in [7]. The combination of antenna structure gives the 3-dB axial ratio of 2.7% and the resonance frequency can also be tuned by changing the length of slots. In [8], the authors used the combination of quadruple strip feed and a 90° hybrid coupler pair to excite the cylindrical DRA. Here, the four orthogonally orientated vertical strips have contributed equal amplitude of power and relative different excitation phases to achieve dual-fed circular polarization. The quadruple strip feed cylindrical DRA exhibits a simulated 3-dB axial ratio bandwidth of 28.9%. The authors discussed three antenna configurations based on using single and dual pair conformal feed associated with the microstrip lines. These conformal feeds are aligned perpendicular to each other for generation of orthogonal modes, which hence shows the circular polarization as reported in [9]. Whereas in [10], the authors have taken the modified Wilkinson power divider associated with the two conformal strips of unequal length that are used to excite the cylindrical DRA for getting the circular polarization waves. For achieving circularly polarized fields in the cylindrical DRA, the authors used dual vertical conformal strips associated with microstrip lines and arranged in a perpendicular fashion to excite hybrid orthogonal mode, as shown in Figure 4.15. The phase quadrature relationships between two linearly polarized signals have been attained by varying corresponding conformal strips heights that have been reported in [11]. Such a combination of antenna geometry along with excitation mechanism achieved 24.6% of axial ration bandwidth in the broadside directions, which is suitable for wireless applications.

In [12], the same authors have used different combinations of unequal parallel conformal strips as the outside of the rectangular DRA for excitation of orthogonal modes. The axial ratio bandwidth of this combination of antenna configuration shows only 9.2% compared to previous designs, as depicted in Figure 4.16.

In [13], Wuo et al. have for the first time investigated a circularly polarized unidirectional DRA with lateral radiation. The conformal strip is made of the combination of microstrip line and coaxial probe in this structure. Here, it requires two sets of circularly polarized fields that have complementary patterns, and the first set of orthogonal mode is obtained using the DRA whereas the second set of orthogonal mode is obtained through the loading effects (i.e., slot antenna loaded with a DRA). The second set has been realized by a cutting cross-slot in the ground plane. Hence, the compact circularly polarized unilateral DRA gives the axial ratio bandwidth of 4.9%, operating in 2.4 GHz for

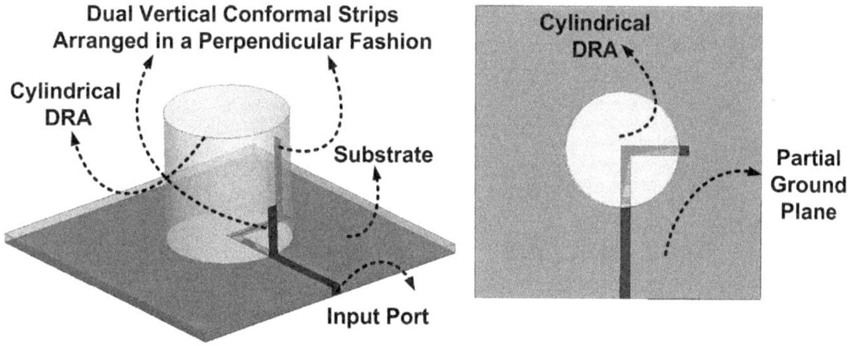

(a) 3D view (b) Top view

Figure 4.15 Dual vertical conformal strips arranged in a perpendicular fashion to a coupled cylindrical DRA for wideband circular polarization application [11].

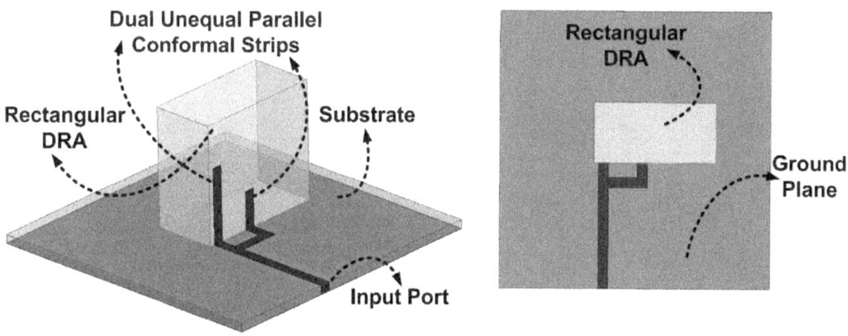

(a) 3D view (b) Top view

Figure 4.16 Dual unequal parallel vertical conformal strips to a coupled rectangular DRA for circular polarization application [12].

wireless local area network applications. A similar type of antenna structure and its results can also be analyzed, as given in Figure 4.17.

In [14], the authors utilized a broadband phase delay line (PDL) for the first time in a cylindrical DRA for dual-function applications. The cylindrical DR is used as the load on the broadband PDL design to improve the impedance matching and phase shift. In this feeding configuration at the same time, the delay line is used to feed the same DRA for the antenna applications. By doing this, three different results can be obtained, such as the phase shift of 90° and 180° PDLs, as well as circularly polarized and differential DRAs. Similar types of antenna configuration are shown in Figure 4.18. The new circularly polarized and differential DRAs can be obtained in a much more compact way and also avoid external 50-ohm load compared with those using a hybrid coupler.

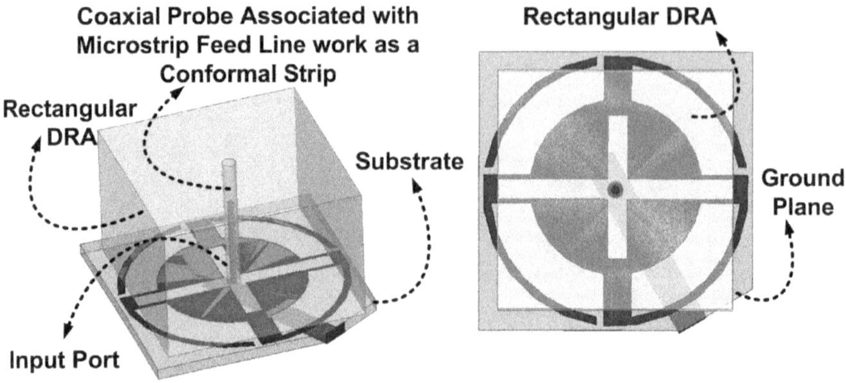

(a) 3D view (b) Top view

Figure 4.17 Coaxial probe associated with microstrip feed used as a conformal strip to a coupled rectangular DRA for circular polarization application [13]. (Modified after [13].)

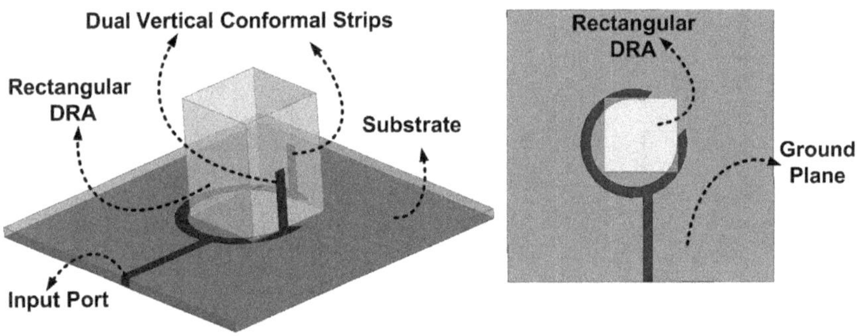

(a) 3D view (b) Top view

Figure 4.18 Dual vertical conformal strips to a coupled rectangular DRA for wideband circular polarization application. (Modified after [14].)

The compact broadband circularly polarized cylindrical DRA with PDL has an overlapping axial ratio and impedance passband of 24.2% in the broadside direction.

A unique conformal H-shaped metal strip has been used for exciting a rectangular DRA, as reported in [15]. Using H-shape conformal metal strip excitation, the degenerate orthogonal $TE^{x}_{\delta 13}$ and $TE^{y}_{1\delta 3}$ modes of higher order have been generated for achieving improved bandwidth and high gain. A singly fed circularly polarized rectangular DRA using a unique H-shaped conformal metal strip shows the 3-dB axial ratio bandwidth of 20% and useful gain of 6.8 dBic in the broadside direction; similar results can also be obtained using the presented antenna structure shown in Figure 4.19.

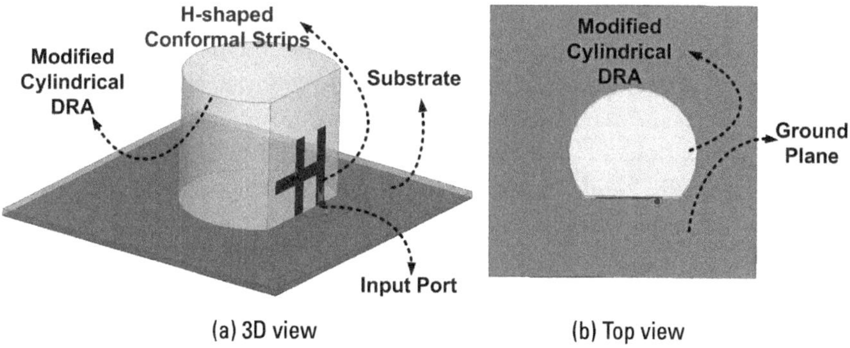

Figure 4.19 H-shaped conformal strips to couple modified a cylindrical DRA for circular polarization application. (Modified after [15].)

4.3.2 Single-Point Conformal-Fed along with Shorted/Open-Ended Stubs on a Single DRA

The combination of conformal strip feeding techniques and shorted or open-ended stub structures are employed to obtain the orthogonal modes in conventional DRAs [16–20]. The excitation of the CP radiation using printing two vertical metal strips on the surface of the cylindrical dielectric resonator is reported in [16] and a similar type of antenna structure has been realized in Figure 4.20. In this antenna configuration, the two identical metal strips are printed on the cylindrical DRA surface at the different position of $\varphi = 45°$ and $\varphi = 225°$ to get the orthogonal modes. It has been provided a narrow 3-dB AR bandwidth of 2.20%.

The single use of a parasitic patch along with conformal strip techniques to excite circular polarization waves in hemispherical DRA is presented in [17].

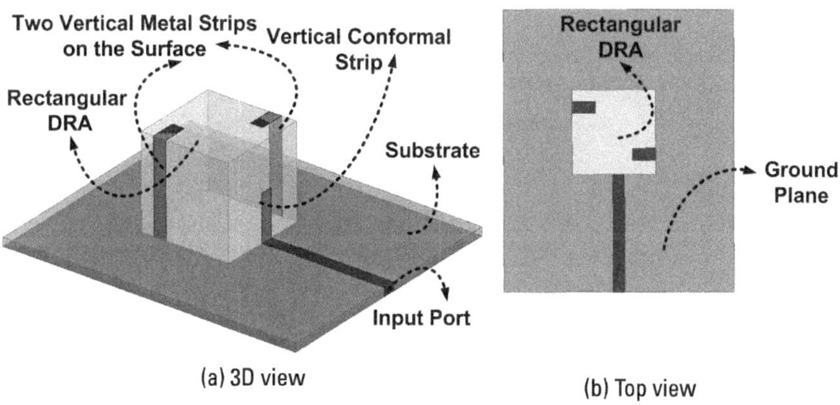

Figure 4.20 Conformal strip coupled rectangular DRA with two vertical metal strips on the surface for small band circular polarization. (Modified after [16].)

Here the Green's function approach is applied for analytical study of DRA and also observes the fundamental TE_{111} mode in a hemispherical DRA. For matching the appropriate boundary conditions, the integral equations of the conformal-strip and parasitic-patch currents are formulated, which is solved by method of moments (MoM). The calculated 3-dB axial ratio bandwidth is found to be 2.4%. Similar antenna results could also be analyzed by using the antenna structure shown in Figure 4.21.

In [18], a rectangular DRA is excited by using a conformal strip technique, which is compatible with a coaxial probe. This configuration only supports the linearly polarized field. For generation of circular polarization, a parasitic patch is attached on a side wall of the DRA to excite a degenerate orthogonal mode. The fundamental mode of TE_{111} can be excited and achieve the 3-dB AR bandwidth of 2.70%. Figure 4.22. depicts a similar type of antenna structure and the same antenna results could be also achieved.

It is also possible to use multiple parasitic strips along with conformal feeding techniques to generate the orthogonal modes in a hemispherical DRA, as reported in [19]. The advantage of multiple parasitic strips is that if it is placed asymmetric to generate the circularly polarized fields, it can be used to tune the operating frequency. Hence this configuration shows the multiple 3-dB AR bandwidths. Similar kinds of results could be also analyzed by utilizing the antenna geometry shown in Figure 4.23.

In [20], the conformal strip is associated with the coaxial probe to excite the rectangular DRA. For getting the orthogonal modes, an E-shaped conformal strip has been applied but it shows the narrowband circular polarization applications. Further, enhancement of the AR bandwidth has been achieved by introducing the parasitic patch at an optimal location on the rectangular DRA. Hence, it excites the $TE^x_{\delta 13}$ and $TE^y_{1\delta 3}$ modes in the DRA to provide circular

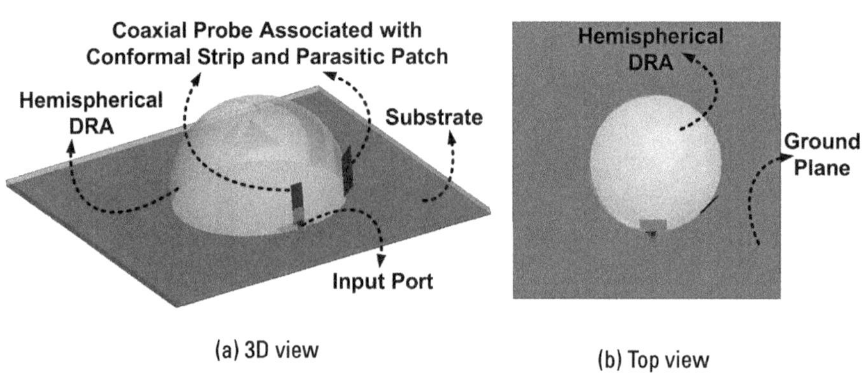

Figure 4.21 Conformal strip coupled hemispherical DRA with a vertical metal strip on the surface for small band circular polarization. (Modified after [17].)

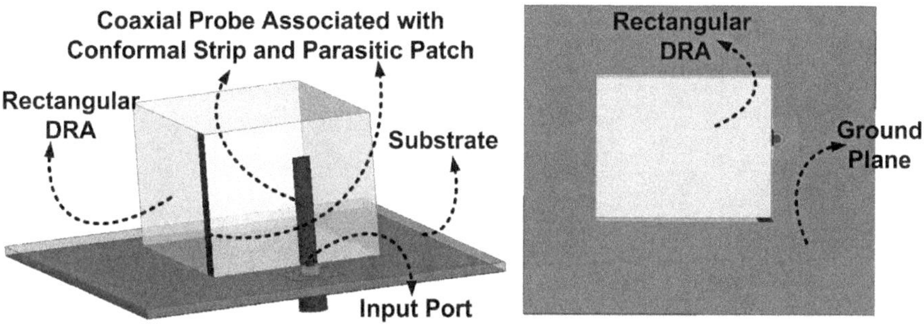

(a) 3D view (b) Top view

Figure 4.22 Conformal strip coupled rectangular DRA with a metal strip on the surface for small band circular polarization. (Modified after [18].)

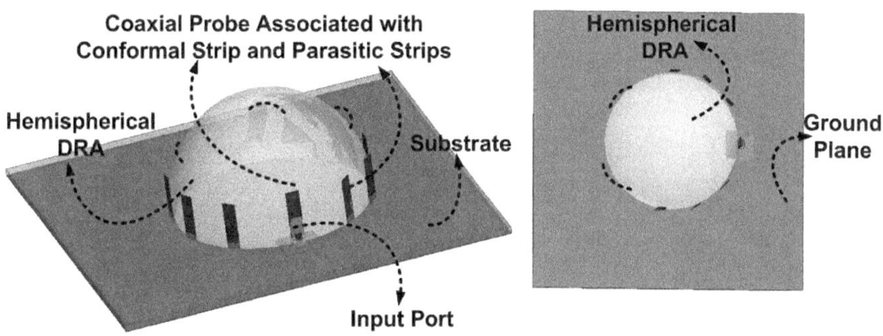

(a) 3D view (b) Top view

Figure 4.23 Conformal strip coupled hemispherical DRA with parasitic strips on the surface for small band circularly polarized applications. (Modified after [19].)

polarization and shows the overlapping bandwidth of 23%. A similar type of antenna structure is shown in Figure 4.24.

4.3.3 Dual-/Multiple-Point Conformal-Fed on a Single DRA

Here, a dual-point conformal strip has been used for generating the orthogonal modes in the appropriate DRA as reported in [21, 22]. The authors in [21] use two vertical strips with quadrature in phase and apply it in a rectangular DRA for obtaining circular polarization, as a similar type of structure shown in Figure 4.25. Here, simultaneously fundamental TE_{111} and higher-order TE_{113} modes are excited to obtain the wideband circular polarization. The result of 4-dB AR bandwidth shows the 36.99% covering 2.6–3.78 GHz in the broadside direction. A broadband axial ratio bandwidth is obtained in the rectangular DRA by applying the dual-feed type, as reported in [22]. The conformal feeding

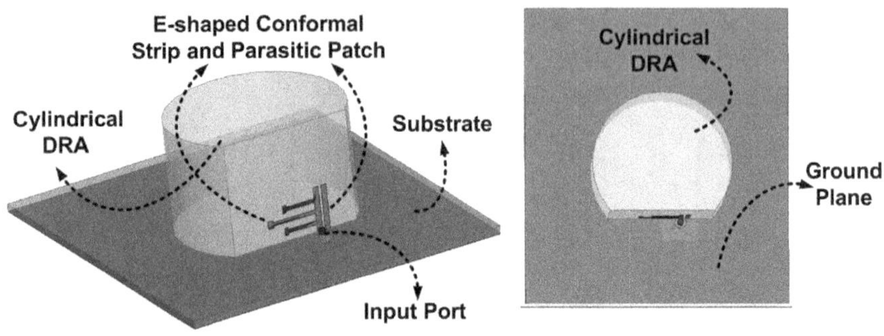

(a) 3D view (b) Top view

Figure 4.24 E-shaped conformal strip coupled cylindrical DRA with a parasitic patch on the surface for narrowband circular polarization. (Modified after [20].)

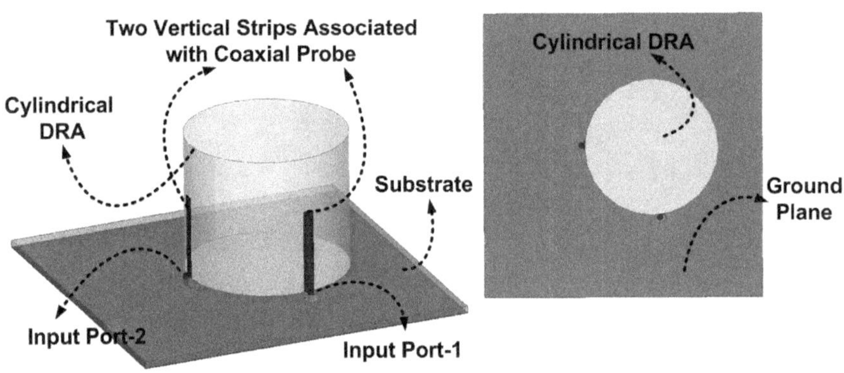

(a) 3D view (b) Top view

Figure 4.25 Dual-point conformal strip coupled cylindrical DRA for circular polarization. (Modified after [21].)

structure comprised 3-dB Wilkinson power divider and a wideband 90° phase shifter for getting an equal power splitting along with consistent 90° phase shift and the good impedance matching over a wide bandwidth range. Such a feeding configuration is responsible for obtaining the broadband AR bandwidth of 41.7% in the broadband direction.

A brief summary of all designs is provided in Table 4.1. Mostly conventional shapes of DRAs are utilized by different authors and excited by a single or dual conformal feeding technique and in some configuration using stubs or metallic parasitic strips. The antenna design and its process can be characterized by seeing each antenna configuration. Furthermore, it is broadly based on four different steps to achieve the orthogonal modes in the antenna geometry using a conformal feeding technique.

Table 4.1

Brief Comparison of Various CP Designs Using Single-Point-Fed Conformal Strip Technique with Conventional Shape of a Single DRA

Ref	DRA Shape, Number of DRs (ε_r)	Conformal Strip Arrangements	IMBW (%)	ARBW (%)	f_c (GHz)	Modes Excited
[6]	Cylindrical, 1 (9.5)	Dual unequal strip lengths	15.80	3.50	1.97	Not mentioned
[7]	Cylindrical, 1 (79)	Conformal strips embedding two pairs of unequal slots in the finite ground plane	13.63	2.7	2.17	Not mentioned
[8]	Cylindrical, 1 (9.5)	Quadruple strip feed utilizing a pair of 90^0 hybrid couplers	34.5	25.9	1.89	Not mentioned
[9]	Rectangular, 1 (9.8)	Dual pair of conformal strips	36.8	14.46	3.95	$TE^x_{\delta 11}$ and $TE^x_{\delta 13}$
[10]	Cylindrical, 1 (9.8)	Two conformal strips of unequal length with modified Wilkinson power divider	35.61	15.21	1.72	Not mentioned
[11]	Cylindrical, 1 (9.8)	Dual vertical microstrip lines arranged in a perpendicular fashion	30.37	24.6	3.13	$HE^x_{11\delta}$ and $HE^y_{11\delta}$
[12]	Rectangular, 1 (9.8)	Two parallel conformal strips of unequal length	23.80	9.2	3.48	TE_{111}
[13]	Cylindrical, 1 (10)	Conformal strip is made of the combination of microstrip line and coaxial probe	9.43	4.9	2.45	$TM_{01\delta}$ and $TE_{01\delta+1}$
[14]	Cylindrical, 1 (10)	Broadband PDL is used for conformal strips	34	24.20	2.64	$HEM_{11\delta}$
[15]	Rectangular, 1 (10)	Unique conformal H-shaped metal strip	23	20	4.03	$TE^x_{\delta 13}$ and $TE^y_{1\delta 3}$
[16]	Cylindrical, 1 (9.5)	Conformal strip and two vertical metal strips on the surface of DRA	25	2.2	6.3	Not mentioned
[17]	Hemispherical, 1 (9.5)	Conformal-strip and parasitic patch on surface of DRA	Not mentioned	2.4	3.56	TE_{111}
[18]	Rectangular, 1 (9.5)	Conformal-strip and parasitic-patch surface of DRA	14	2.7	3.4	TE_{111} and TE_{113}
[19]	Hemispherical, 1 (9.5)	Conformal-strip and multiple parasitic strips on surface of DRA	2–4	2–4	3.7	TE_{111}
[20]	Rectangular, 1 (10)	E-shaped conformal strip and parasitic patch	23	22.01	4.01	$TE^x_{\delta 13}$ and $TE^y_{1\delta 3}$
[21]	Rectangular, 1 (10.8)	Dual vertical conformal strips	32.8	36.99	4.01	TE_{111} and TE_{113}
[22]	Rectangular, 1 (12)	Wilkinson power divider and a wideband 90° phase shifter associated with conformal strips	50.8	41.7	6.04	Not mentioned

IMBW = fractional –10-dB impedance bandwidth, ARBW = fractional 3-dB axial ratio bandwidth, ε_r = dielectric constant of DRA, f_c = center frequency with regard to axial ratio plot.

1. The design parameters of DRAs have been chosen based on the resonance frequency by using different formula for different convention shape of dielectric resonator, which were given in Chapter 2;
2. For the generation of pure circular polarization, the path difference must be 0.5λ or have a 90° phase shift between two different linearly polarized signals, as given in (4.1);
3. Generally, conformal strip will only provide a phase shift of 45° or a path difference of $\lambda/8$ distance, so it requires an additional 45° phase shift or a $\lambda/8$ path difference to fulfill condition of circular polarization, which are obtained by arranging the position of DRAs or directly using the stubs or metallic parasitic strips.;
4. So, generation of circular polarization is totally based on the 90° phase shift or $\lambda/4$ path difference between the two signals, which can be obtained by arranging the different combinations, like using stubs, parasitic elements, and partial ground plane along with the conformal technique in the conventional shape of DRAs.

4.4 Conformal Coupling in a Modified DRA for CP Configuration

Initially, basic geometrical shapes of dielectric resonators, such as hemispherical, cylindrical, and rectangular, were studied for the generation of circularly polarized fields by using the conformal strip technique. In this section, a modified shape of dielectric resonators has been considered for the generation of circular polarizations in different antenna design configurations. It is very much a difficult process to make a modified shape of dielectric resonators, but it helps to maximize the coupling as well as formation of orthogonal modes. Additionally, the process can be split into two classes for comprehension grounds: single-point conformal fed with modified shape of DR and dual-point conformal fed with modified shape of DR for circular polarization.

4.4.1 Single-Point Conformal-Fed with a Modified DRA

This section provides clear direction for using the modified shapes of dielectric resonators for generation of orthogonal modes with the help of the conformal technique as presented [23–28]. In [23], a new comb-shaped DRA is used for circularly polarized applications. Here, a coaxial probe connected to a conformal strip for exciting a comb DRA and the longitudinal slots are deliberated oriented with respect to the feed to produce two orthogonal components with quadrature phase shift. The slots will reduce the effective permittivity of the cylindrical DRA and hence achieve wider impedance bandwidth. This structure shows the covering AR bandwidth of 4%. Similar results can be obtained by using the information in Figure 4.26.

Conformal spiral wire has been used as a feed to excite the DRA and obtain orthogonal mode to get the circular polarization, as reported in [24]. This configuration of the DRA structure shows the 3-dB AR bandwidth of 3.4%. Figure 4.27 shows a similar kind of antenna structure and the antenna results could be also verified by using this antenna geometry.

The authors in [25] used conformal strips attached with an underlaid quadrature coupler to energize a hollow rectangular DRA. The idea is used to realize a circularly polarized field in a hollow rectangular DRA. Because the coupler is located beneath the DRA, it does not increase the complete size of the antenna, making the system compact. A pair of conducting strips were placed on two adjacent side walls of the DRA to excite the degenerate TE^{x}_{111} and TE^{y}_{111} orthogonal modes, producing circular polarizations. Such a combination of conformal strips gives AR bandwidths of 10.54%. The authors in [26] used an equilateral triangular DRA excited by the conformal spiral loop for obtaining circular polarizations. The conformal spiral loop was placed on two side walls of the TDRA to create the phase excitation and ensure the generation of the orthogonal modes. The presented antenna structure shows the dual band AR bandwidths of 1.6% and 1.8%, respectively.

In [27], the authors have used a stacked rectangular DRA with different volumes for generation of wideband circularly polarized broadside radiation characteristics, as shown in Figure 4.28. Here wide impedance bandwidths and AR bandwidths came from three different factors: stepped-shaped conformal strip associated with microstrip line, a stacked rectangular DRA with different volumes, and a vertical type of partial ground plane. The orthogonal $TE^{x}_{\delta 11}$

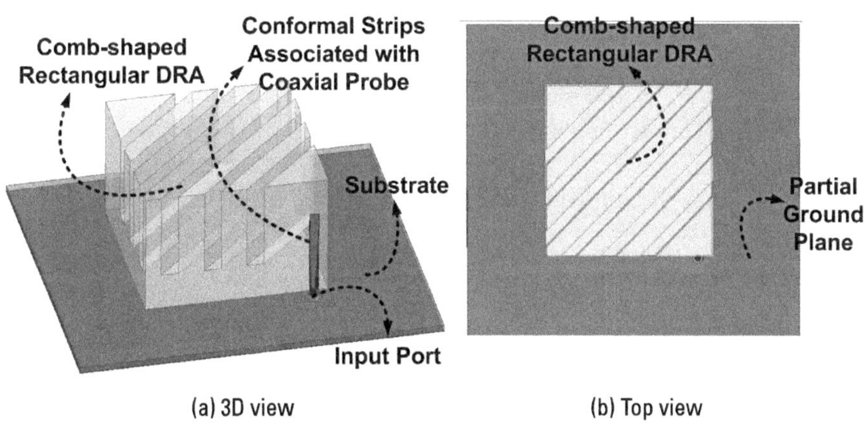

Figure 4.26 Conformal strip coupled comb-shaped rectangular DRA for narrowband circular polarization applications. (Modified after [23].)

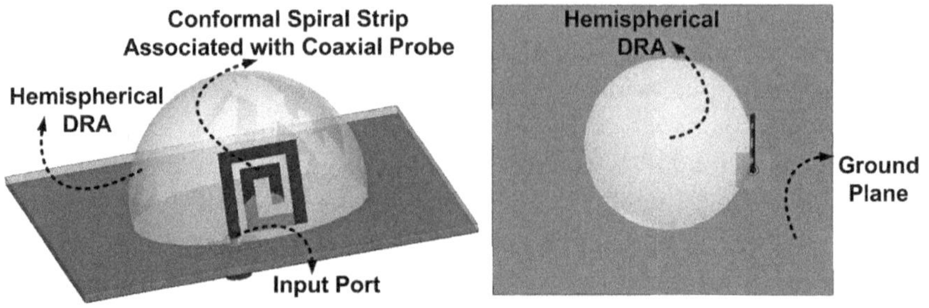

(a) 3D view (b) Top view

Figure 4.27 Conformal spiral strip coupled hemispherical DRA for narrowband circular polarization applications. (Modified after [24].)

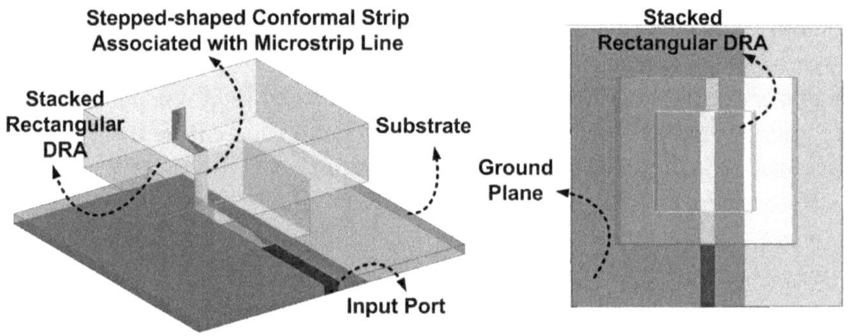

(a) 3D view (b) Top view

Figure 4.28 Stepped-shaped conformal strip coupled stacked rectangular DRA for wideband circular polarization applications [27].

and $TE^{y}_{1\delta 1}$ modes are mainly responsible in the stacked rectangular DRA for generation of circularly polarized fields. The stacked rectangular DRA gives AR bandwidths of 11.53% in broadside direction.

A V-shaped DRA loaded with circular patches and those patches attached with the coaxial probe has been reported in [28]. Such a modification of the DRA shows the AR bandwidths of 4% at a centered frequency of 8.65 GHz.

4.4.2 Dual-Point Conformal Fed with a Modified Single DRA

This section shows that the multiple port can also be used to generate the orthogonal modes in a modified shape of a DRA. In [29], a stair-shaped rectangular DRA was excited by using a single- and dual-section Wilkinson power divider with a 90° phase shifter to get circular polarizations. The presented antenna configuration shows the generation of triband circular polarizations (i.e.,

L5-, L1-, and S-bands). By using the dual-section Wilkinson power divider (WPD) with a wideband 90° phase shifter you get the circular polarization radiations in the L5-band and L1-band, respectively, while circular polarization is obtained in the S-band by using a single-section WPD with a wideband 90° phase shifter in triband DRA. The TE^y_{111}, TE^y_{113}, and TE^y_{112} modes are responsible for broadside circularly polarized radiation patterns of a triband DRA. It shows the 3-dB AR bandwidths of 1.6%, 2.0%, and 2.1% for the L5-, L1-, and S-bands, respectively. Similar results can also be analyzed by utilizing the antenna geometry shown in Figure 4.29.

All the designs based on a modified shape DRA are summarized in Table 4.2, which was excited by the conformal technique. To get the same amplitude and a phase shift of 90°, the modified shapes of DRAs are most suitable and help to get a CP condition in the DRAs. It is observed from Table 4.2 that the antenna design can be divided into four different steps for obtaining the CP fields in a modified shape of DRA, conformal feed, and ground plane:

1. The dimensions of a modified DRA have been chosen based on the optimizing process and use of the frequency range.
2. For the generation of pure circular polarization, the path difference must be $\lambda/2$ or have a 90° phase shift between two different linearly polarized signals, as given in (4.1).
3. Mostly the modified ground plane and conformal strips will provide a phase shift of 45° or a path difference of $\lambda/8$ distance, and the remaining 45° phase shift or $\lambda/8$ path difference is given by a modified shape of DRA.
4. So the combination of modified shapes and conformal strips along with ground plane gives a 90° phase shift or a $\lambda/4$ path difference

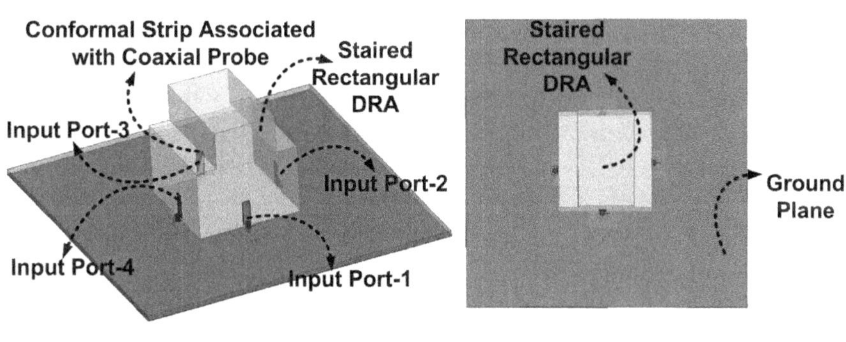

Figure 4.29 Conformal strip coupled with staired rectangular DRA for wideband circular polarization applications. (Modified after [29].)

Table 4.2
Brief Comparison of Various CP Designs Using Single-/Dual-Point-Fed Conformal Techniques with Modified Shapes of DRA

Ref	DRA Shape, Number of DRs (ε_r)	Conformal Strip Arrangements	IMBW (%)	ARBW (%)	f_c (GHz)	Modes Excited
[23]	Comb-shaped-cylindrical, 1 (10)	Vertical conformal	9	4	4.55	Not mentioned
[24]	Modified hemi-spherical, 1 (9.5)	Conformal wire	12	3.9	3.8	Not mentioned
[25]	Hollow rectangular, 1 (10)	Underlaid quadrature coupler associated with conformal	24.95	33.8	2.45	TE^x_{111} and TE^y_{111}
[26]	Equilateral triangular, 1 (10)	Dual-face spiral loop	12.37, 4.8	1.6, 1.8	7.43, 8.62	Not mentioned
[27]	Stacked rectangular, 1 (10)	Stepped-shaped conformal strip	54.84	11.53	2.34	$TE^x_{\delta 11}$ and $TE^y_{1\delta 1}$
[28]	V-shaped, 1 (10)	Circular patches associated with coaxial probe	25	4	8.65	Not mentioned
[29]	Stair shaped rectangular, 1 (38)	Dual sections WPD with a 90° phase shifter	15, 20, 16	1.6, 2.0, 2.1	1.176, 1.575, 2.49	$HEM_{11\delta}$

IMBW = fractional –10-dB impedance bandwidth, ARBW = fractional 3-dB axial ratio bandwidth, ε_r = dielectric constant of DRA, f_c = center frequency with regard to axial ratio plot.

between the two signals. Hence, it gives the circularly polarized fields the respective antenna designs, as shown in Table 4.2.

4.5 Conformal Coupling in Multiple DRAs for CP Configuration

Several design examples of multiple DRAs are discussed in this section for the generation of orthogonal modes by using conformal technique. Such types of designs offer comparatively lower permittivity of the materials, hence it shows good impedance matching as well as AR bandwidth. Here, the conformal strip is either associated with the microstrip feed line or coaxial probe to couple the maximum energy into dielectric resonators and also plays an important role for obtaining the circularly polarized fields. Multiple DRAs could also be separated into two different classes for clarity: single-point conformal-fed multiple DRAs and dual-point conformal-fed multiple DRAs.

4.5.1 Single-Point Conformal-Fed Multiple DRA

In [30] the author used conformal strips to excite a three-layer hemispherical DRA along with parasitic patches for generation of the orthogonal modes.

Wide impedance and AR bandwidths could be achieved through the use of the three layers with different permittivity of a hemispherical DRA, conformal strip, and feed locations (through patch). The wider impedance bandwidth can be realized with proper selection of permittivity in all three different layers. The dimensions of the strip and patch as well as their locations at different dielectric layer interfaces show vital insight into realizing wide impedance and AR bandwidths. The conformal strip-fed three-layer hemispherical DRA can be obtained by 3-dB AR bandwidths of 5.03%. The three-layer hemispherical DRA naturally shows that the radiation pattern is broadside because the TE_{111} mode is excited. A stair-shaped DRA, which is made of two rectangular dielectric resonators, is joined together to offer wideband circular polarization fields for wireless local area network applications, as reported in [31]. The excitation of multiple orthogonal modes depends on the stair-shaped DRA and the trapezoidal patch is used as a conformal mechanism that shows a wideband AR bandwidth. In this structure, the quasi-TE^x_{111}, quasi-TE^y_{111}, TE^x_{121}, and TE^y_{211} modes are responsible for getting the wide 3-dB AR bandwidth of 22% and the impedance bandwidth of 37%.

In [32], a rectangular and cubical-shaped dielectric resonator having optimum gap is excited by using a combination of conformal strip associated with the inverted L-shaped microstrip feed line, as shown in Figure 4.30. The combinations of two different feeds (i.e., conformal-strip and microstrip-feed) excites the TE_{111} orthogonal degenerate modes in two DRs to obtain the circularly polarized fields. The gap between the rectangular and cubical DRs and conformal strips decide the AR bandwidth in broadside direction. The inverted sigmoid-shaped DRA is used for dual-band circular polarization applications as reported in [33]. Here, a conformal strip is used for the excitation of a sigmoid-shaped

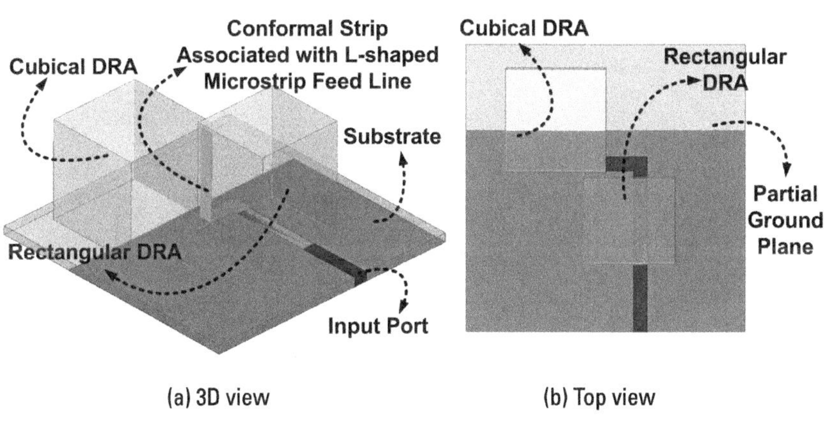

Figure 4.30 Conformal strip coupled rectangular and cubical DRA for wideband circular polarization applications [32].

DRA, but it shows the only single narrow AR bandwidth. Therefore, to achieve a dual-band AR bandwidth response, a metallic strip is placed at the corner of the DR from the center of antenna structure. The $HEM_{11\delta}$ and $HEM_{22\delta}$ modes are mainly responsible for providing circularly polarized radiations in the first and fourth bands, respectively. Hence, it shows the lower and upper band 3-dB AR bandwidth of 19.98% and 3.07% in broadside direction.

4.5.2 Dual-/Multiple-Point Conformal-Fed Multiple DRA

A differentially inserted compact dual-band that is circularly polarized has been observed in a multiple cylindrical DRA as reported in [34]. The inserted differential conformal feeding structure comprised of a pair of arc-shaped strips connecting with coaxial probes plays a significant role in reducing the size of the antenna. Such a configuration excites the fundamental mode of HEM_{111} of the cylindrical DRA and also shows an additional resonant mode to form the second band. The upper and lower bands of the cylindrical DRA can be independently controlled by the DR and the strips. It shows the 3 dB AR bandwidth of 8.06% and 6.38% for the lower and upper bands, respectively. Similar types of antenna results could also be identified by using the presented antenna geometry shown in Figure 4.31.

Table 4.3 summarizes all the designs based on the multiple DRAs that are excited by different combinations of the conformal strip technique. You use the different conformal strips either connected with the microstrip feed line or coaxial probe to excite the multiple DRAs for generation of circular polarization radiations. Therefore, the process can be divided into four different phases for obtaining the CP radiation patterns:

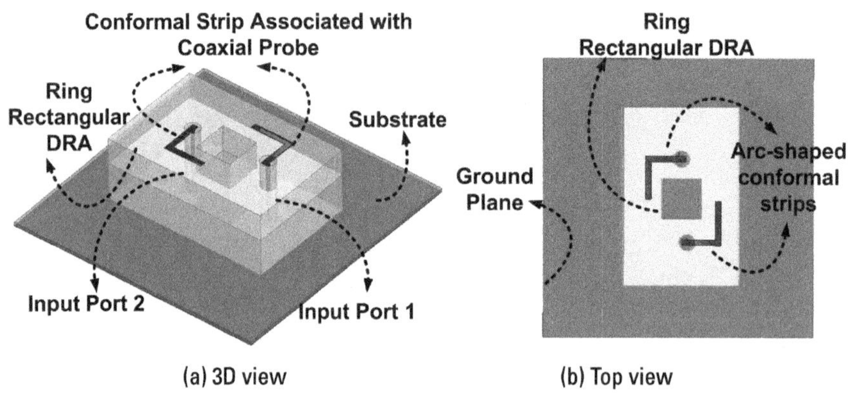

Figure 4.31 Arc-shaped conformal strips coupled ring rectangular DRA for dual band circular polarization applications. (Modified after [34].)

Table 4.3

Brief Comparison of Various CP Designs Using Single-/Dual-Point-Fed Conformal Techniques with Multiple DRAs

Ref	DRA Shape, Number of DRs (ε_r)	Microstrip Shape	IMBW (%)	ARBW (%)	f_c (GHz)	Modes Excited
[30]	Three-layer HDRA, 3 (3, 1, 9)	Conformal strip and parasitic atch	49.34	8.04	3.35	TE_{111}
[31]	Stair-shaped DRA, 2 (10.2)	Trapezoidal patch is associated with conformal strip	37	22	5.75	quasi-TE^x_{111}, quasi-TE^y_{111} and TE^x_{121}, TE^y_{211}
[32]	Rectangular and cubical, 2 (9.8)	Conformal-strip and inverted L-shaped microstrip feed line	27.36	23.00	5.65	TE_{111}
[33]	Inverted-sigmoid, 2 (12.8)	Conformal strip and metallic strip	30.07, 7.98	19.98, 3.07	5.96, 11.95	$HEM_{11\delta}$ and $HEM_{22\delta}$
[34]	Ring cylindrical, 2 (9.9)	Arc-shaped strips connecting with coaxial probes	30.4, 18.5	8.06, 6.38	3.5, 4.9	HEM_{111}

IMBW = fractional –10-dB impedance bandwidth, ARBW = fractional 3-dB axial ratio bandwidth, ε_r = dielectric constant of DRA, f_c = center frequency with regard to axial ratio plot.

1. The dimensions of multiple DRAs have been chosen based on the optimizing process and uses of the frequencies.
2. For the generation of pure circular polarization, the path difference must be $\lambda/2$ or have a 90° phase shift between two different linearly polarized signals, as given in (4.1).
3. Mostly the conformal strips along with ground plane provide a phase shift of 45° or a path difference of $\lambda/8$ distance, and the other 45° phase shift or $\lambda/8$ path difference is taken by the multiple DRAs.
4. So the combination of the multiple DRAs and different arrangements of the conformal strips along with ground plane provide a 90° phase shift or a $\lambda/4$ path difference between the two signals. Hence, it shows the circularly polarized waves in the respective design antenna.

4.6 Conformal Strip Coupling in MIMO CP-DRA Configuration

Over the last few years, MIMO has gained the attention of researchers worldwide due to its features, such as large signal capacity, link reliability, high data rate, and stability without increasing the channel bandwidth. The MIMO technology offers negligible effects of cochannel and adjacent channel interferences, and equal power levels of transmitter and receiver multiple uncorrelated signals. A DRA provides a better alternative in terms of the generation of different orthogonal modes, high-impedance BW, and high gain as compared to microstrip antennas. Therefore, circularly polarized based MIMO DRAs have been reported in [35–39] for various wireless applications.

A wideband circularly polarized DRA MIMO hybrid technique has been studied in [35] for worldwide interoperability for microwave access (WiMAX) applications. The hybrid technique is used and places a parasitic patch at an optimized distance beside the conformal metal strip of the two identical rectangular DRAs. Therefore, the parasitic patch is mainly responsible for the generation of the degeneration of two orthogonal modes (i.e., $TE^{x}_{\delta 13}$ and $TE^{y}_{1\delta 3}$) and obtained circular polarization. Here, isolation between the radiators is also intensified by inserting the parasitic patch and incorporating the diagonal position of DRAs.

The mutual coupling is significantly suppressed by placing the DRAs diagonally at an optimized position. The presented MIMO DRA is a hybrid of the parasitic patch and diagonal position of radiating elements, which offers AR bandwidth of 20.82%. Similar types of MIMO DRA results could also be identified by using the presented antenna configuration as shown in Figure 4.32.

In [36], a two-port MIMO DRA is proposed to achieve circularly polarized radiations. In this structure, by placing the dielectric resonator without separation along with increasing the length to width ratio helps to enhance the impedance bandwidth. Due to no separation between the dielectric resonators, it offers isolation between the two ports of the MIMO DRA at more than 15 dB. The authors have also demonstrated the performance of MIMO parameters, such as diversity gain. ECC, mean effective gain, channel capacity loss, and total active reflection coefficient are found in the acceptable range. The $TM_{21\delta}$ mode is responsible for the generation of circular polarization in the MIMO DRA and gives an AR bandwidth of 4.06%. A dual-band circularly polarized

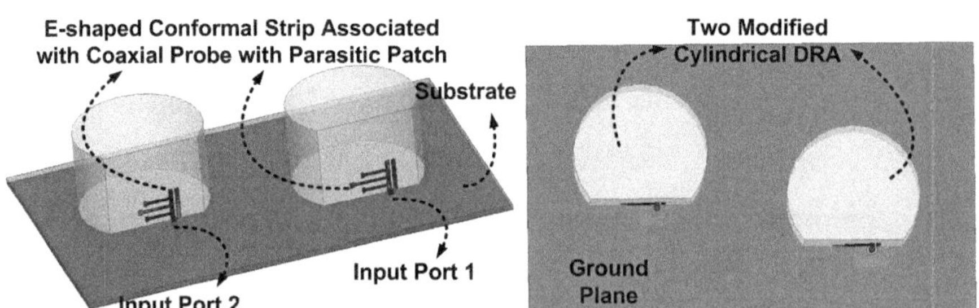

(a) 3D view (b) Top view

Figure 4.32 E-shaped conformal strip coupled modified cylindrical DRA for MIMO circular polarization applications. (Modified after [35].)

MIMO DRA for sub-6 or as midband applications has been reported in [39]. The presented antenna structure consists of a two-ring cylindrical DRA, which is excited by two arc-shaped microstrip feed lines and four conformal probes. Here, the defect ground structure is made by cutting a rectangular slot in the ground to increase isolation between these ports. The conformal probes are responsible for exciting the $HE_{11\delta}$ and $HE_{12\delta}$ modes having different resonant frequencies of the MIMO DRA. These HE modes help to produce the orthogonal by tuning the conformal probes and hence show dual-band circularly polarized fields. The multiband MIMO DRAs show an AR bandwidth of 5% and 2% in the broadside directions.

A summary based on a circularly polarized MIMO DRA is given in Table 4.4, which is excited by different arrangements of the conformal techniques. Using the conformal strip along with different shapes of DRAs helps to produce a degenerate orthogonal mode in antenna configuration. The process of a circularly polarized MIMO DRA is dependent on the use of multiple DRAs, conformal strips, and different ground planes. Thus, for obtaining circularly polarized radiation in MIMO, the process can be also separated into four different parts :

1. The dimensions of different shapes of DRAs have been chosen based on the optimizing process and uses of the frequency bands.

Table 4.4

Brief Comparison of MIMO CP Designs Using Conformal Techniques with DRAs

Ref	DRA Shape, Number of DRs (ε_r)	Conformal Strip Arrangements	IMBW (%)	ARBW (%)	f_c (GHz)	Modes Excited
[35]	Rectangular, 2 (10)	Two symmetric E-shaped conformal metal strip	38.51	20.82	3.99	$TE^x_{\delta 13}$ and $TE^y_{1\delta 3}$
[36]	Half-cylindrical and rectangular, 2-2 (12.8)	Vertical conformal strip	35.80	4.55	7.88	$TM_{21\delta}$
[37]	Rectangular, 1 (10)	Vertical conformal strip with inverted Z-shaped slot	13.79, 5.55	5.12, 1.0	5.85, 7.10	TE^x_{111}, TE^y_{111} and TE^x_{121}, TE^y_{121}
[38]	Ring cylindrical DRA, 2 (9.8)	Two L type feed networks, conformal probes	21.2, 22.2	5.66, 4.25	3.35, 3.52	$HE_{11\delta}$
[39]	Ring cylindrical DRA, 2 (10)	Vertical conformal strip with DGS	19, 9.4	5, 2	3.3, 5.5	$HE_{11\delta}$ and $HE_{12\delta}$

IMBW = fractional –10-dB impedance bandwidth, ARBW = fractional 3-dB axial ratio bandwidth, ε_r = dielectric constant of DRA, f_c = center frequency with regard to axial ratio plot.

2. For the generation of exact circular polarization waves, the path difference should have a 90° phase shift between two linearly polarized signals, as given in (4.1).
3. In MIMO configuration, the different combination of conformal strips and ground plane are used to provide a path difference of $\lambda/8$ and another path difference of $\lambda/8$ is obtained by different shapes of DRAs.
4. Therefore, the final combination of the MIMO DRAs, conformal feeding technique, and ground plane give a $\lambda/4$ path difference between the two signals. Hence, it gives circularly polarized fields in the respective MIMO DRA design.

4.7 Conformal Coupling Array and Filtering CP-DRA Configuration

4.7.1 Conformal Strip-Fed Circularly Polarized DRA Array

With the fast growth of fifth generation (5G) wireless communication systems, the millimeter-wave (mm-wave) antenna array has received more attention in research and industrial applications. Moreover, the mm-wave has a large data transmission capacity but also has large losses, which include dielectric loss, atmospheric path loss, metallic loss of transmission lines, and radiation loss as well as multipath interferences and polarization mismatch. To avoid these losses, circularly polarized DRA array antennas in mm-wave bands have been widely sought out. In [40], the conformal technique was used to obtain circularly polarized fields in DRA array configuration. Here, a dual Marchand balun network is employed for excitation of 2 × 2 array antenna and found that the proposed array DRA shows a pair of degenerate orthogonal modes (i.e., TM_{101}, and TM_{011}). These modes can be split by introducing perturbations on a rectangular DRA and after that it is more suitable for designing a circularly polarized DRA array antenna. A circularly polarized DRA array utilizing a dual Marchand balun network was proposed with a 3-dB AR bandwidth that is 90 MHz (5.32–5.41 GHz). A similar type of circularly polarized DRA array could be also realized by using the antenna configuration shown in Figure 4.33.

4.7.2 Conformal Strip-Fed Circularly Polarized DRA Filter

The next-generation mobile communication network, such as 5G or 6G, will greatly improve communication efficiency at sub-6-GHz and mm-wave frequency bands, although the circuits are becoming more compact and multifunctional at these bands but also suffer from electromagnetic interference and multicoupling problems. In response, the RF front-end system with filtering

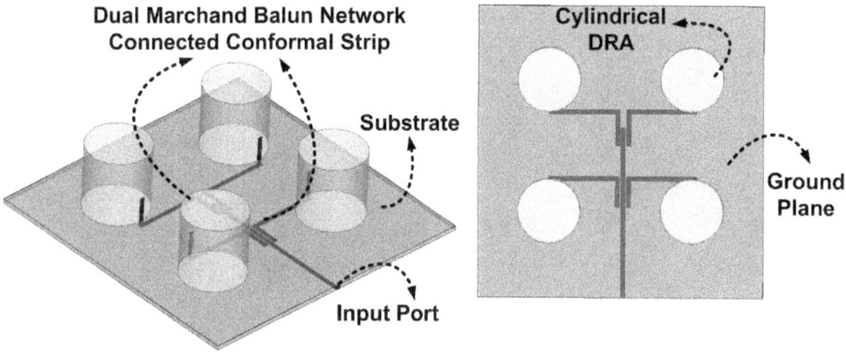

(a) 3D view (b) Top view

Figure 4.33 Dual Marchand balun network conformal strips coupled cylindrical DRA for array circular polarization applications. (Modified after [40].)

functionality rejects the interference and noise while simultaneously offering low cost, compact size, high integration, simple design, and multifunctionality. The linearly and circularly polarized filtering DRA as well as its equivalent horizontal magnetic dipoles are reported in [41]. The orthogonal degenerate $HEM_{11\delta}$ mode is obtained by perturbing the elliptical cross section and two notches to a cylindrical DRA and hence it gives circularly polarized fields. Here, a V-shaped conformal strip is used instead of a rectangular strip because the rectangular strip makes the angle 180° while the V-shaped strip gives the angle 120°, therefore it substantially improves the 3-dB AR bandwidth. A circularly polarized filtering DRA gives a 3-dB AR bandwidth of 2% (2.44–2.49 GHz) and 5% (2.35–2.47 GHz), respectively. Figure 4.34 shows similar types of antenna structures of circularly polarized filtering DRAs and all the near- and far-field results could also be realized.

Table 4.5 gives a complete summary of circularly polarized array and filtering DRAs, which were excited by different shapes of conformal strips either associated with a microstrip feed or coaxial probe. In these designs, the conformal strips have been placed in a such a way that give a quarter-wavelength for providing the orthogonal degenerate modes as well as arrays and filtering applications in DRAs. Hence, these design processes are also divided into four different sections for obtaining a circularly polarized field:

1. The dimensions of different shapes of DRAs have been chosen based on the optimizing process and uses of the frequency bands.
2. For the generation of an exact circular polarization wave, the path difference should be $\lambda/2$ or have a 90° phase shift between two linearly polarized signals, as given in (4.1).

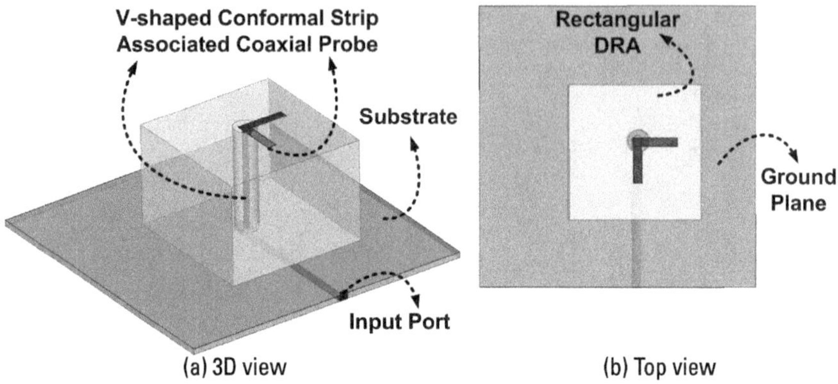

Figure 4.34 V-shaped conformal strip coupled rectangular DRA for filtering circular polarization applications. (Modified after [41].)

Table 4.5
Brief Comparison of Array and Filtering CP Designs Using Conformal Techniques with DRAs

Ref	Antenna Configuration	DRA Shape, Number of DRs (ε_r)	Conformal Strip Arrangements	IMBW (%)	ARBW (%)	f_c (GHz)	Modes Excited
[40]	Array	Rectangular, 1×4, 2×2 (72)	Dual Marchand balun network connected conformal strip	9.6	1.7	5.40	TM_{101}, and TM_{011}
[41]	Filter	Cylindrical, 1 (10)	V-shaped strip associated conformal strip	4.1	4.9	2.44	$HEM_{11\delta}$

IMBW = fractional −10-dB impedance bandwidth, ARBW = fractional 3-dB axial ratio bandwidth, ε_r = dielectric constant of DRA, f_c = center frequency with regard to axial ratio plot.

3. In an array or filtering configuration, a conformal strip is placed in such a way that it gives quarter-wavelength transformers to the DRA for obtaining circularly polarized fields.
4. Therefore, the final array/filter DRA configuration is mostly dependent on the shape and positions of the conformal strips, dielectric resonators, and ground plane. Hence, it provides a $\lambda/4$ path difference between the two signals. Therefore, it shows the circularly polarized fields in the respective array and filtering DRAs.

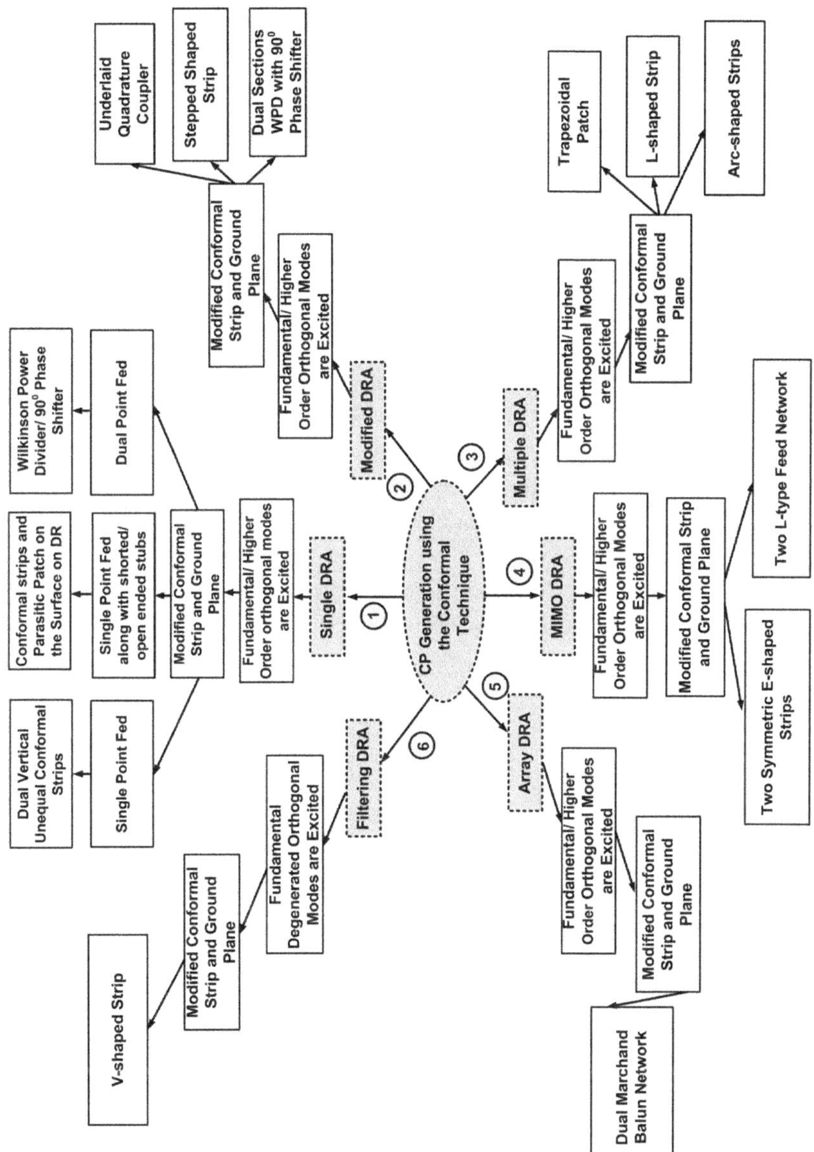

Figure 4.35 A complete summary of the conformal technique for generation of circular polarization in DRAs.

4.8 Summary

This chapter presented the details of the conformal techniques and generation of circularly polarized fields in DRAs for various wireless applications. A detailed analysis of different shapes of DRAs has been utilized based on the requirements of their respective design configurations. The circularly polarized fields can be obtained by combinations of conformal techniques, which are associated with either microstrip feed lines or coaxial probes along with using some extra elements such as parasitic metallic patch power dividers, stubs, strips, simple dividers, and quadrature couplers. Due the widespread availability of conformal techniques, researchers were able to design diversified combinations of circularly polarized DRAs, such as single-band, dual-band, triple-band, MIMO, array, and filtering applications. The first and second sections covered the field distribution and design methodology of conformal techniques to generate the circularly polarized field in DRAs. Then generation of circularly polarized radiations in a single DRA by using a conformal strip along with or without shorted/open-ended stubs or parasitic metallic plates/strips was described. Similarly, modified shaped and multiple DRAs with single or double conformal feeds were also described for circular polarization. The conformal technique was also used to develop a circularly polarized based MIMO, array, and filtering DRAs configurations. A final summary of the conformal technique is shown in Figure 4.35. Tables were provided at the end of each section for better understanding the key parameters.

References

[1] Leung, K. W., "Dielectric Resonator Antenna Excited by a Conformal strip," *IEEE Antennas and Propagation Society International Symposium*, 1999, pp. 2812–2815.

[2] Leung, K. W., "Conformal Strip Excitation of Dielectric Resonator Antenna," *IEEE Transactions on Antennas and Propagation*, Vol. 48, No. 6, 2000, pp. 961–967.

[3] Petosa, A., *Dielectric Resonator Antenna Handbook*, Norwood, MA: Artech House, 2007.

[4] Patel, P., B. Mukherjee, and J. Mukherjee, "Wideband Circularly Polarized Rectangular Dielectric Resonator Antennas Using Square-Shaped Slots," *IEEE Transactions on Antennas and Propagation*, Vol. 15, 2016, pp. 1309–1312.

[5] Fakhte, S., H. Oraizi, R. Karimian, and R. Fakhte, "A New Wideband Circularly Polarized Stair-Shaped Dielectric Resonator Antenna," *IEEE Transactions on Antennas and Propagation*, Vol. 63, No. 4, 2015, pp. 1828–1832.

[6] Wong, W. C., and K. W. Leung, "Circularly Polarized Dielectric Resonator Antenna Excited by Dual Conformal Strips of Unequal Lengths," *Microwave and Optical Technology Letters*, Vol. 29, No. 5, 2011, pp. 348–350.

[7] Huang, C. Y., and C. W. Ling, "Frequency-adjustable Circularly Polarised Dielectric Resonator Antenna with Slotted Ground Plane," *Electronics Letters,* Vol. 39, No. 11, 2003, pp. 1030–1031.

[8] Khoo, K. W., Y. X. Guo, and L. C. Ong, "Wideband Circularly Polarized Dielectric Resonator Antenna," *IEEE Transactions on Antennas and Propagation,* Vol. 55, No. 7, 2007, pp. 1929–1932.

[9] Chowdhury, R., and R. K. Chaudhary, "Wideband Circularly Polarized Rectangular DRA Fed with Dual Pair of Right-Angled Microstrip Line," *International Journal of Microwave and Millimeter-Wave Computer-Aided Engineering,* Vol. 26, No. 8, 2016, pp. 713–723.

[10] Kumari, R., and R. K. Gangwar, "Circularly Polarized Cylindrical Dielectric Resonator Antenna Excited by Dual Conformal Strips Along with Modified Wilkinson Power Divider For High Gain Application," *Microwave and Optical Technology Letters,* Vol. 59, No. 4, 2017, pp. 908–913.

[11] Chowdhury, R., N. Mishra, M. M. Sani, and R. K. Chaudhary, "Analysis of a Wideband Circularly Polarized Cylindrical Dielectric Resonator Antenna with Broadside Radiation Coupled with Simple Microstrip Feeding," *IEEE Access,* Vol. 5, 2017, pp. 19478–19485.

[12] Chowdhury, R., and R. K. Chaudhary, "Circularly Polarized Rectangular Dielectric Resonator Antenna Fed with Unequal Parallel Microstrip Lines for Wi-MAX Applications," *Microwave and Optical Technology Letters,* Vol. 59, No. 5, 2017, pp. 1182–1188.

[13] Guo, L., and K. W. Leung, "Compact Unilateral Circularly Polarized Dielectric Resonator Antenna," *IEEE Transactions on Antennas and Propagation,* Vol. 66, No. 2, 2018, pp. 668–674.

[14] Sun , Y. X., K. W. Leung, and J. F. Mao, "Dualfunction Dielectric Resonator as Antenna and Phase-Delay-Line Load: Designs of Compact Circularly Polarized/Differential Antennas," *IEEE Transactions on Antennas and Propagation,* Vol. 66, No. 1, 2018, pp. 414–419.

[15] Illahi, U., J. Iqbal, M. I. Sulaiman, M. M. Alam, M. M. Su'ud, and M. H. Jamaluddin, "Singly-Fed Rectangular Dielectric Resonator Antenna with a Wide Circular Polarization Bandwidth and Beamwidth for WiMAX/Satellite Applications," *IEEE Access,* Vol. 7, 2019, pp. 66206–66214, 2019.

[16] Lee, M. T., K. M. Luk, E. K. N. Yung, and K. W. Leung, "Microstrip-Line Feed Circularly Polarized Cylindrical Dielectric Resonator Antenna," *Microwave and Optical Technology Letters,* Vol. 24, No. 3, pp. 206–207, 2000.

[17] Leung, K. W., and H. K. Ng, "Theory and Experiment of Circularly Polarized Dielectric Resonator Antenna with a Parasitic Patch," *IEEE Transactions on Antennas and Propagation,* Vol. 51, No. 3, 2003, pp. 405–411.

[18] Li, B., and K. W. Leung, "Strip-Fed Rectangular Dielectric Resonator Antennas With/Without a Parasitic Patch," *IEEE Transactions on Antennas and Propagation,* Vol. 53, No. 7, 2005, pp. 2200–2207.

[19] Ng, H. K., and K. W. Leung, "Frequency Tuning of the Linearly and Circularly Polarized Dielectric Resonator Antennas Using Multiple Parasitic Strips," *IEEE Transactions on Antennas and Propagation,* Vol. 54, No. 1, 2006, pp. 225–230.

[20] Iqbal, J., U. Illahi, M. I. Sulaiman, M. M. Alam, M. M. Su'ud, M. N. M. Yasin, and M. H. Jamaluddin, "Bandwidth Enhancement and Generation of CP by Using Parasitic Patch on Rectangular DRA for Wireless Applications," *IEEE Access*, Vol. 7, 2019, pp. 94365–94372.

[21] Li, B., C. X. Hao, and X. Q. Sheng, "A Dual-Mode Quadrature-Fed Wideband Circularly Polarized Dielectric Resonator Antenna," *IEEE Antennas and Wireless Propagation Letters,* Vol. 8, 2009, pp. 1036–1038.

[22] Han, R. C., S. S. Zhong, and J. Liu, "Design of Broadband Circularly Polarized Dielectric Resonator Antenna Using Improved Feed Network," M*icrowave and Optical Technology Letters,* Vol. 56, No. 9, 2014, pp. 2191–2195.

[23] Chu, L. C. Y., D. Guha and Y. M. M. Antar, "Comb-shaped Circularly Polarised Dielectric Resonator Antenna," *Electronics Letters*, Vol. 42, No. 14, 2006.

[24] Khamas, S. K., "Circularly Polarized Dielectric Resonator Antenna Excited by a Conformal Wire," *IEEE Antennas and Wireless Propagation Letters,* Vol. 7, 2008, pp. 240–242.

[25] Lim, E. H., K. W. Leung, and X. S. Fang, "The Compact Circularly-Polarized Hollow Rectangular Dielectric Resonator Antenna with an Underlaid Quadrature Coupler," *IEEE Transactions on Antennas and Propagation*, Vol. 59, No. 1, 2011, pp. 288–293.

[26] Dash, S. K. K., T. Khan, and B. K. Kanaujia, "Circularly Polarized Dual Facet Spiral Fed Compact Triangular Dielectric Resonator Antenna for Sensing Applications," *IEEE Sensors Letters*, Vol. 2, No. 1, 2018.

[27] Kumar, R., and R. K. Chaudhary, "Stacked Rectangular Dielectric Resonator Antenna with Different Volumes for Wideband Circular Polarization Coupled With Step-Shaped Conformal Strip," *International Journal of Microwave and Millimeter-Wave Computer-Aided Engineering*, Vol. 29, No. 6, 2019, pp. 1–8..

[28] Singhwal, S. S., B. K. Kanaujia, A. Singh, and J. Kishor, "Circularly Polarized V-shaped Dielectric Resonator Antenna," *International Journal of Microwave and Millimeter-Wave Computer-Aided Engineering*, Vol. 29, No. 9, 2019, pp. 1–10.

[29] Chaudhary,P., D. K. Ghodgaonkar, S. Gupta, R. Jyoti, M. B. Mahajan, "Compact Circularly Polarized Triband Staired Rectangular Dielectric Resonator Antenna Using Single and Dual Sections WPD with Phase Shifter For Navigation Satellite Applications," M*icrowave and Optical Technology Letters,* Vol. 62, No. 5, 2020, pp. 2047–2058.

[30] Kakade, A. B., and M. S. Kumbhar, "Wideband Circularly Polarized Conformal Strip Fed Three Layer Hemispherical Dielectric Resonator Antenna with Parasitic Patch," M*icrowave and Optical Technology Letters,* Vol. 56, No. 1, 2014, pp. 72–77.

[31] Fakhte, S., H. Oraizi, R. Karimian, and R. Fakhte, "A New Wideband Circularly Polarized Stair-Shaped Dielectric Resonator Antenna," *IEEE Transactions on Antennas and Propagation*, Vol. 63, No. 4, 2015, pp. 1828–1832.

[32] Kumar, R., and R. K. Chaudhary, "A Wideband Circularly Polarized Dielectric Resonator Antenna Excited with Conformal-Strip and Inverted L-Shaped Microstrip Feed Line For WLAN/WI-MAX Applications," M*icrowave and Optical Technology Letters,* Vol. 58, No. 10, 2016, pp. 2525–2531.

[33] Varshney, G., S. Gotra, V. S. Pandey, and R. S. Yaduvanshi, "Inverted-Sigmoid Shaped Multiband Dielectric Resonator Antenna with Dual-Band Circular Polarization," *IEEE Transactions on Antennas and Propagation*, Vol. 66, No. 4, 2018, pp. 2067–2072.

[34] Tong, C. W., H. Tang, W. Qin, W. W. Yang, X. F. Zhang, and J. X. Chen, "Differentially Inserted-Fed Compact Dual-Band Circularly Polarized Dielectric Resonator Antenna," *IEEE Antennas and Wireless Propagation Letters*, Vol. 18, No. 12, 2019, pp. 2498–2502.

[35] Iqbal, J., U. Illahi, M. I. Sulaiman, M. M. Alam, M. M. Su'ud, and M. N. M. Yasin, "Mutual Coupling Reduction Using Hybrid Technique in Wideband Circularly Polarized MIMO Antenna for WiMAX Applications," *IEEE Access*, Vol. 7, 2019, pp. 40951–40958.

[36] Varshney, G., R. Singh, V. S. Pandey, and R. S. Yaduvanshi, "Circularly Polarized Two-Port MIMO Dielectric Resonator Antenna," *Progress in Electromagnetics Research M*, Vol. 91, 2020, pp. 19–28.

[37] Kumar, V., "Rectangular DR-based Dual-band CP-MIMO Antenna with Inverted Z-shaped Slot," *International Journal of Electronics*, Vol. 107, No. 10, 2020., pp. 1559–1573.

[38] Singhwal, S. S., B. K. Kanaujia, A. Singh, J. Kishor and L. Matekovits, "Multiple Input Multiple Output Dielectric Resonator Antenna with Circular Polarized Adaptability for 5G Applications," *Journal of Electromagnetic Waves and Applications*, Vol. 34, No. 9, 2020, pp. 1180–1194.

[39] Singhwal, S. S., B. K. Kanaujia, A. Singh, J. Kishor and L. Matekovits, "Dual-band Circularly Polarized MIMO DRA for sub-6 GHz Applications," *International Journal of Microwave and Millimeter-Wave Computer-Aided Engineering*, Vol. 30, No. 10, 2020, pp. 1–12.

[40] Yang, L. L., Y. L. Li, W. W. Yang, Y. J. Yang, J. X. Chen, "A Low-Profile Circularly Polarized Dielectric Patch Antenna and Array," *International Journal of Microwave and Millimeter-Wave Computer-Aided Engineering*, Vol. 30, No. 3, 2019, pp. 1–11.

[41] Liu, Y. T., K. W. Leung, J. Ren, and Y. X. Sun, "Linearly and Circularly Polarized Filtering Dielectric Resonator Antennas," *IEEE Transactions on Antennas and Propagation*, Vol. 67, No. 6, 2019, pp. 3629–3640.

5

Coplanar Waveguide-Fed Dielectric Resonator Antenna for Circular Polarization

This chapter scrutinizes a few circularly polarized DRA designs reported in the literature using coplanar waveguide feeding approaches. A brief introduction to coplanar coupling techniques will be presented to build a basic foundation. Then, reported research ideas are surveyed along with a comparison between them. Finally, a design example is presented in which a circularly polarized antenna is designed and analyzed using this approach.

5.1 Introduction to Coplanar Waveguide Feeding Technique

In the previous chapter, the designing of a circularly polarized DRA was based on the planar and conformal microstrip coupling technique. However, there is another interesting excitation mechanism that is available to couple energy to DRAs, which is CPW feeding. The coplanar waveguide technique is basically a combination of the microstrip technique (presented in Chapter 3) and the aperture coupling technique (Chapter 6). The coplanar waveguide technique involves the concept of planar transmission lines and it is mainly a combination of the ground plane and the microstrip feeding, which are present on the same side of the substrate and separated by slots [1, 2]. Some examples of CPW coupling are shown in Figure 5.1.

The feeding of dielectric resonators through a shorted CPW provides an electric monopole equivalence similar to the coaxial probe technique (discussed

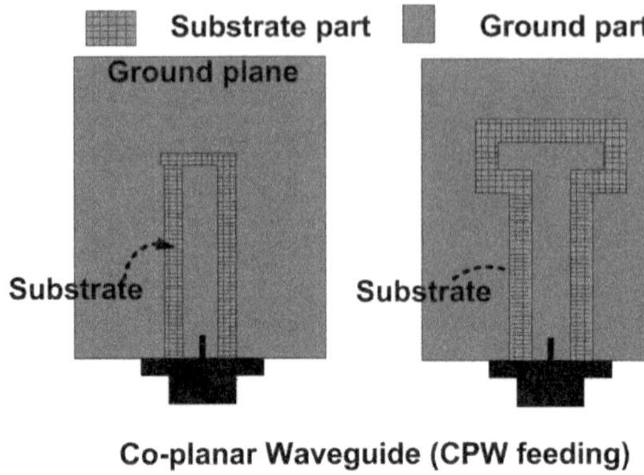

Figure 5.1 Examples of coplanar waveguide coupling methods.

in Chapter 7), and thus CPW-based feeding has the advantage of simple fabrication and easy integration to the antenna and substrate without involving a drilling process. There is another advantage associated with its electric monopole property, which is that interference is reduced to the field distribution of the excited modes [3–6] using a CPW feed.

There are some applications in antenna technology that need to manage the ground plane and feed position in the same plane, so the system can be easily integrated into the MMICs where this technique is useful. The coplanar waveguide technique also provides less dispersion, low radiation loss, and design flexibility required for antenna engineers [7, 8].

Yet another advantage of CPW feeding is that the radiation from the feed structure is negligible because it is excited in the odd mode of a coupled slot line. The modes of a DRA depend on the shape and location of the CPW feed line; therefore, by changing the location of the CPW feed, the modes of a DRA can be changed. For example, in the case of a cylindrical DRA, if a CPW feed line is in the center, it generates $TM_{01\delta}$ mode (Figure 5.2(b)) while edge positions generate the $HE_{11\delta}$ mode (Figure 5.2(a)) as shown in Figure 5.2.

Therefore, the CPW could be an alternative option to fulfill the demands of the antenna designers and research in CP technology. This coupling technique is not explored for realizing CP designs and therefore few research articles fall under this category of feeding and to be more specific, three or four research articles have been reported based on the CPW fed technique for realizing CP waves.

5.2 CP Generation Techniques in DRAs Using CPW Feeding

This section inspects designs reported using CPW feeding techniques. An example of a CPW-fed circularly polarized DRA was reported in [9], in which a

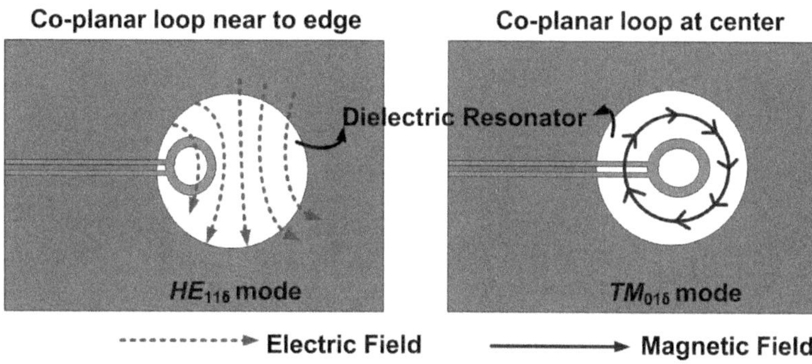

Figure 5.2 CPW coupling to a cylindrical DRA for excitation of (a) HE11δ mode, and (b) TM01δ mode.

wideband planar inverted-*F*-shaped DRA was designed for circular polarization radiation. A conducting patch was designed on top of the dielectric resonator substrate and a slot was introduced in the middle of the radiating element. With a nano-sized DRA having a thickness of 3.5 mm, the proposed antenna provides a –10-dB input impedance bandwidth of 255 MHz (2,300–2,555 MHz) or about 10.4% centered at 2.45 GHz in the lower mode, whereas in the upper mode, it reaches up to 835 MHz (5,595–6,430 MHz) or about 14.4% with respect to 5.8 GHz.

Another example of a circularly polarized rectangular DRA is reported in [10] that was excited by a CPW coupled folded slot in which two orthogonal modes were generated at two frequencies with a 90° phase difference at the design frequency f_d= 2.49 GHz to produce circular polarization. The rectangular DRA was rotated with respect to the *x*-axis to generate two orthogonal modes in space domain.

The dimensions of the CPW coupled folded slot were optimized to generate a 90° phase difference at the design frequency f_d and to maximize coupling to the orthogonal modes of the DRA. A similar modified CPW fed antenna structure is shown in Figure 5.3 adapted from the design presented in [10]. In this design, the asymmetric placement of the DRA on the modified CPW feeding is responsible for the generation of orthogonal modes with a 90° phase difference.

There are two articles in which a wideband CP was reported [11, 12], which are now discussed. The authors in [11] investigated for the first time a wideband CP in a DRA by using a modified CPW feed, as shown in Figure 5.4. A half-split cylindrical dielectric resonator having two layers with different permittivity and a new modified CPW feed has been utilized for achieving wideband CP in this designed antenna. With this technique, it offers -10 dB impedance bandwidths of 25.94% and 3-dB axial ratio bandwidth of 17.34% in boresight direction. The position of the DRA has been tuned to maximize

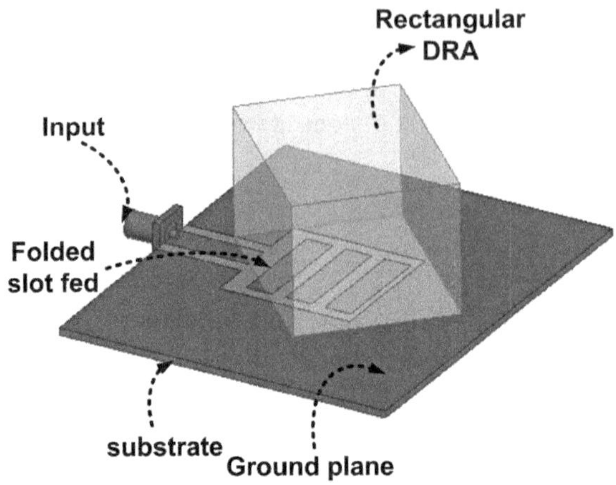

Figure 5.3 Design of a CPW-fed rectangular DRA for CP generation. (Modified after [10].)

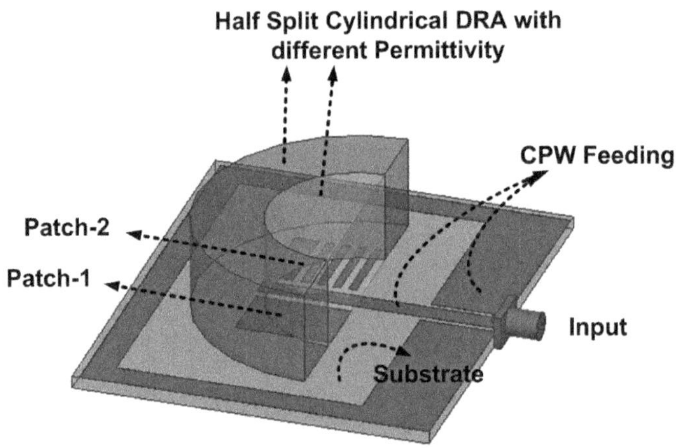

Figure 5.4 Example of a circularly polarized DRA with CPW feeding [11].

the coupling between the modified CPW feed and the DRA for obtaining optimum impedance bandwidth as well as excitation of orthogonal modes in this antenna configuration. The half-split DRA is excited in its $TM_{11\delta}$ mode, which is responsible for the generation of orthogonal fields in the reported antenna.

The orthogonal modes are excited by using the two patches (one is slotted) along with the CPW-fed line. Initially,the first rectangular patch (Patch-1) is placed below the DRA. This is the main region for placing the rectangular patch to reduce the excess tangential electric field and increase the gain of antenna. The addition of a rectangular patch helps to improve the matching of the presented antenna, while it also provides the orthogonal modes in the DRA. Next, another rectangular patch (Patch-2) is added below the DRA. By introducing another rectangular patch, it shows better matching compared to the matching when a single patch was there. The two rectangular patches are mainly responsible for the matching. Then, Patch-2 is modified that gives rise to more electromagnetic coupling to the DRA and also excites the orthogonal modes in the proper way to achieve better circularly polarized performance. The presented antenna provides LHCP in broadside direction.

In another example, the same authors reported a different wideband bidirectional circularly polarized antenna using CPW feeding in [12]. The geometry of the antenna is shown in Figure 5.5. By modifying the rectangular slot as an asymmetrical slot, the orthogonal modes with approximate equal magnitude were generated in the rectangular DRA across a narrow frequency range. The *L*-shaped feed generates two orthogonal modes with approximate equal magnitude across the wide frequency range.

It is worth to mentioning that the *L*-shaped feed line with an asymmetric (deformed) rectangular-slot resonator is capable of generating two orthogonal modes with approximate equal magnitude (i.e., $TE^{x}_{\delta 11}$ and $TE^{y}_{1\delta 1}$) across the wideband in the DRA to achieve wideband circularly polarized radiation in both directions (broadside and backside). A pair of orthogonal $TE^{x}_{\delta 11}$ and $TE^{y}_{1\delta 1}$ modes are generated in the rectangular hybrid DR antenna for the circularly polarized fields. The reported circularly polarized hybrid DRA operates across a wide impedance bandwidth of 80.51% and provides a wideband 3-dB axial ratio bandwidth of 43.75% in the broadside and the backside directions. The proposed antenna has bidirectional circularly polarized radiation and shows right-handed circularly polarized in +*Z*-direction and left-handed circularly polarized

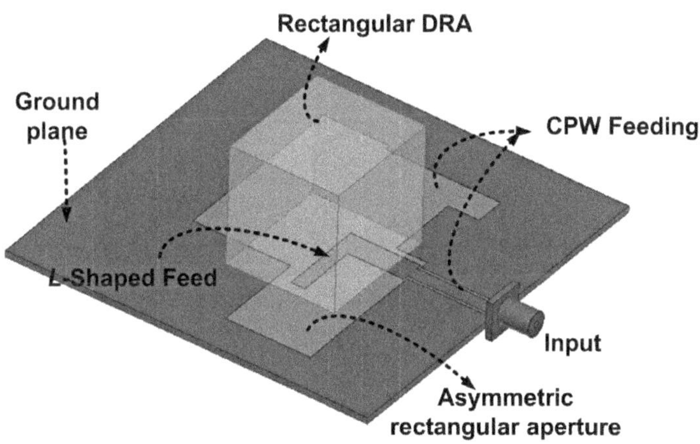

Figure 5.5 Example of a circularly polarized DRA with CPW feeding with L-shaped feed and modified rectangular aperture [12].

in -*Z*-direction; therefore, it is utilized in the potential applications in various wireless systems to cover a wide-frequency range. A summary of designs reported using coplanar waveguide feed is given in Table 5.1.

5.3 Design Example for a CPW-Fed Circularly Polarized DRA

A design example is now presented for CP generation in DRA using the coplanar waveguide technique. A circularly polarized rectangular dielectric resonator antenna is designed using a CPW feeding approach as shown in Figure 5.6. A rectangular DRA having a dielectric constant of 9.8 (alumina) is used, which has equal lateral dimensions of $l = 20$ mm, $w = 20$ mm, and $h = 19$ mm. These dimensions are optimized when maximum axial ratio bandwidth is obtained.

Table 5.1
Brief Comparison of Various CP Designs using CPW Feeding Techniques in DRAs

Ref	DRA Shape (ε_r)	Feed Arrangement	IMBW (%)	ARBW (%)	f_c (GHz)	Modes Excited
[10]	Rectangular (10)	CPW coupled folded slot	11.4	3.6	2.5	NM
[11]	Two half-split cylindrical DRA (10.2. 2.32)	CPW feeding with slotted patch	25.94	17.34	4.90	$TM_{11\delta}$
[12]	Rectangular DRA (9.8)	CPW feeding with modified aperture	80.51	43.75	2.24	$TE^{x}_{\delta 11}$ and $TE^{y}_{1\delta 1}$

IMBW = fractional –10-dB impedance bandwidth, ARBW = fractional 3-dB axial ratio bandwidth, NM = not mentioned, ε_r = dielectric constant of DRA, f_c = center frequency with regard to axial ratio plot.

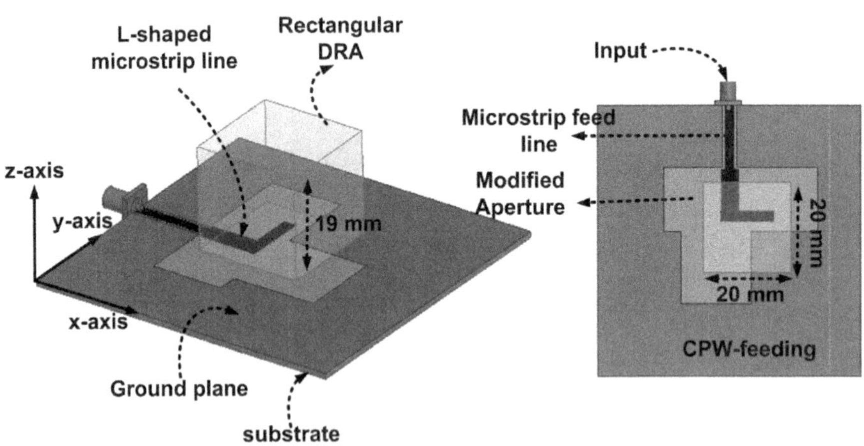

Figure 5.6 A design example showing a rectangular DRA fed with a CPW feeding for CP generation. (Modified after [12].)

The DRA is placed above an *L*-shaped microstrip feed line along with an asymmetric rectangular modified aperture. The asymmetric slot with *L*-shaped feed line is designed on FR4 glass epoxy substrate ($\varepsilon_r = 4.4$, $\tan\delta = 0.025$). The rectangular slot is converted into an asymmetric shape in order to generate two orthogonal modes inside the rectangular DRA.

5.3.1 Antenna Design Methodology and Discussion

The design methodology of the considered design example is elaborated on in this section. The antenna considered is broadly divided into four different configurations (subantenna structures) called Sub-Ant. 1 to Sub-Ant. 4, in which the Sub-Ant. 4 is the proposed antenna. This is done to understand the working principle of circularly polarized radiation.

Initially, the Sub-Ant. 1 is designed using a CPW-feed mechanism in which a large rectangular slot is created, and the rectangular DR is excited using simple straight microstrip feed line as shown in Figure 5.7(a). The corresponding input reflection coefficient and axial ratio plot is shown in Figure 5.8. It has been observed that the rectangular DRA is not properly coupled at this antenna stage. Next, Sub-Ant. 2 is obtained by modifying the straight feed line into an *L*-shaped feed line as shown in Figure 5.7(b). From Figure 5.8 it can be seen that the input reflection plot now comes below –10 dB as the rectangular DRA is now properly coupled. However, the axial ratio is still above 3 dB as shown in Figure 5.8(b).

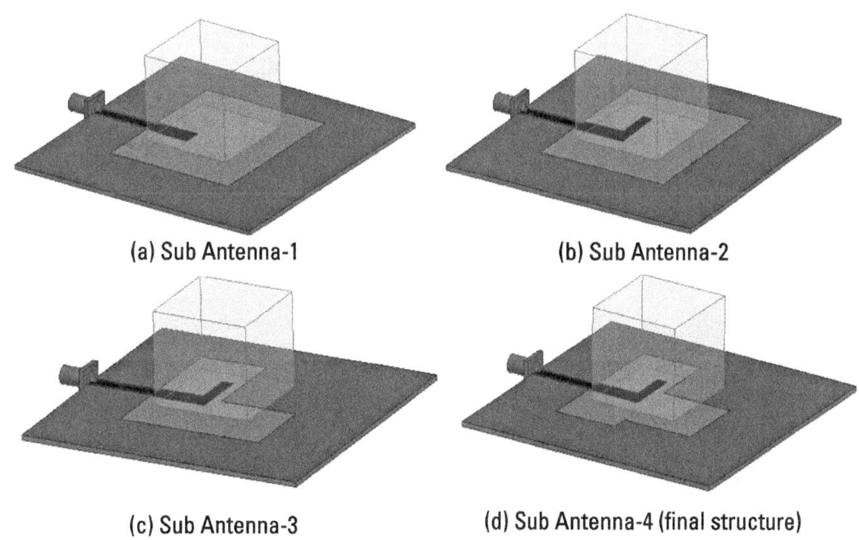

Figure 5.7 Design evolution of the proposed design example antenna.

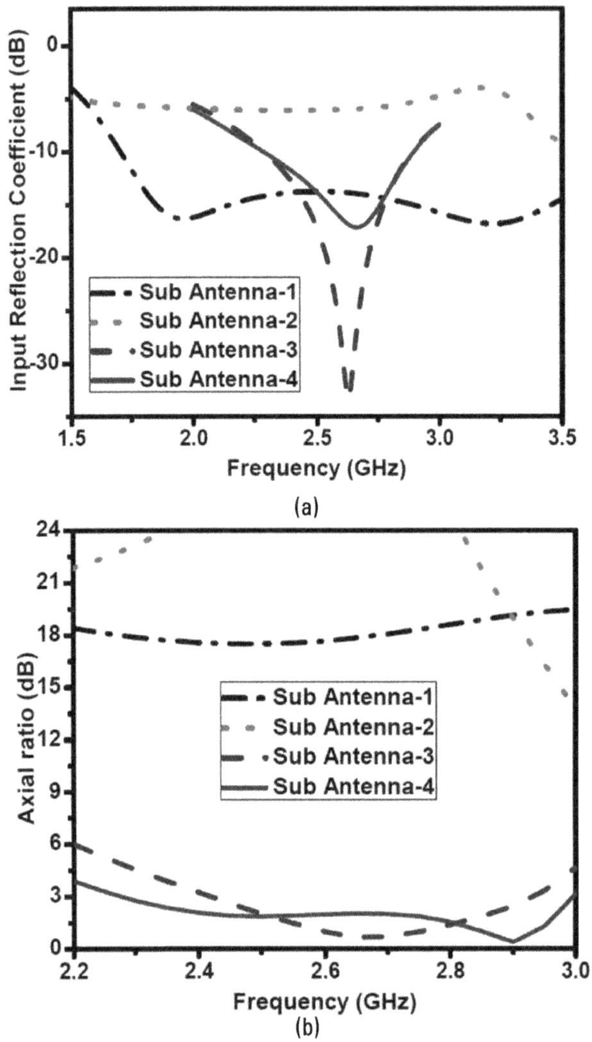

Figure 5.8 Simulated (a) input reflection coefficient and (b) axial ratio at different antenna stages of the proposed design example.

Both Sub-Ant. 1 and Sub-Ant. 2 are not generating two orthogonal modes for circularly polarized radiation. To improve the axial ratio below 3 dB, the Sub-Ant. 3 is formed by modifying the rectangular slot geometry as shown in Figure 5.7(c) with the same feed line as the Sub-Ant. 2.

By modifying the rectangular slot as an asymmetric slot, the orthogonal modes with approximate equal magnitude are generated in the rectangular DRA across the narrow frequency range. This can be confirmed from the axial ratio plot shown in Figure 5.8(b), where the axial ratio comes below 3 dB. The

L-shaped feed and asymmetrical rectangular slot helps in proper coupling and generates two orthogonal modes with approximate equal magnitude across the wide frequency range. Finally, Sub-Ant. 4 is formed by further modifying the slot as shown in Figure 5.7(d). The Sub-Ant. 4 shows the simulated –10 dB impedance bandwidth of around 24.2% (2.28–2.91 GHz) and the usable 3-dB axial ratio bandwidth in the broadside of the same 24.2% (2.28–2.91 GHz). A pair of orthogonal $TE^{x}_{\delta 11}$ and $TE^{y}_{1\delta 1}$ modes are generated in the rectangular dielectric resonator antenna for the circularly polarized fields.

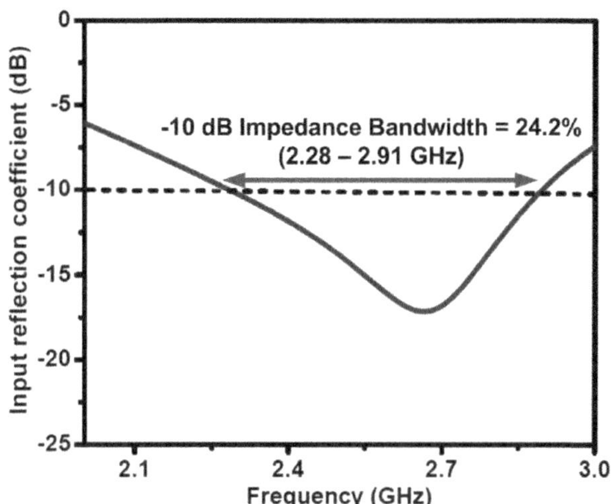

(a)

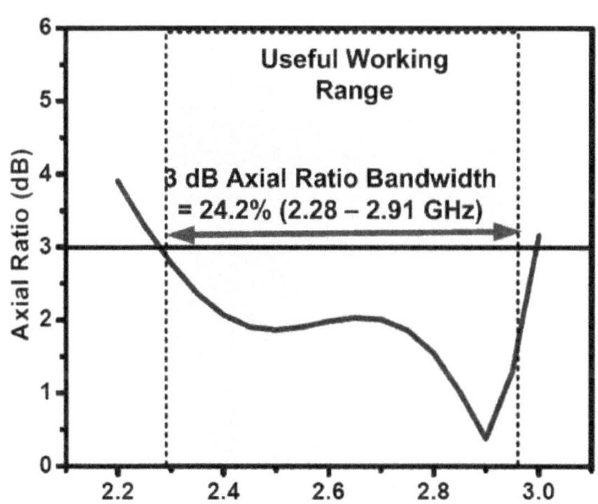

(b)

Figure 5.9 Simulated (a) input reflection coefficient and (b) axial ratio plot of the presented design examples.

It is worth mentioning here that the *L*-shaped feed line with asymmetrical (deformed) rectangular aperture that is combined and behaving like a CPW feeding is capable of generating two orthogonal modes with approximate equal magnitude and a 90° phase difference.

The corresponding simulated input reflection coefficient and axial ratio plots are shown separately in Figure 5.9 for the design example under consideration. It is observed that a –10-dB impedance bandwidth of 24.2% and a usable 3-dB axial ratio bandwidth of 24.2% have been obtained, respectively.

Normalized radiation patterns are plotted on an *xz*-plane in Figure 5.10. It has been perceived that right-handed circularly polarized (RHCP) waves are leading components as compared to left-handed circularly polarized (LHCP) waves along broadside direction. Hence, this proposed antenna design radiates in RHCP mode. It has been found that RHCP fields are stronger then LHCP fields in broadside direction approximately by 20 dB (2.7 GHz) and 30 dB (2.9 GHz) respectively on an *xz*-plane. Figure 5.11 shows the electric field distribution at different antenna stages where linearly polarized behavior has been noticed for Sub-Ant. 2 and circular polarized behavior for Sub-Ant. 3 and Sub-Ant. 4. Also, the counterclockwise motion of electric field confirms the RHCP mode of circular polarization.

5.4 Summary

A brief investigation of circular polarization analysis in DRAs using the coplanar waveguide feeding method was presented in this chapter. Although few

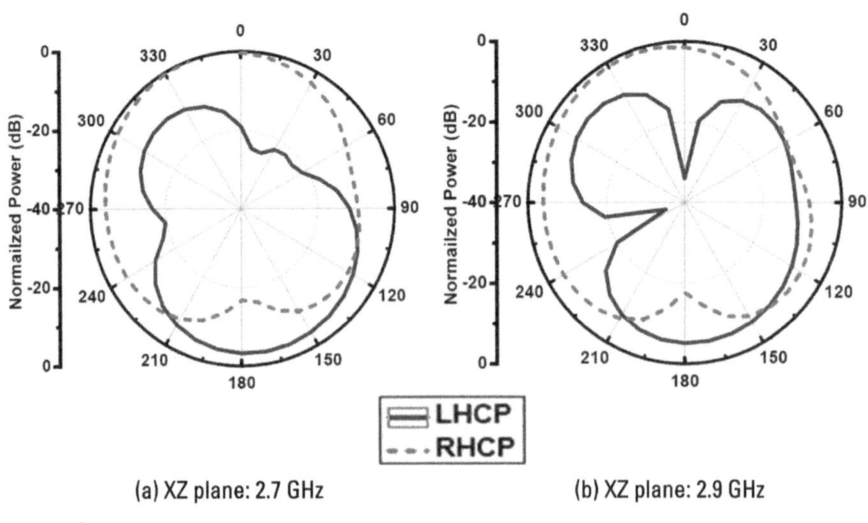

(a) XZ plane: 2.7 GHz (b) XZ plane: 2.9 GHz

Figure 5.10 Simulated radiation pattern of the proposed design example in xz-plane at 2.7 GHz and 2.9 GHz.

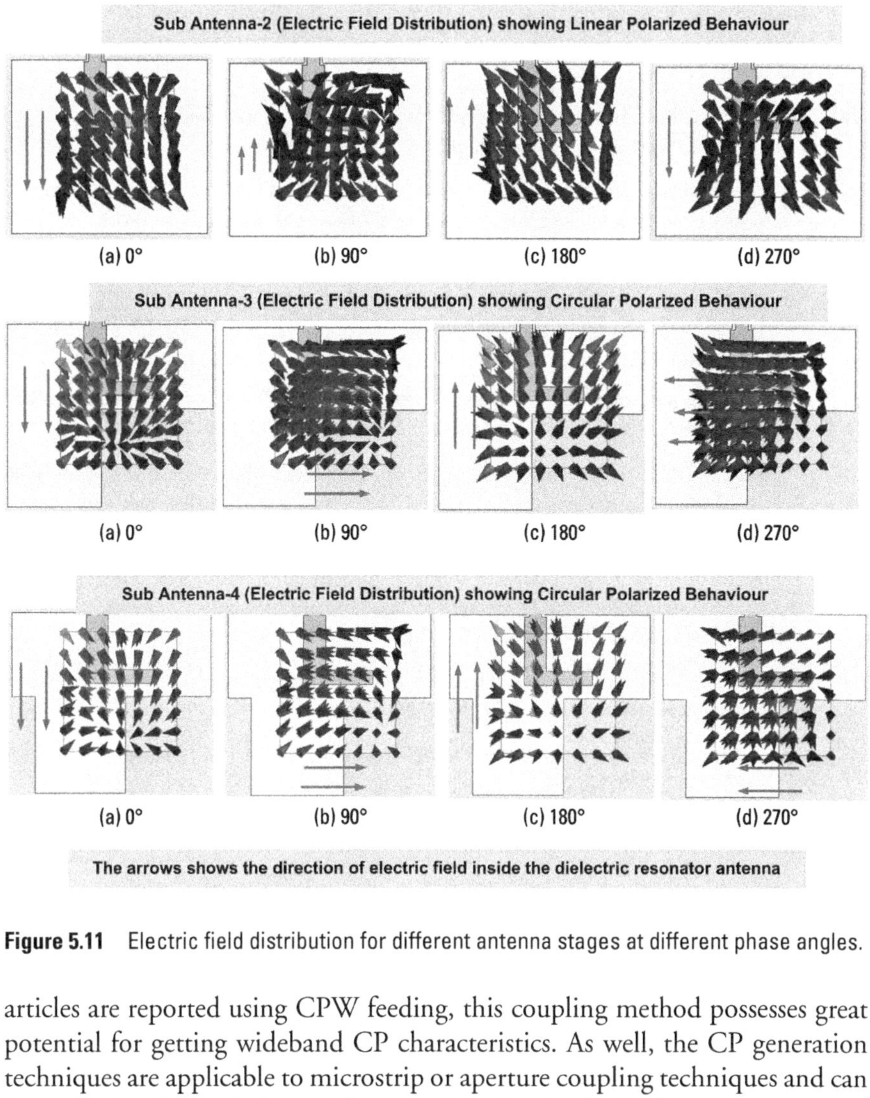

Figure 5.11 Electric field distribution for different antenna stages at different phase angles.

articles are reported using CPW feeding, this coupling method possesses great potential for getting wideband CP characteristics. As well, the CP generation techniques are applicable to microstrip or aperture coupling techniques and can be applied to CPW feeding with some alteration in the feeding.

References

[1] Wu, J. Y., C. Y. Huang, and K. L. Wong, "Low-profile, Very High-permittivity Dielectric Resonator Antenna Excited by a Coplanar Waveguide," *Microwave and Optical Technology Letters*, Vol. 22, No. 2, 1999, pp. 96–97..

[2] Deng, S. M., T. W. Chen, and H. H. Kan, "A CPW-Fed Rectangular Dielectric Resonator Antenna," *Asia-Pacific Microwave Conference*, Vol. 2, 2001, pp. 954–957.

[3] Kranenburg, R. A., S. A. Long, and J. T. Williams, "Coplanar Waveguide Excitation of Dielectric Resonator Antennas," *IEEE Transactions on Antennas and Propagation,* Vol. 39, No. 1, 1991, pp. 119–122.

[4] Ghosh, B., Y. M. Antar, A. Petosa, and A. Ittipiboon, "Feed Configurations of CPW Fed DRA," *IEEE Antennas and Propagation Society Symposium,* Vol. 2, 2004, pp. 1347–1350.

[5] Gao, Y., Z. Feng, and L. Zhang, "Compact CPW-Fed Dielectric Resonator Antenna with Dual Polarization," *IEEE Antennas and Wireless Propagation Letters,* Vol. 10, 2011, pp. 544–547.

[6] Gao, Y., B. Ooi, W. B. Ewe, and A. P. Popov, "A Compact Wideband Hybrid Dielectric Resonator Antenna," *IEEE Microwave and Wireless Components Letters,* Vol. 16, No. 4, 2006, pp. 227–229.

[7] Ghosh, B., Y. M. Antar, A. Petosa, and A. Ittipiboon, "CPW Feed to Rectangular DRA," *Microwave and Optical Technology Letters,* Vol. 45, No. 3, 2005, pp. 210–216.

[8] Al Salameh, M. S., Y. M. Antar, and G. Seguin, "Coplanar-Waveguide-Fed Slot-Coupled Rectangular Dielectric Resonator Antenna," *IEEE Transactions on Antennas and Propagation,* Vol. 50, No. 10, 2002, pp. 1415–1419.

[9] Tseng, C. F., C. L. Huang, and C. H. Hsu, "A Wideband Planar Inverted-F Dielectric Resonator Antenna for RFID System Applications," *Microwave and Optical Technology Letters,* Vol. 48, No. 7, 2006, pp. 1302–1305.

[10] Curry, C., "Novel Size-Reduced Circularly Polarized Antennas," Master's Thesis, Royal Military College, Kingston, Canada, July 2000.

[11] Kumar, R., and R. K. Chaudhary, "A New Modified CPW Fed Wideband Circularly Polarized Half-Split Cylindrical Dielectric Resonator Antenna with Different Permittivity of Two Layers in Radial Direction," *International Journal of RF and Microwave Computer-Aided Engineering,* Vol. 27, No. 3, 2017.

[12] Kumar, R., and R. K. Chaudhary, "Wideband Circularly Polarized Hybrid Dielectric Resonator Antenna with Bi-Directional Radiation Characteristics for Various Wireless Applications," *International Journal of RF and Microwave Computer-Aided Engineering,* Vol. 29, No. 9, 2019.

6

Aperture-Fed Dielectric Resonator Antennas for Circular Polarization

This chapter examines various DRA designs fed by aperture coupling techniques for the generation of circular polarization and improving its performance. A brief introduction to the aperture coupling technique will be presented to build a basic foundation. Then, different aperture coupling methods are surveyed along with a comparison between them in each method. There is also a brief review of circularly polarized DRA designs in other antenna types, such as MIMO antennas. Filtering antennas are also included in this chapter.

6.1 Introduction to the Aperture Coupling Technique

The aperture coupling or slot coupling method is one of the most common techniques being used in DRA technology for various applications, such as wideband, circular polarization, and multiband. In this method, a DRA is excited through an aperture (of different shapes) cut in the ground plane above which the DRA is positioned. The aperture is fed by an open-ended microstrip line (simple or modified) or waveguide [1–3] that is present on the other side of the ground plane (below substrate). This arrangement, where the radiating element and feeding network are separated by a ground plane in between, helps in minimizing the spurious radiation in the system. Figure 6.1 shows the general arrangement of aperture coupling techniques.

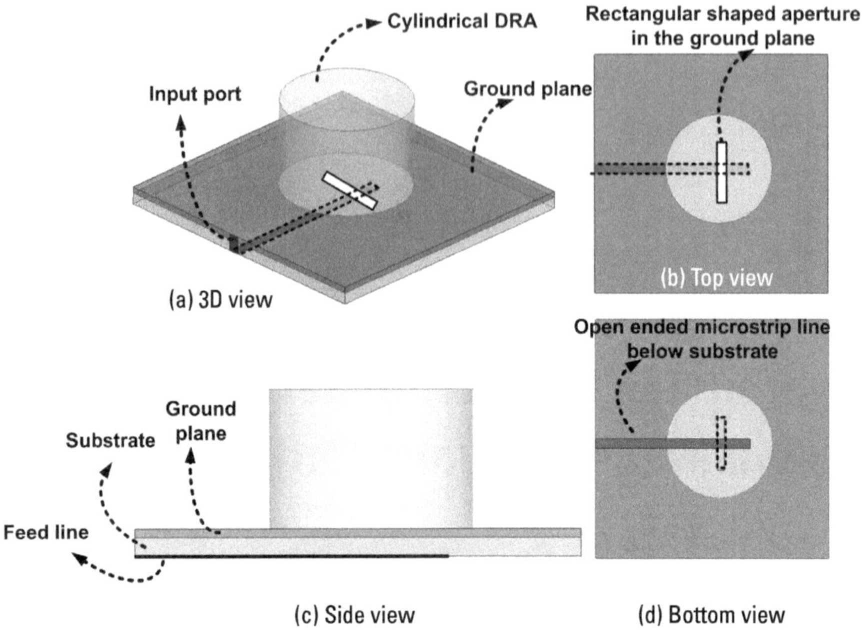

Figure 6.1 General arrangement of aperture coupling technique in a DRA.

Aperture coupling resembles a horizontal equivalent magnetic dipole and behaves as a magnetic current source (like a loop) that can be conveniently incorporated on any substrate and makes its implementation easier. Therefore, in order to excite any particular mode in a DRA using this coupling method, the aperture should be placed in a location of maximum magnetic field intensity. Thus, knowledge of internal field configuration (particularly magnetic field distribution) of the mode in the DRA is prerequisite critera. The electric field is sketched in Figure 6.2 within a rectangular aperture and direction of equivalent magnetic current source is also given as parallel to the slot length.

The aperture length must be large enough to achieve proper coupling to the dielectric resonator, but large apertures may resonate at the same frequency as the DRA, resulting in backlobe radiation. Thus, there is a trade-off between the backlobe radiation and proper coupling to the dielectric resonator with respect to aperture length [4]. A suitable starting values of slot length (L_s) and width (W_s) can be calculated by using (6.1) and (6.2) [4]:

$$L_s = \frac{0.4\lambda_0}{\sqrt{\varepsilon_e}} \tag{6.1}$$

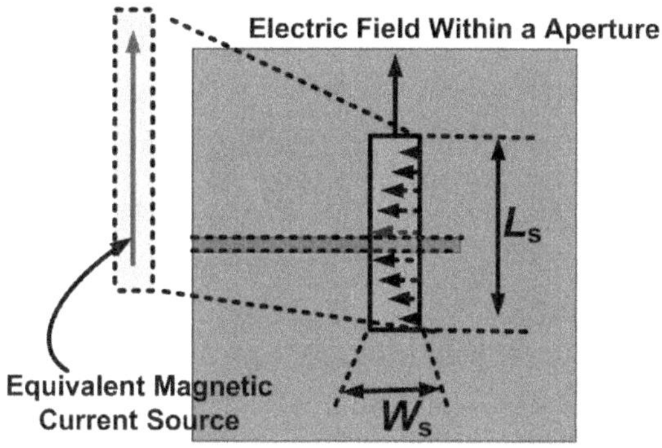

Figure 6.2 Electric field and magnetic current sources for a rectangular aperture.

$$W_s = 0.2L_s \tag{6.2}$$

Depending on the required aperture shape and dimensions, it can be adapted into various shapes. Various types of aperture coupling have been reported in the literature and some of most popular shapes are shown collectively in Figure 6.3.

6.2 CP Generation Using Rectangular Slot Coupling

This section investigates various antenna designs reported using aperture coupling technique methods starting from the simplest rectangular slot. It was mentioned in Chapter 1 that two conditions are needed for CP: excitation of orthogonal modes with equal magnitude and maintaining a ±90° phase difference between them. Several examples of rectangular slot coupling for CP generation are presented in the following section.

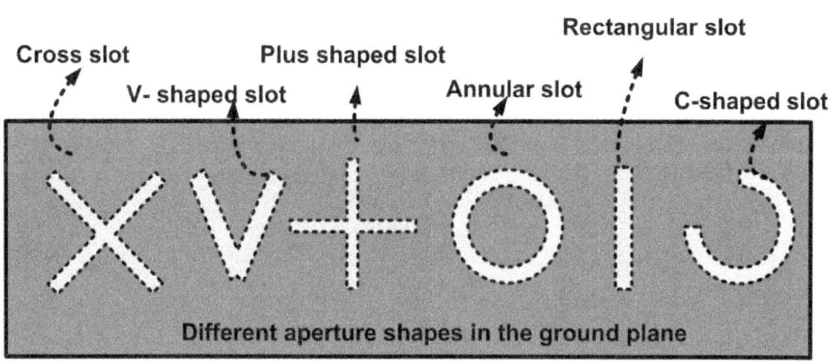

Figure 6.3 Different aperture shapes used in a DRA for circular polarization generation.

6.2.1 Dual-Point Slot-Fed DRAs for Circular Polarization

The easiest way to achieve circular polarization using this method is displayed in Figure 6.4(a). In order to excite orthogonal modes in a rectangular DRA, two rectangular slots are present at orthogonal positions fed by a modified microstrip line arrangement exciting both rectangular slots individually. The electric field orientation in both slots will excite orthogonal TE^{x}_{111} and TE^{y}_{111} modes in the DRA. The input power is split into two current paths that excites the fundamental TE_{111} mode in the rectangular DRA through both orthogonal slots. The length of the microstrip feed lines and slot dimensions are adjusted to achieve the desired magnitude and phase quadrature needed between the orthogonal modes. An ARBW of 16.52% has been reported using this simple approach. This is an example of dual feeding (feeding at orthogonal positions).

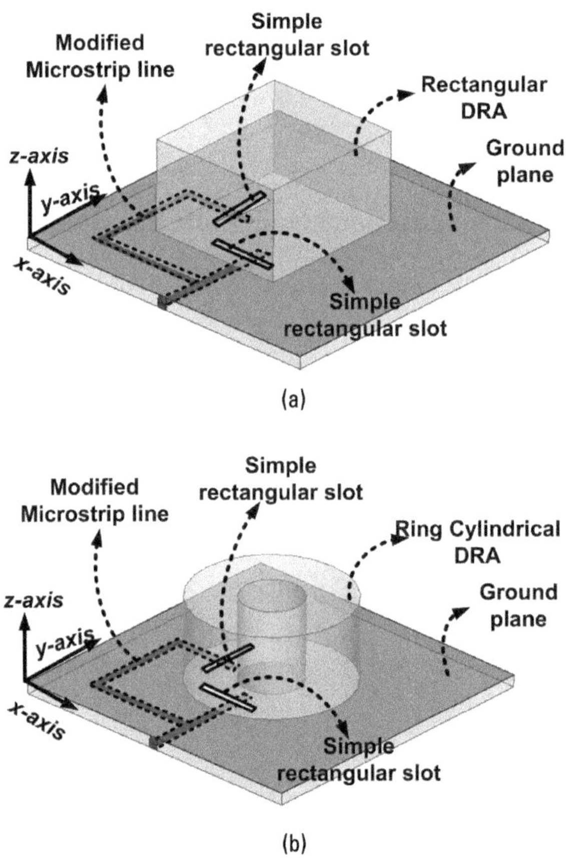

Figure 6.4 Example of a dual-point slot fed DRA for circular polarization: (a) rectangular DRA [5], and (b) cylindrical ring DRA [6].

This simple aperture coupling method is also applicable to cylindrical and hemispherical DRAs [5, 6].

However, the magnitude and phase difference conditions are very sensitive to feed dimensions. These two rectangular slots can also be fed from different ports and a power splitting network, such as hybrid coupler and Wilkinson splitter, can be used to divide the power equally in phase quadrature at the cost of increased system complexity. Therefore, most of the current research on circularly polarized antennas are more focused on single-point-fed DRAs, which are discussed in the next examples.

6.2.2 Single-Point Slot-Fed Conventional DRAs for Circular Polarization

A simpler approach for getting CP radiation from DRAs is borrowed from microstrip antennas [7]. This approach uses single-point feeding to excite the appropriate modes in DRA. The generation of CP using single-point-fed generally requires excitation of degenerate orthogonal modes in the structure and through some additional adjustments, the resonances of these modes are separated. Two examples of single-point rectangular slot-fed DRAs modified from [8, 9] are shown in Figure 6.5. In the first example (Figure 6.5(a)), a rectangular DRA with slightly unequal axial dimensions is fed along the diagonal at the center position through a rectangular aperture to excite degenerate TE^{x}_{111} and TE^{y}_{111} modes. This arrangement provides a low ARBW of 2%. In the second example (Figure 6.5(b)), 3.4% of an ARBW was obtained where the patch dimensions and slots have been used for generating orthogonal modes with phase quadrature.

However, the obtained axial ratio was narrowband when using simple rectangular slot in conventional DRAs. To obtain wide axial ratio bandwidth, various researchers have reported extra modifications in DRA system using the

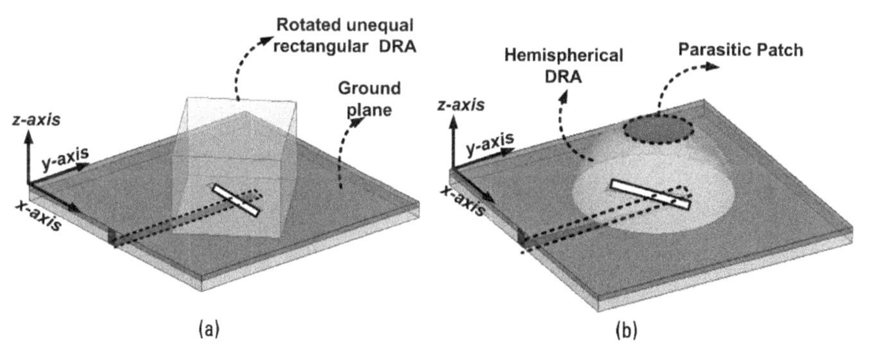

Figure 6.5 Example of a single-point slot-fed DRA for circular polarization: (a) rectangular DRA (modified after [8]), and (b) hemispherical DRA (modified after [9]).

same rectangular slot by introducing additional dielectric resonators organized in some specific arrangement, which will be discussed now.

6.2.3 Single-Point-Fed Multielement Conventional DRAs for CP

In the previous examples, single-element DRAs have been used. In the next examples [10–13], multielement (more than one DRA) is used for improving the axial ratio bandwidth.

In the first example, three rectangular DRAs have been arranged in a stair-shaped form fed through a common rectangular aperture in the ground plane. A similar structure is shown in Figure 6.10 modified after [10]. The overall arrangement is rotated 45° with respect to its sides and its length to width ratio is optimized for getting optimum performance. At a length to width ratio of 1.9 of each DRA, an axial ratio bandwidth of 10.6% was obtained.

Another example of multielement DRA design for CP generation coupled through a simple rectangular slot is shown in Figure 6.7 modified after [11]. The radiating part consists of four thin rectangular DRAs (15 mm × 5 mm × 1.27 mm) and each DRA is rotated at an angle of 30° relative to its lower DRA [11]. The whole DRA arrangement is fed through a slot coupling fed by a simple microstrip line. This arrangement reported an axial ratio bandwidth of 6% (9.55 to 10.15 GHz) due to the merging of TE^{x}_{111} and TE^{y}_{111} modes in the structure. In this example, the rectangular slot excites the four rotated layers of RDRAs simultaneously, resulting in two orthogonal components of electric field, and the phase difference is obtained by optimizing the angle of rotation.

The authors in [12] used a fan-blade-shaped DRA obtained by the combination of a pair of half-split cylindrical DRAs and one rectangular DRA.

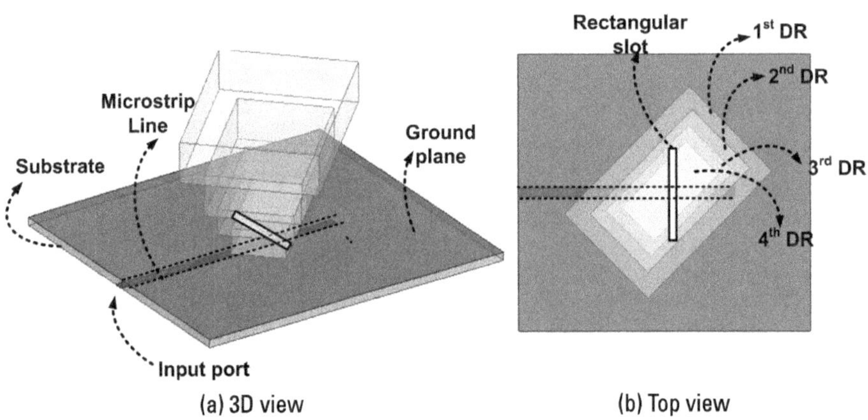

Figure 6.6 A multielement (3-DRA) arrangement excited by a microstrip line fed slot coupled for circular polarization. (Modified after [10].)

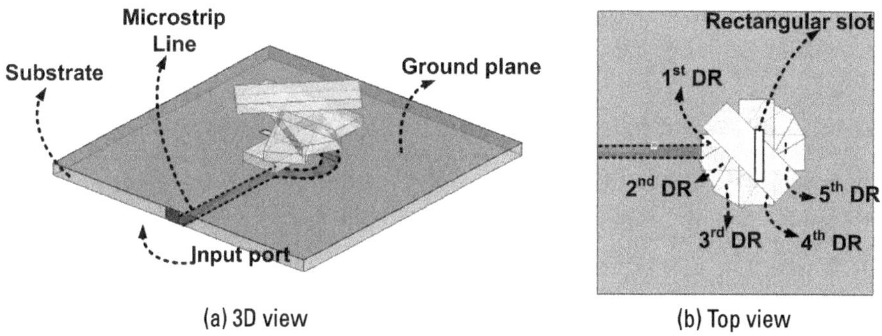

Figure 6.7 Example of a circularly polarized multielement DRA using simple rectangular slot coupling. (Modified after [11].)

This multielement DRA arrangement is excited by a simple rectangular slot to achieve dual-band circular polarization. This configuration offers 3-dB axial ratio bandwidth for lower and upper bands as 13.33% and 7.81%, respectively. There is one more flexibility that can be added in these approaches by modifying the shape of the microstrip line that is used to feed the slot as shown in Figure 6.8, which is slightly modified after [13]. In this example, two rectangular DRAs are stacked and form upper and bottom layers in which the upper layer is rotated by an angle of 48° (after optimization) and again excited by a simple rectangular slot cut in the ground plane [13]. Unlike in previous examples, in this case the slot is fed through a modified microstrip line network, which is dual offset for better impedance matching. In this method the slot coupling results in simultaneous excitation of TE_{111} and TE_{113} modes in the

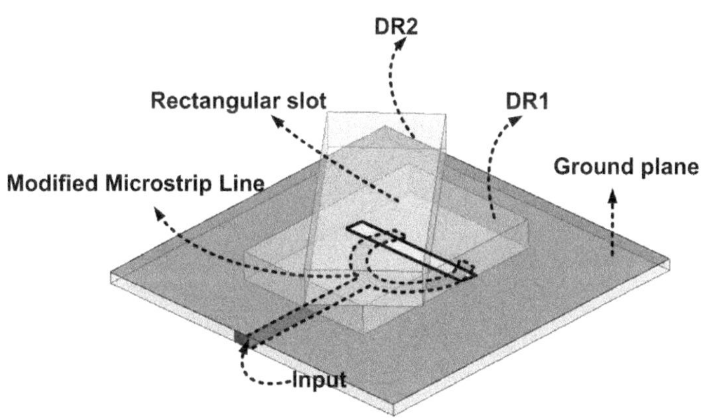

Figure 6.8 Example of a modified microstrip line slot coupled stair-shaped dielectric resonator antenna. (Modified after [13].)

dual rectangular DRA due to their specific placement and hence, a wide axial ratio bandwidth of 18.2% was obtained.

6.2.4 Single-Point-Fed Nonconventional DRAs for Circular Polarization

In previously discussed CP designs, conventional DRAs were used. There are some examples also reported in which the shape of the DRA has been changed in such a way that, through a simple rectangular slot coupling, orthogonal modes can be excited in the structure resulting in CP fields. One such design is shown in Figure 6.9 modified after [14]. The DRA is itself deformed in a cross-shaped form and is considered as two linearly polarized DRAs in orthogonal arrangement [14]. This nonconventional DRA is then excited through a slot at the center position to excite the same modes in both rectangular DRAs with equal magnitude. However, in this case, a low axial ratio bandwidth of 4% was achieved.

A summary of various designs reported using simple rectangular slot coupling is given in Table 6.1. The design requirements for CP generation using a simple rectangular slot can be arrived at by the following steps:

- The rectangular slot length can be approximated using (6.1);
- The rectangular slot width can be approximated using (6.2);
- The slot location and microstrip feed length (usually $\lambda/4$ at center frequency) has to be estimated to excite the required degenerate modes;
- CP can be obtained by detuning the degenerate modes by means of some other design aspects, such as modifying the DRA geometry, placing the DRA asymmetric with regard to slot, or using parasitic elements.

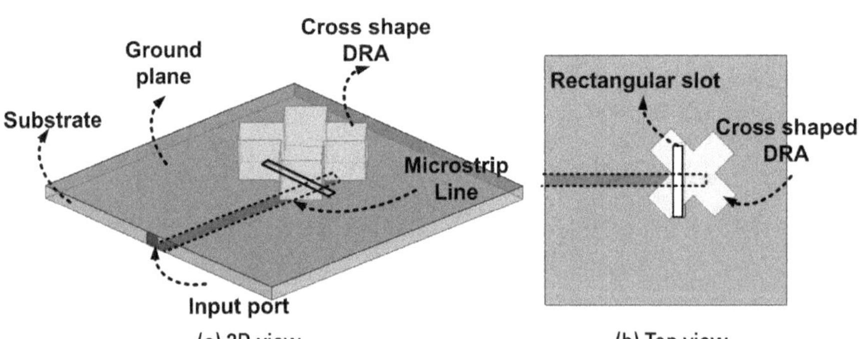

Figure 6.9 Example of a cross-shaped dielectric resonator antenna excited by a microstrip coupled rectangular slot feeding for circular polarization generation. (Modified after [14].)

Table 6.1
Brief Comparison of Various CP Designs Using Rectangular Slot Coupling Techniques

Ref	DRA Shape, Number of DRs (ε_r)	Aperture Shape	IMBW (%)	ARBW (%)	f_c (GHz)	Modes Excited
[5]	Rectangular, 1 (9.8)	Dual rectangular slot at orthogonal positions	17.87	16.52	3.57	TE^x_{111} and TE^y_{111}
[6]	Cylindrical ring, 1 (9.8)	Dual rectangular slot at orthogonal positions	13.12	10	3.70	$HEM_{11\delta}$
[8]	Rectangular,1 (40)	Single rectangular slot	~7.4	~2	NM	TE^x_{111} and TE^y_{111}
[9]	Hemispherical, 1 (9.5)	Single rectangular slot with parasitic patch	16	3.4	3.5	TE_{111}
[10]	Rectangular, 3 (12)	Single rectangular slot	36.6	10.6	9.92	Not mentioned
[11]	Rectangular, 4 (10.2)	Single rectangular slot	21	6	9.85	TE^x_{111} and TE^y_{111}
[12]	Semi cylindrical, 2 and rectangular, 1 (12.8)	Single rectangular slot	17.47 and 28.94	13.33 and 7.81	3.52, 4.73	TE_{111}, TE_{211} and TE_{311}
[13]	Rectangular, 2 (9.8)	Single rectangular slot	31	18.2	5.5	TE_{111} and TE_{113}
[14]	Cross-shaped, 1 (10.8)	Single rectangular slot	9.2	4	11	TE^x_{111} and TE^y_{111}

IMBW = fractional −10-dB impedance bandwidth, ARBW = fractional 3-dB axial ratio bandwidth, ε_r = dielectric constant of DRA, f_c = center frequency with regard to axial ratio plot.

6.3 CP Generation Using Cross Slot Coupling

This section presents the most popular aperture coupling technique in dielectric resonator antennas for circular polarization generation, which is the cross slot coupling method. In this method, two rectangular slots of unequal lengths are arranged in a cross-shaped (or a rotated plus shaped) form with unequal arm lengths to excite the required orthogonal modes in a DRA with phase quadrature relationship. This method is also applicable to cylindrical, rectangular, and hemispherical DRAs.

The first circularly polarized DRA designs using this technique were reported in [15], as shown in Figure 6.10. where a cylindrical DRA (ε_r = 79, radius = 14.72 mm and height = 5.1 mm) was excited using a cross slot shaped aperture with arm lengths of 12 mm and 13 mm, respectively (arm width kept the same = 1 mm), which is fed though a simple microstrip line across the cross slot center below the substrate. As a result, two near-degenerate fundamental $HE_{11\delta}$ modes are excited, which results in 3.9% of ARBW. Since then, various CP designs have been investigated using this method [16, 17].

In the next example [18], previous design concepts were altered by setting the two arms of the cross slot at different positions with electrically large dimensions to make the slot modes independent from DRA modes.

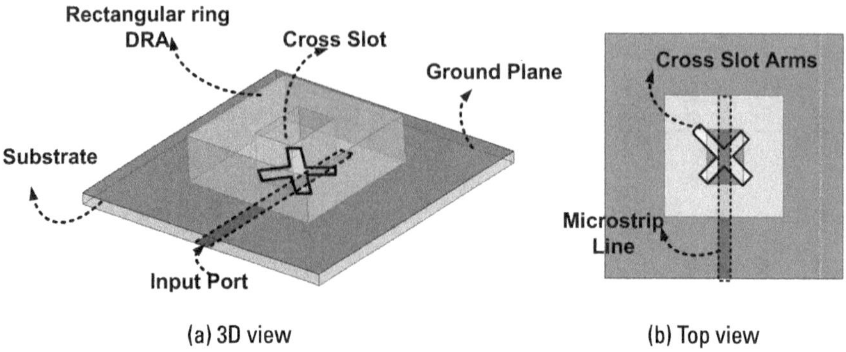

Figure 6.10 Example of a cylindrical DRA coupled by a cross slot through a simple microstrip line for circular polarization generation. (Modified after [15].)

In this case, a microstrip line is fed at an offset position across the slots, unlike previous examples as shown in Figure 6.11(a). With a cylindrical DRA (ε_r = 9.8, radius = 15.2 mm and height = 5.1 mm), an axial ratio bandwidth of 4.7% was obtained. In [19], a large aspect ratio rectangular DRA (3:2:1) was used to obtain dual-band circularly polarized fields using cross slot coupling as shown in Figure 6.11(b). Due to the large aspect ratio of rectangular DRA (ε_r = 10, dimensions = 18 × 58 × 10 mm^3), cross slot helps in exciting lower-order TE_{111} modes and higher-order TE_{121} and TE_{131} modes simultaneously. The lower band axial ratio obtained was 19.8% (due to TE_{111} and TE_{121} modes), whereas the upper band axial ratio obtained was 6.2% (due to the TE_{131} mode), respectively.

There was another method reported in [20] for getting wide axial ratio bandwidth using cross slot coupling techniques and parasitic metallic plates without increasing the footprint and height of antenna. The modified antenna layout is shown in Figure 6.12 in which a square cross section DRA (ε_r = 10, dimensions = 28 × 28 × 19 mm^3) was coupled through a cross slot (two orthogonal slots of different lengths and the same widths) to excite CP fields

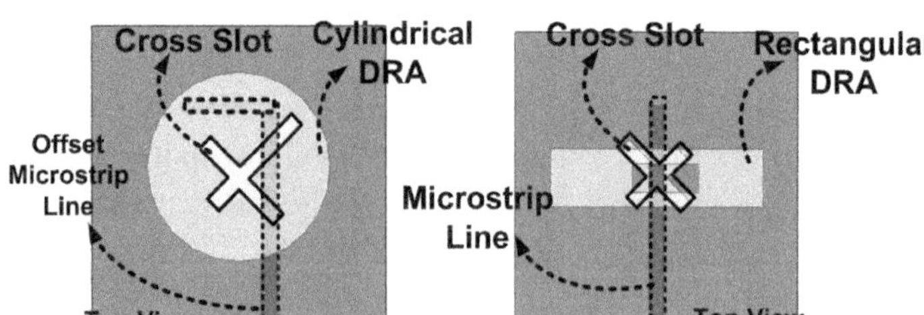

Figure 6.11 Examples of cross slot fed designs: (a) cylindrical DRA coupled by a offset microstrip line (modified after [18]), and (b) rectangular DRA with large aspect ratio (modified after [19]).

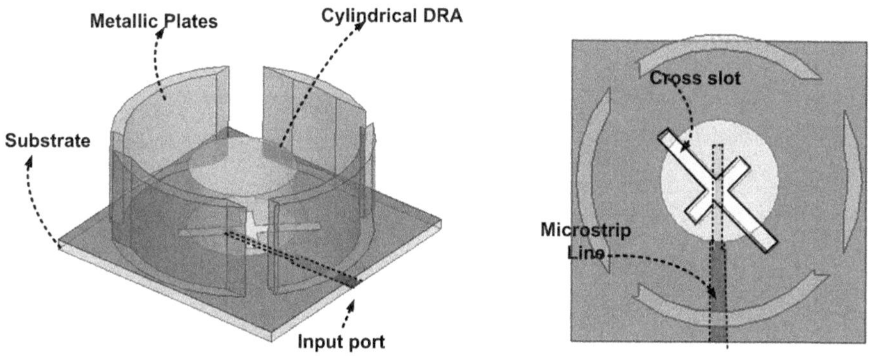

Figure 6.12 Example of circularly polarized rectangular DRA coupled by a cross slot through a simple microstrip line with parasitic vertical plates for large axial ratio bandwidth. (Modified after [20].)

in the structure. To further enhance the axial ratio bandwidth, four rectangular metallic plates (of the same dimensions) were sequentially arranged about the z-axis surrounding the rectangular DRA to induce orthogonal currents for creating additional axial ratio passband. Using this approach, axial ratio from conventional cross-slot fed DRAs increased from ~10% to ~49.5%.

In the previously mentioned examples, a single-element DRA was used for getting CP and improving its performance. In the next two examples, more than one DRA are used for getting better CP performance, which is now discussed.

In [21], two rectangular DRAs are arranged in T-shape forms and are called stacked rectangular DRAs, as shown in Figure 6.13. These are made up of two layers and excited by an asymmetrical cross slot in the ground plane. By stacking two layers of rectangular DR and using slot excitation, the designed antenna generates three sets of degenerate orthogonal modes (i.e., quasi-TE_{111}, quasi-TE_{113}, and quasi-TE_{115}) and slot mode.

By combining these modes, wide dual-band CP behavior was achieved in this antenna. The lower band is achieved by a combination of slot and quasi-TE_{111} modes, while quasi-TE_{113}, and quasi-TE_{115} modes are responsible for generating the upper band. This design reported 3-dB axial ratio bandwidth for the first and second bands to be 9.70% and 20.0%, respectively, in broadside direction. There are some circularly polarized designs that have been reported with enhanced gain also using cross slot coupling techniques [22, 23]. One such example is shown in Figure 6.14 where a two-layer rectangular DRA of different dielectric constant was used and excited through a microstrip line coupled cross slot of unequal slot lengths to excite two near-degenerate orthogonal modes of equal amplitude and a 90° phase difference needed for CP radiation. High gain was accomplished in the structure by exciting higher-order modes in which separation between a set of equivalent magnetic dipoles was adjusted.

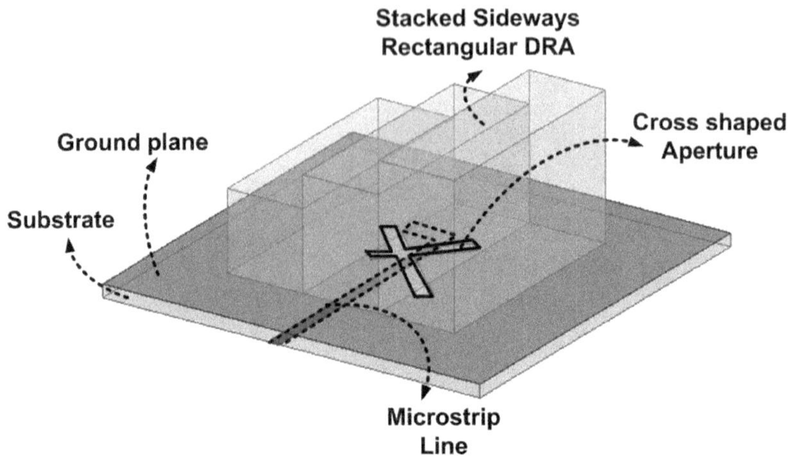

Figure 6.13 Dual-rectangular DRA coupled by cross slot for dual-band circular polarization generation. (Modified after [21].)

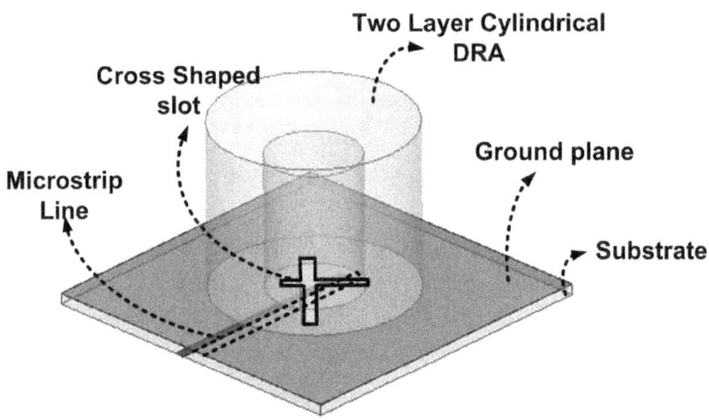

Figure 6.14 Two-layer rectangular DRA coupled by cross slot for circular polarization with enhanced gain. (Modified after [22].)

Additionally, it was surrounded by another dielectric medium. This design [22] achieves 9.5% of axial ratio bandwidth along with high gain of 11 dBic in the working band.

In the next two examples, a cross slot coupling fed through a modified microstrip line technique is examined. In [24], the authors used plus-shaped slots coupled with a microstrip circular ring feeding structure to excite the rectangular DRA shown in Figure 6.15. By means of this organization of feed and slot, the designed antenna is able to show a –10-dB impedance bandwidths of 20.79% and a 3-dB AR-bandwidth of 12.09% in broadside direction. In another example shown in Figure 6.16, a plus-shaped micrsotrip line was used

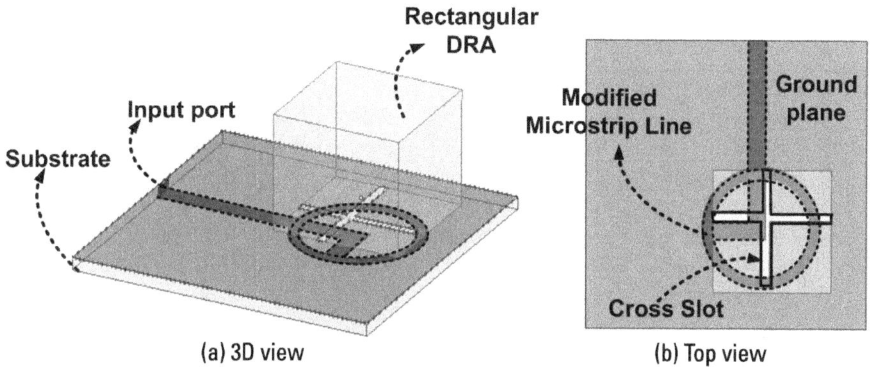

Figure 6.15 A cross slot coupled rectangular DRA fed by a modified circular ring microstrip line for circular polarization [24].

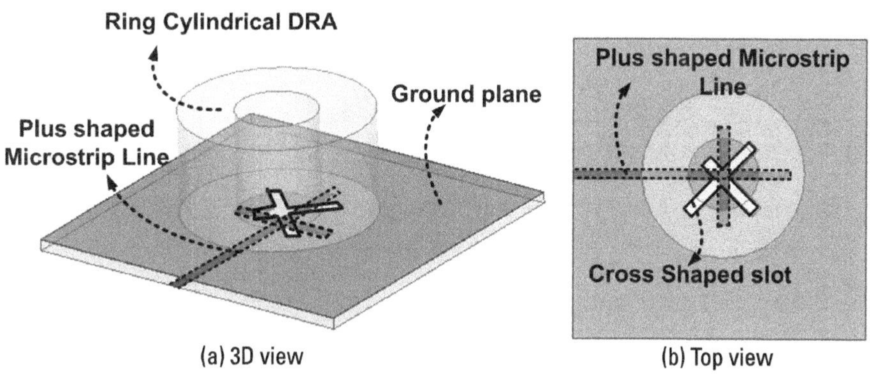

Figure 6.16 A cross slot coupled rectangular DRA fed by a modified plus-shaped microstrip line for dual-band circular polarization. (Modified after [25].)

across the cross slot for getting dual-band circular polarization. Due to this modified micrsotrip line and cross slot arrangement, two pairs of degenerate orthogonal TE_{111} and TE_{113} modes are excited simultaneously.

A comparison is shown in Table 6.2, where various CP designs reported using cross slot coupling techniques are illustrated. The design requirements for CP generation using a cross slot can be arrived at by the following steps:

- Initially, two rectangular slots are etched in the ground plane in cross slot or plus-shaped arrangement so that both slots have a 90° spatial separation between them.
- This cross-slot arrangement helps in exciting the orthogonal modes in the DRA.

Table 6.2
Brief Comparison of Various CP Designs Using Cross Slot Coupling Techniques

Ref	DRA Shape, Number of DRs (ε_r)	Aperture Shape	IMBW (%)	ARBW (%)	f_c (GHz)	Modes Excited
[15]	Cylindrical, 1 (79)	Cross slot fed by simple microstrip line	10.1	3.9	2.05	$HEM_{11\delta}$
[16]	Rectangular, 1 (12)	Cross slot fed by simple microstrip line	14.0, 12.8	3.3, 3.7	2, 2.84	TE_{111} and TE_{113}
[17]	Bowtie shaped (10.2)	Cross slot fed by simple microstrip line	43.8	7.4	8.1	NM
[18]	Cylindrical, 1 (9.8)	Cross slot fed by offset fed microstrip line	30	4.7	5.72	$HEM_{11\delta}$
[19]	Rectangular, 1 (10)	Cross slot fed by simple microstrip line	24.3, 8.5	19.8, 6.2	3.24, 4.17	TE_{111}, TE_{121} and TE_{131}
[20]	Rectangular, 1 (10)	Cross slot fed by simple microstrip line	46.9	49.5	2.19	TE^x_{111} and TE^y_{111}
[21]	Rectangular, 2 (9.8)	Cross slot fed by simple microstrip line	12.2, 21.7	9.7, 20	1.78, 2.43	*Quasi*-TE_{111}, TE_{113} and TE_{115}
[22]	Rectangular, 2 (10,3.5)	Cross slot fed by simple microstrip line	21	9.5	10.92	$TE_{11,11}$ and $TE_{11,13}$
[24]	Rectangular, 1 (9.8)	Cross slot fed by circular ring microstrip	20.79	12.09	3.39	$TE_{11\delta}$
[25]	Rectangular, 1 (20.5)	Cross slot fed by plus-shaped microstrip line	11.4, 8.4	2.1, 2.2	1.26, 1.55	TE_{111} and TE_{113}

IMBW = fractional –10-dB impedance bandwidth, ARBW = fractional 3-dB axial ratio bandwidth, ε_r = dielectric constant of DRA, f_c = center frequency with regard to axial ratio plot.

- The initial values of length and width can be determined by using (6.1) and (6.2), respectively (usually the width of both slots have equal value).
- The remaining conditions of the phase difference and equal magnitude can be obtained by optimizing the length of both slots.
- The microstrip network below the substrate can also be adjusted to improve the CP performance in various shapes.

6.4 CP Generation Using the Annular Slot Coupling Technique

Another shape of aperture used for getting circularly polarized fields—annular-shaped apertures—is examined in this section. This shape is also potentially suitable for exciting CP fields in a DRA structure. It consists of a circular ring-shaped slot in the ground plane over which a DRA is positioned, where the amount of coupling is controlled by the width and relative position of microstrip line and slot. This design was first experimentally analyzed in 1998 [26] using a simple microstrip line. A circularly polarized design using an annular slot was reported in [27] with a backing cavity (inserted beneath the substrate) to avoid any unwanted backlobe radiation. In order to generate CP radiation, the annular slot was perturbed by introducing opposing stubs (located at 45° and 225°) with respect to the *x*-axis as shown in Figure 6.17. With this feed configuration, an axial ratio bandwidth of 3.4% was obtained whereas a large impedance bandwidth of 18% was obtained, which is much wider when obtained from a rectangular slot. Other designs using annular slot feeding have been reported in [28, 29].

An offset microstrip line coupled annular slot was used in [30] by cutting the slot at two positions, A and B, in succession as demonstrated in Figure 6.18. The current distribution was found to be of the same amplitude at these points with a $\lambda/4$ path difference. By using this approach, two field arrangements with orthogonal polarization were obtained with a 90° phase difference to get CP fields. With a cylindrical DRA ($\varepsilon_r = 79$, radius = 15 mm, height = 4.5 mm), a narrow axial ratio bandwidth of 3.47% was accomplished.

In [31], a wide axial ratio bandwidth of 14% was realized using a C-shaped slot in which the DRA geometry was modified in a fractal spidron shape as shown in Figure 6.19. A comparison is shown in Table 6.3, where various CP designs reported using annular slot coupling are presented.

6.5 Different Shapes of DRAs Coupled with Modified Slot

There are some additional aperture shapes that are also reported in the literature. We have put these into a miscellaneous-shaped category that can be employed for CP generation with some examples discussed in this section in the next few designs.

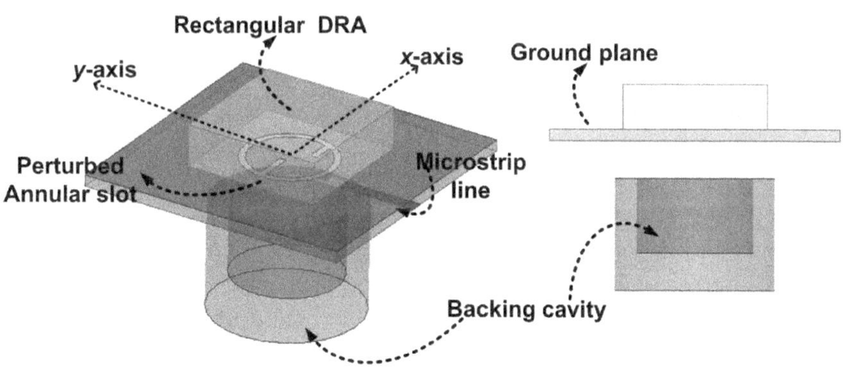

Figure 6.17 Example of a circularly polarized dielectric resonator antenna coupled by a perturbed annular slot with backing cavity: (a) 3D-view, and (b) side view. (Modified after [27].)

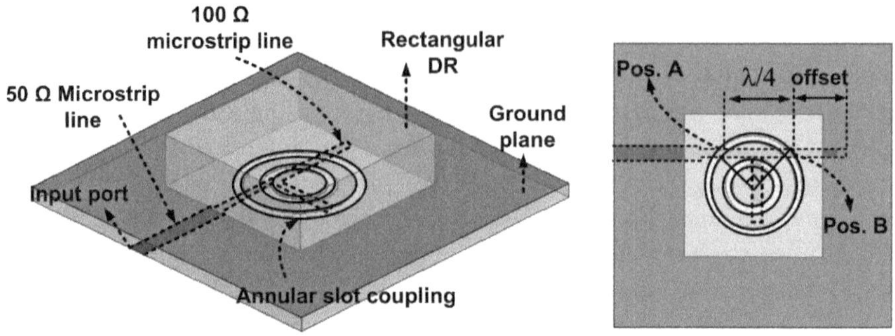

Figure 6.18 A cylindrical DRA coupled with annular slot fed by an offset microstrip line for circular polarization generation. (Modified after [30].)

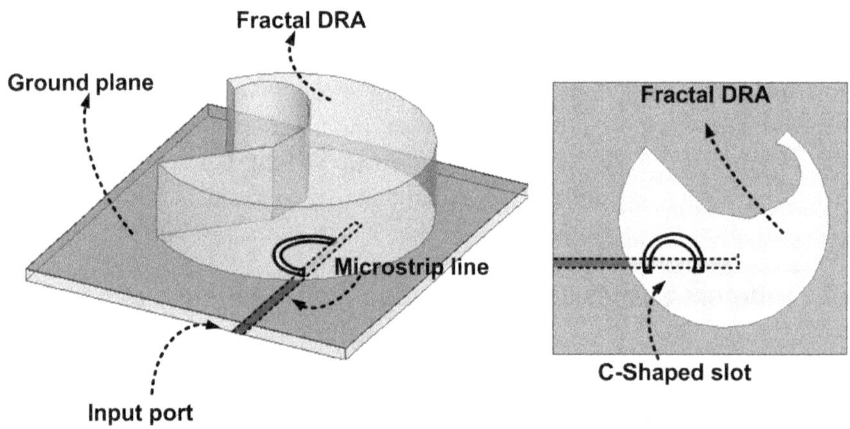

Figure 6.19 A fractal-shaped DRA coupled with a C-shaped slot fed by a microstrip line for wideband circular polarization generation. (Modified after [31].)

A stair-shaped slot was proposed in [32] for exciting multiple orthogonal modes in a multielement DRA arrangement so that wideband circular polarization could be generated as given in Figure 6.20. By using stair shapes for excitation, horizontal slots and vertical slots align the *E*-field along orthogonal directions. Also, the excitation of two pairs of orthogonal higher-order modes leads to a large axial ratio bandwidth due to stair-shape aperture. A very large axial ratio bandwidth of 41.01% was achieved in the structure.

An alternative aperture coupling is also possible by introducing parasitic apertures for getting improved CP performance. The authors in [33] designed a dual-band CP by using modified shapes of cylindrical DRs (i.e., two notches are truncated from the cylindrical DR and excited through a rectangular and parasitic arc-shaped slots in the ground plane. A similar structure is shown in

Table 6.3
Brief Comparison of Various CP Designs Using the Annular Slot Coupling Technique

Ref	DRA Shape, Number of DRs (ε_r)	Aperture Shape	IMBW (%)	ARBW (%)	f_c (GHz)	Modes Excited
[27]	Cylindrical, 1 (9.5)	Perturbed annular slot with backing cavity	18	3.4	1.96	NM
[28]	Hemispherical, 1 (9.5)	Annular feeding with parasitic slot	15	4.5	~3.76	TE_{111}
[29]	Cylindrical, 1 (79)	C-shaped slot	7.24	1.63	1.35	$HEM_{11\delta}$
[30]	Cylindrical, 1 (79)	Annular slot fed by offset microstrip line	5.8	~3.5	2.47	$HEM_{11\delta}$
[31]	Fractal shaped. 1 (10)	C-shaped slot	37.29	14.01	5.35	NM

IMBW = fractional –10-dB impedance bandwidth, ARBW = fractional 3-dB axial ratio bandwidth, ε_r = dielectric constant of DRA, f_c = center frequency with regard to axial ratio plot.

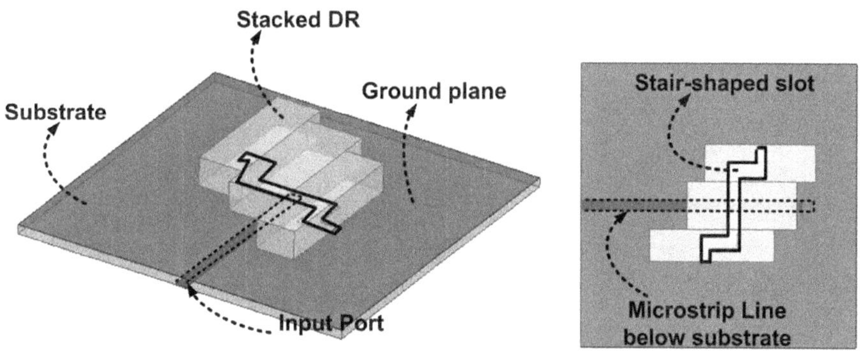

Figure 6.20 Example of a stair-shaped slot excitation in a multielement DRA arrangement for wideband circular polarization. (Modified after [32].)

in Figure 6.21 after [33]. Due to this arrangment, HE_{111} and $HE_{11\delta}$ (2<δ<3) modes were excited. Here, a pair of extra arc-shaped slots were utilized for improving the impedance bandwidths and axial ratio bandwidths. This designed antenna obtained 15.8% and 5.02% of axial ratio bandwidth for lower and upper bands in boresight direction. There are other examples where the aperture shape was modified in alphabetical shapes, such as in *F*-shaped [34] and *L*-shaped [35] apertures.

For example, CP was excited in a rectangular DRA by using an *F*-shaped slot [34] with partial ground plane as given in Figure 6.22(a) where an axial ratio bandwidth of 14.28% was obtained, whereas in [35] an axial ratio bandwidth of 17.59% of was reported by using an *L*-shaped slot as shown in Figure 6.22(b).

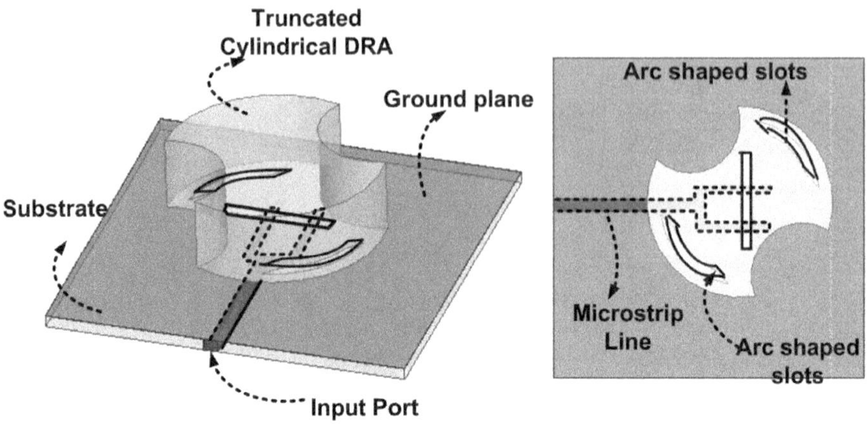

Figure 6.21 A truncated cylindrical DRA excited by a rectangular slot and parasitic apertures for dual-band circular polarization. (Modified after [33].)

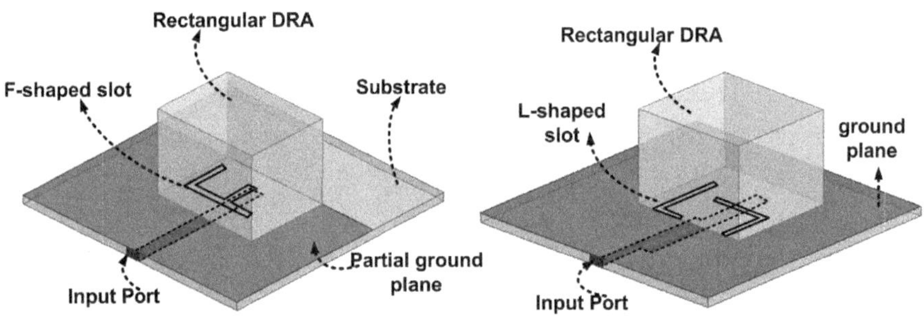

Figure 6.22 Example of a circularly polarized DRA design with an (a) F-shaped aperture [34], and (b) an L-shaped aperture [35].

In the previous examples, a broadside axial ratio was of interest while an omnidirectional cylindrical DRA (excited in $TM_{01\delta}$ mode) was also designed for circular polarization in [36]. A list of various miscellaneous aperture shapes is collectively compared in Table 6.4. Figure 6.23 summarizes the discussion about aperture coupling methods for exciting CP fields in a dielectric resonator antenna in a flowchart form for quick understanding.

6.6 Design Example of a Cross-Slot DRA for CP Generation

Among all aperture fed methods, the cross-slot coupling technique is the most popular method and is widely used for CP generation. Hence, a design example is now discussed for CP generation in DRA using the cross-slot coupling meth-

Table 6.4
Brief Comparison of Various CP Designs Using Miscellaneous Slot Shape Coupling Techniques

Ref	DRA Shape, Number of DRs (ε_r)	Aperture Shape	IMBW (%)	ARBW (%)	f_c (GHz)	Modes Excited
[32]	Semicylindrical, 2 and rectangular, 1 (12.8)	Stair-shaped slot	49.67	41.01	5.53	TE_{111}, TE_{121} and TE_{131}
[33]	Truncated cylindrical, 1, (9.8)	Rectangle and parasitic arc shaped slot	26.25, 11.17	15.8, 5.02	5.19, 9.17	HE_{111} & $HE_{11\delta}(2<\delta<3)$
[34]	Rectangular, 1 (9.8)	F-shaped slot	50.55	14.28	5.32	TE_{111}, TE_{121} and TE_{113}
[35]	Rectangular,1 (9.8)	Dual *L*-shaped slot-fed stair-shaped microstrip	26.84	17.59	3.24	TE^x_{111} and TE^y_{111}
[36]	Cylindrical, 1 (6.85)	Spiral-shaped slots	~8.6	6.2	2.43	$TM_{01\delta}$

IMBW = fractional −10-dB impedance bandwidth, ARBW = fractional 3-dB axial ratio bandwidth, ε_r = dielectric constant of DRA, f_c = center frequency with regard to axial ratio plot.

od. A circularly polarized cylindrical dielectric resonator antenna is designed using the cross slot coupling approach as shown in Figure 6.24. A cylindrical DRA having a dielectric constant of 9.8 (alumina) is used with a radius and height equal to 10 mm and 17.5 mm, respectively. Two rectangular slots, called slot arm-1 and slot arm-2 (having the same width = 0.9 mm) with different slot lengths (16 mm and 11.5 mm, respectively) are etched on the ground plane in a cross-shaped form. The initial values of individual rectangular slot lengths can be chosen according to (6.1), which comes out to be 13 mm at the center frequency of 3.5 GHz. Thus, initially two slot arms have 13 mm lengths and are then organized in cross-slot form to excite orthogonal field components. After the generation of orthogonal modes due to a cross-shaped arrangement, the length of both slot arms is optimized in order to get the requisite phase difference conditions. The optimized lengths come out to be 11.5 and 16 mm in this design. The reasonable width of both slots can be calculated from (6.2) as the starting point. Initially the orthogonal mode generation is illustrated using a cross slot, which is simple to visualize.

This cross slot helps in exciting the fundamental $HEM_{11\delta}$ mode in a dielectric resonator antenna. The schematic direction of an excited electric field is shown in Figure 6.25. If only slot arm-1 is present, then it would excite a field pattern as shown in Figure 6.25(a). Considering slot arm-2 only, the electric field pattern developed inside cylindrical dielectric resonator antenna (CDRA) is displayed in Figure 6.25(b). Thus, these two field patterns generated by incorporating individual slots are orthogonal to each other. To simultaneously excite these two electric field patterns, both slots are connected simultaneously in a cross-shaped form and are excited by the same microstrip line crossing through its center (beneath the substrate). Hence, two field patterns are generated inside

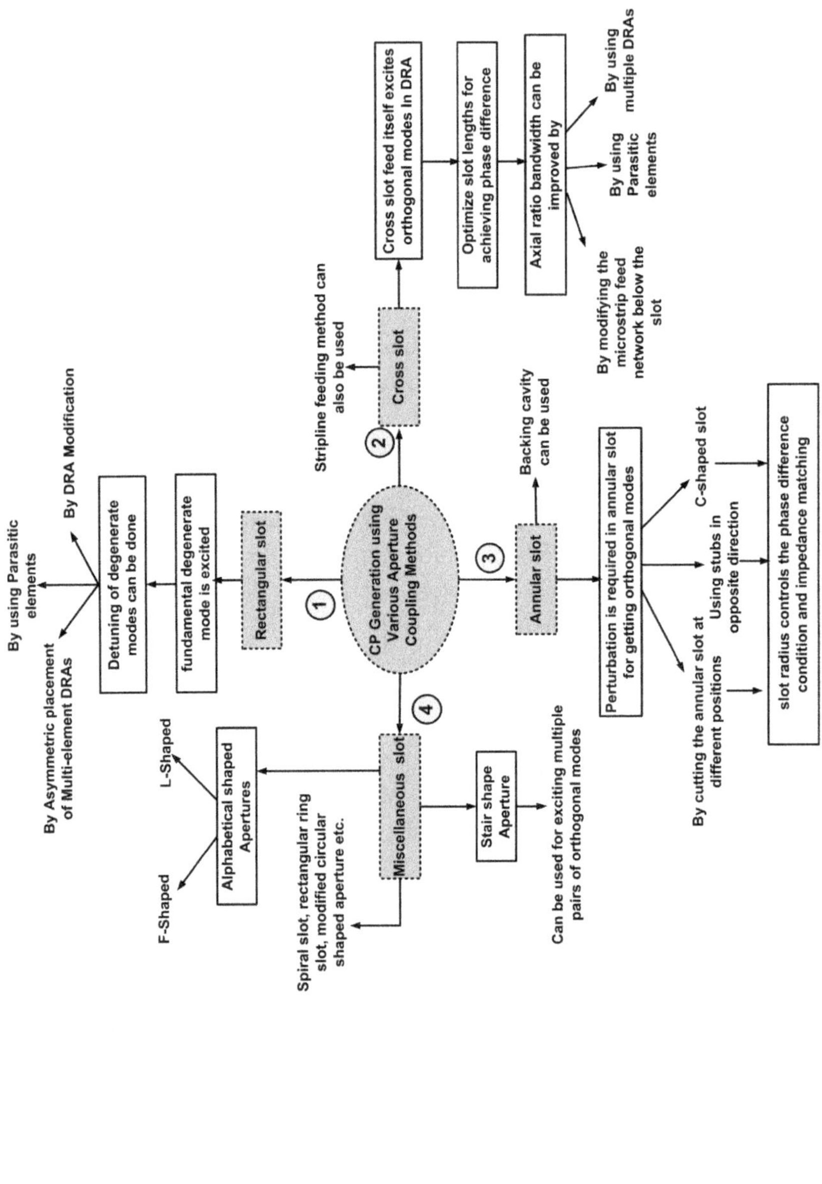

Figure 6.23 An overview of aperture coupling methods in DRA for CP generation.

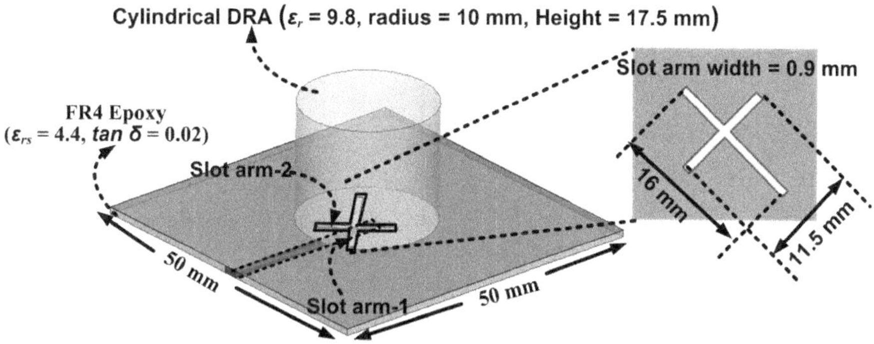

Figure 6.24 A design example showing a cylindrical DRA fed with a cross slot for circular polarization generation.

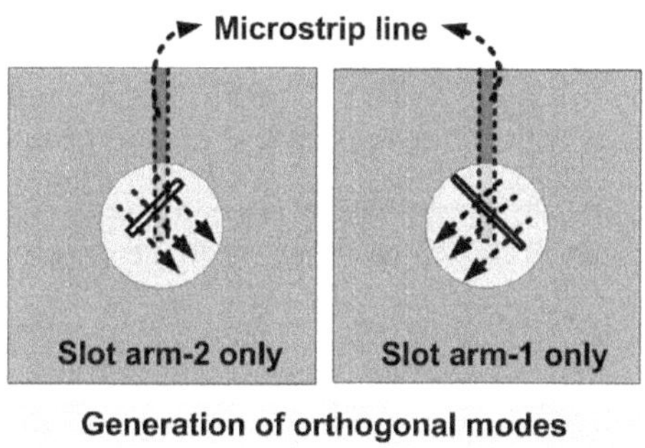

Figure 6.25 Schematic of E-field direction in DRA when (a) only slot arm-2 is used, and (b) only slot arm-1 is used.

cylindrical DRA in an orthogonal arrangement with the help of the geometry of the cross slot. This is how easily orthogonal modes are obtained using cross-slot coupling.

Figure 6.26 shows the simulated electric field distribution inside a cylindrical DRA with its 3D view and top view to confirm the excitation of orthogonal $HEM_{11\delta}$ modes. The required phase quadrature relationship between the orthogonal modes is obtained by varying their relative slot lengths. For this purpose, an axial ratio plot is shown in Figure 6.27 for different values of the length of slot arm-1 when the length of slot arm-2 is fixed at 11.5 mm to show the sensitivity of slot length on axial ratio.

The corresponding simulated input reflection coefficient and axial ratio plots are shown in Figure 6.28 for the design example under consideration. It is observed that a -10-dB impedance bandwidth of 23.6% and a 3-dB axial ratio bandwidth of 7.3% has been obtained, respectively.

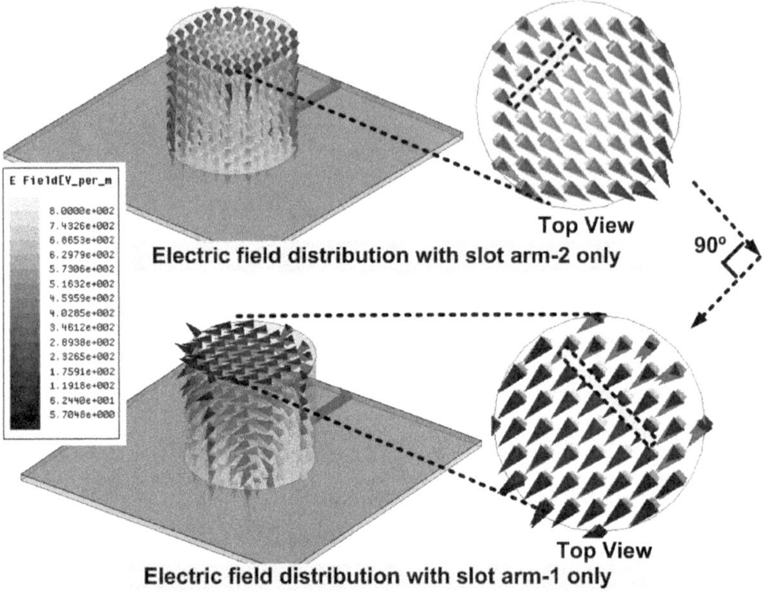

Figure 6.26 Simulated electric field distribution in proposed cylindrical DRA due to the presence of individual slots with 3D and top view showing orthogonal conditions between them.

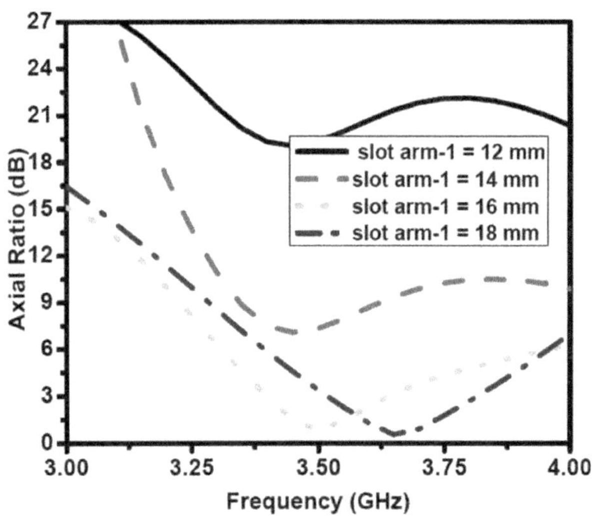

Figure 6.27 Simulated axial ratio variation with varying slot arm-2 length.

Normalized radiation patterns are plotted on the *xz*-plane in Figure 6.29. It has been perceived that LHCP waves are leading component as compared to RHCP waves along broadside direction. Hence, this proposed antenna design

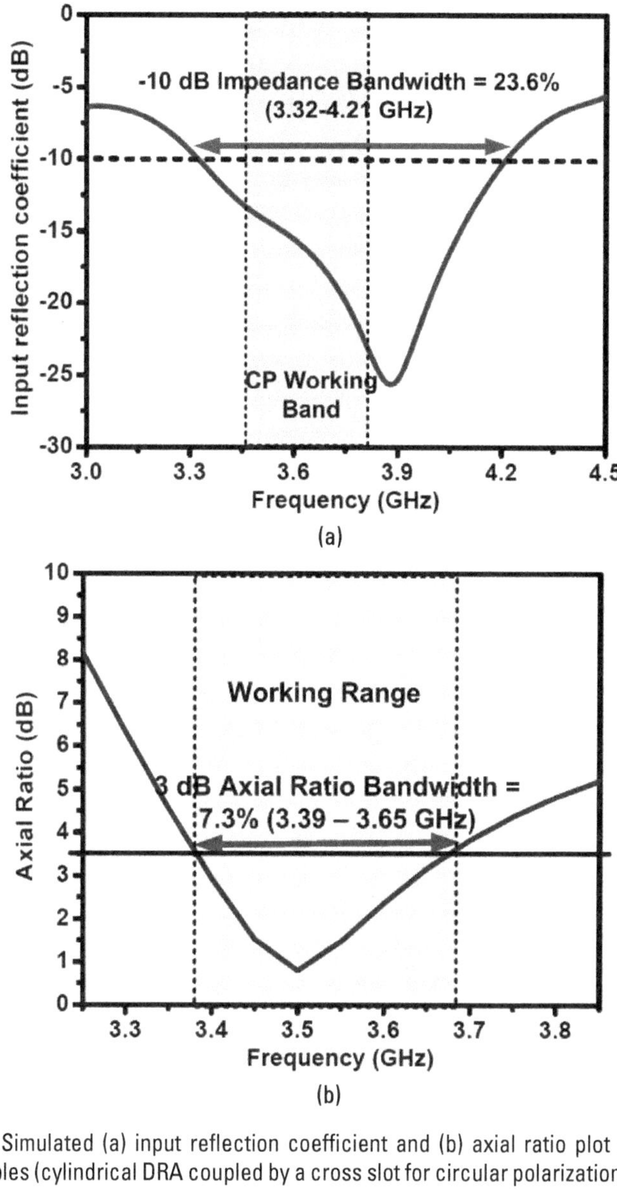

Figure 6.28 Simulated (a) input reflection coefficient and (b) axial ratio plot of presented design examples (cylindrical DRA coupled by a cross slot for circular polarization generation).

radiates in LHCP mode. It has been found that LHCP fields are stronger than RHCP fields in broadside direction approximately by 18 dB (3.4 GHz) and 26 dB (3.5 GHz), respectively, on an *xz*-plane.

In Figure 6.30, the electric field distribution is demonstrated inside cylindrical DRA for different phase angles (0°, 90°, 180°, and 270°) at 3.5 GHz when only a single slot is present. It has been observed that the orientation of a

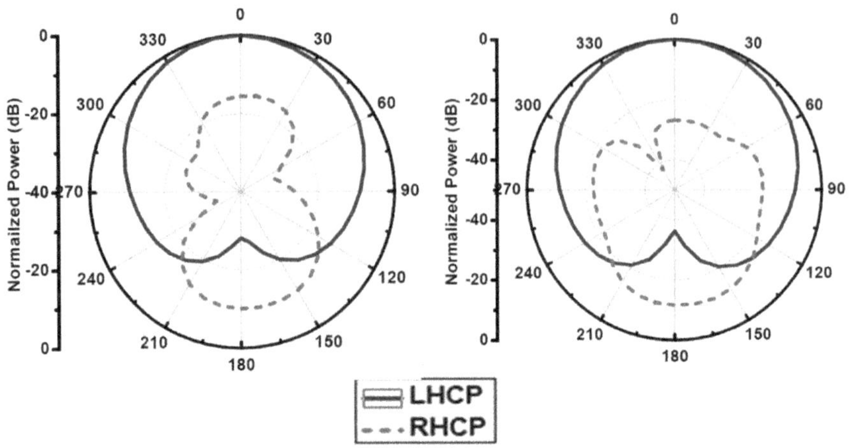

(a) XZ plane: 3.4 GHz (a) XZ plane: 3.5 GHz

Figure 6.29 Example of a cylindrical DRA coupled by a cross slot through a simple microstrip line for circular polarization generation.

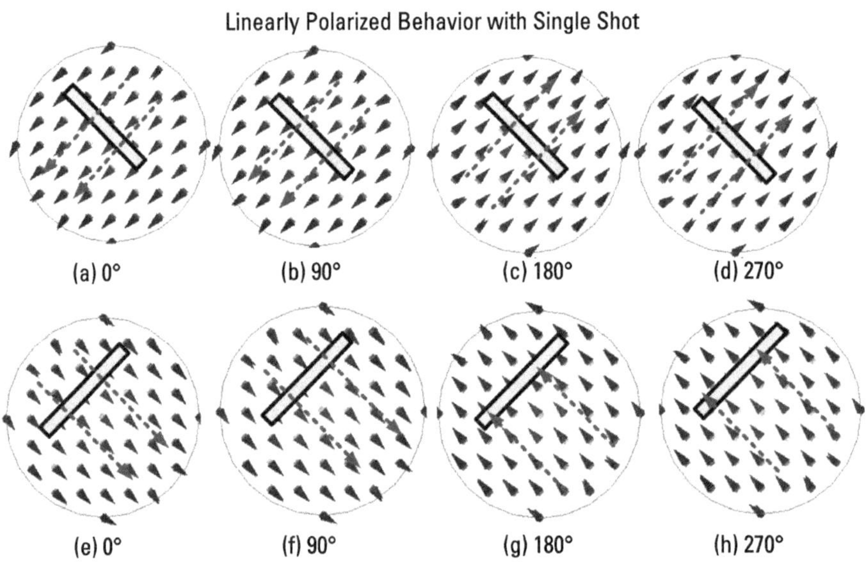

Figure 6.30 Electric field distribution inside a cylindrical DRA with a single slot only for different phase angles showing linearly polarized behavior at 3.5 GHz.

net electric field is confined in a single direction and varies in a linear polarized manner. But when both slots are arranged in a cross-slot shape, the net electric field is found to be varying in a clockwise circular manner, which reveals its cir-

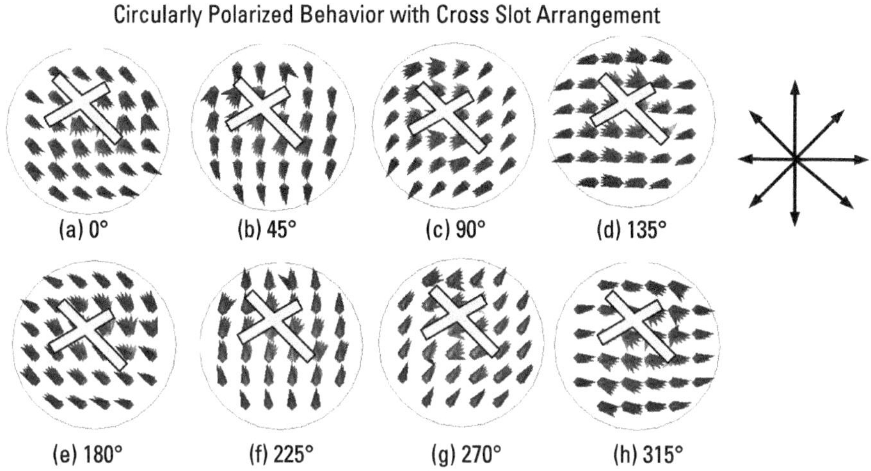

Figure 6.31 Electric field distribution inside a cylindrical DRA with a cross slot for different phase angles showing circularly polarized behavior at 3.5 GHz.

cularly polarized behavior that is depicted in Figure 6.31. As well, the clockwise motion of the electric field confirms the LHCP mode of circular polarization.

6.7 Circular Polarization in Different Applications of DRAs

In this section, examples of aperture coupling techniques in MIMO [37,38], arrays, and filtering antenna are discussed for generation of circular polarization.

There is an example of MIMO antenna design [38] based on dielectric resonators where circular polarization was realized using aperture coupling techniques as shown in Figure 6.32 after a slight modification in the geometry of [38]. In [38], a circular slot aperture was modified by removing a pair of arc-shaped metal parts from opposite boundaries. Due to this modification in slot shape, it excited degenerate orthogonal HE^{x}_{111} & $HE^{y}_{11\delta}$ modes in the structure responsible for circular polarization. The modified circular apertures were mirror images of each other in order to realize dual-sense CP fields. In the same design, a pair of slits were etched on the ground plane in order to improve isolation in the system. Thus, various aperture shapes can be utilized in MIMO technology to also get CP.

Furthermore, the aperture coupling method has also been utilized in dielectric resonator based array antennas for realizing CP fields with enhanced gain [39, 40]. As an example, a stair-shaped slot was used in [39] to excite $HEM_{11\delta}$ and $TE_{01\delta}$ modes simultaneously in dual-cylindrical DRAs as given in Figure 6.33. The horizontal rectangular slot aligns the E-field along the x-direction, whereas the two vertical slots align the E-field along the y-direction,

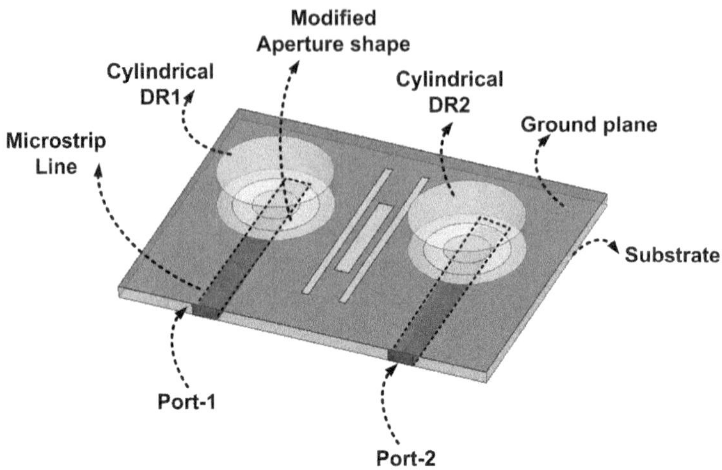

Figure 6.32 A Dielectric resonator-based MIMO antenna with polarization diversity coupled using a modified circular aperture shape for CP generation. (Modified after [38].)

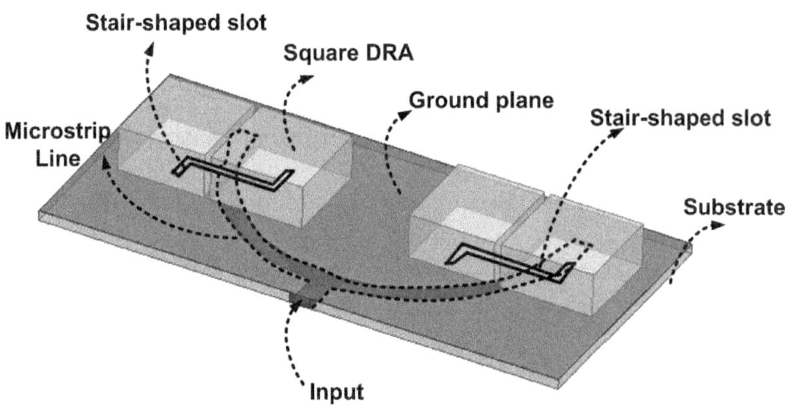

Figure 6.33 A dielectric resonator-based array antenna with polarization diversity coupled using a stair-shaped aperture for wideband CP. (Modified after [39].)

and hence orthogonal components were introduced. A wide axial ratio bandwidth of 23.5% was achieved using this feeding arrangement (Figure 6.33) with enhanced gain (due to the use of array concept). Similarly, a circularly polarized design was reported in [40] using the cross-slot method with high gain through the use of parasitic elements. As a final CP design, a circularly polarized filtering dielectric resonator antenna was also designed using aperture coupling techniques and can be found in [41]. This design achieved a 3.55% axial ratio along with high selectivity and out-of-band signal suppression. To conclude, these antennas are listed in Table 6.5.

Table 6.5
Several CP Designs Using Aperture Coupling in Other Antenna Types

Ref	Antenna Type	Aperture Shape	IMBW (%)	ARBW (%)	f_c (GHz)	Modes Excited
[37]	MIMO antenna	Cross ring shaped	25.9	18.7	3.47	quasi-TE_{111} and TE_{131}
[38]	MIMO antenna	Modified circular aperture	~15	~6.2	5.77	HE^x_{111} & $HE^y_{11\delta}$
[39]	Array antenna	Stair-shaped aperture	30.9	23.5	5.1	$HEM_{11\delta}$ & $TE_{01\delta}$
[40]	Array antenna	Cross slot	17.8	2.2 GHz BW	26.3	$HEM_{11\delta}$
[41]	Filtering antenna	Ring rectangular slot fed by bandpass filter line	6.13	3.55	7.79	quasi $TE^x_{\delta 11}$ and $TE^y_{1\delta 1}$

IMBW = fractional –10-dB impedance bandwidth, ARBW = fractional 3-dB axial ratio bandwidth, ε_r = dielectric constant of DRA, f_c = center frequency with regard to axial ratio plot.

Table 6.6
A Brief Discussion of Some Popular Aperture-Fed Methods for CP Generation

Slot Shape	Advantages	Disadvantages
Rectangular slot 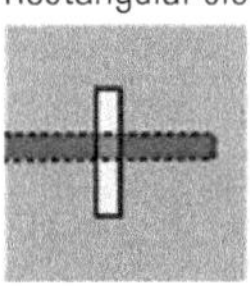	It is the simples slot shape, and it provides lower probability of backlobe radiation as compared to other annular and cross-slot shapes.	This feeding scheme cannot excite the orthogonal modes in the structure by itself (i.e., it requires some modifications in the DRA geometry). Also, independent optimization of slot dimensions is not possible and provides low bandwidth using single conventional DRAs.
Cross slot 	In this slot shape, the DRA itself need not be modified. The slot geometry itself is capable of exciting two orthogonal modes in the DRA and perturbation may not be required. Unlike rectangular slot, it allows independent optimization of slot dimensions for getting CP parameters.	Usually, this slot shape also provides low bandwidth just like rectangular slots. As well, there is a possibility of high backlobe radiation that can be avoided by restricting the slot dimensions or by using the stripline feeding method.
Annular slot 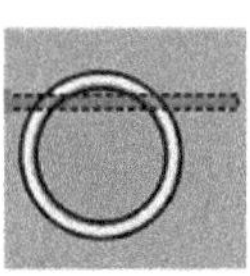	In this case, CP generation generally depends on the geometry and relative positioning of microstrip feed networks beneath the DRA. Moreover, additional feeding circuitry beneath the slot is also possible. This slot can provide a wider impedance bandwidth than a simple rectangular slot.	Like a rectangular slot, in this case perturbation in DRA geometry is also required to excite CP fields As well, there is a higher possibility of backlobe radiation using this technique, just like in cross slot coupling. In this method, usually a backing cavity is introduced beneath the slots, which increases the design complexity. Both impedance matching and axial ratio depend on slot radius.
Stair-shaped slot 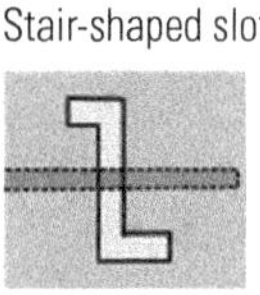	By using this slot shape, a wider axial ratio bandwidth can be realized as compared to simple rectangular slot and cross slot. This is because it helps in the excitation of multiple pairs of orthogonal modes.	This particular slot shape can provide wide axial ratio bandwidth but at the cost of increased antenna size as multiple DRAs are required. It also suffers from high backlobe radiation.

6.8 Comparison between Various Aperture-Fed Methods

The previous section illustrated several aperture coupling methods proposed for getting CP waves and improving their performance for various applications (dual-band, wideband, etc.). Next, the advantages and disadvantageous offered by some popular aperture-fed methods are detailed in Table 6.6. This comparison table will provide a basic foundation on the use of specific methods in any particular application.

6.9 Summary

This chapter discussed several circularly polarized designs (narrowband, dual-band, and wideband) using several shapes of aperture couplings to excite the requisite modes in dielectric resonator antennas. This chapter started with the simplest rectangular slot followed by cross slot, annular slot, stair-shaped slot, spiral slots, and so on. A list of CP-DRA designs were grouped with similar aperture feedings and detailed in tables for a quick glimpse of their CP performance. It is concluded from this chapter that in order to achieve wideband or multiband CP, aperture coupling techniques provide more flexibility in choosing the shapes of DRAs and the geometry of microstrip feeding networks as compared to other coupling techniques. As well, the usage of aperture coupling is also extended to MIMO, array, filtering antennas, and so on, as surveyed in the last section.

References

[1] Kishk, A. A., et al., "Slot-Excitation of Dielectric Disk Radiator," *IEEE Transactions on Antennas and Propagation*, Vol. 43, No. 2, 1995, pp. 198–201.

[2] Leung, K. W., and K. K. So, "Rectangular Waveguide Excitation of Dielectric Resonator Antennas," *IEEE Transactions on Antennas and Propagation*, Vol. 51, No. 9, 2003, pp. 2477–2481.

[3] Eashrah, I., et Al., "Theory and Implementation of Dielectric Resonator Antenna Excited by a Waveguide Slot," *IEEE Transactions on Antennas and Propagation*, Vol. 53, No. 1, 2005, pp. 483–494.

[4] Petosa, A., *Dielectric Resonator Antenna Handbook*, Norwood, MA: Artech House, 2007.

[5] Kumar, R. and R. K. Chaudhary, "Modified Microstrip-Line-Fed Rectangular Dielectric Resonator Antenna Coupled with Slotted Ground Plane for Wideband Circular Polarization," *Microwave and Optical Technology Letters*, Vol. 58, No. 1, 2016, pp. 206–210.

[6] Kumar, R. and R. K. Chaudhary, "Slot Coupled Wideband Circularly Polarized Cylindrical Ring DRA for Wi-MAX Applications," *IEEE Asia Pacific Microwave Conference*, 2016.

[7] Kumar, G., and K. P. Ray, *Broadband Microstrip Antennas*, Norwood, MA: Artech House, 2003.

[8] Oliver, M. B., R. K. Mongia, and Y. M. M. Antar, "A New Broadband Circularly Polarized Dielectric Resonator Antenna," *IEEE Antennas and Propagation Society International Symposium*, Vol. 1, 1995.

[9] Leung, K. W., and H. K. Nge, "The Slot-Coupled Hemispherical Dielectric Resonator Antenna with a Parasitic Patch: Applications to the Circularly Polarized Antenna and Wideband Antenna," *IEEE Transactions on Antennas and Propagation*, Vol. 53, No. 5, 2005, pp. 1762–1769.

[10] Chair, R., S. L. S. Yang, A. A. Kishk, K. F. Lee, and K. M. Luk, "Aperture Fed Wideband Circularly Polarized Rectangular Stair Shaped Dielectric Resonator Antenna," *IEEE Transactions on Antennas and Propagation*, Vol. 54, No. 9, 2006, pp. 1350–1352.

[11] Fakhte, S., H. Oraizi, and R. Karimian, "A Novel Low-Cost Circularly Polarized Rotated Stacked Dielectric Resonator Antenna," *IEEE Antennas and Wireless Propagation Letters*, Vol. 13, 2014, pp. 722–725.

[12] Varshney, G., V. S. Pandey, and R. S. Yaduvanshi, "Dual-band Fan-Blade-Shaped Circularly Polarised Dielectric Resonator Antenna," *IET Microwave, Antennas and Propagation*, Vol. 11, 2017, pp. 1868–1871.

[13] Wang, K. X., and H. Wong, "A Circularly Polarized Antenna by Using Rotated-Stair Dielectric Resonator," *IEEE Antennas and Wireless Propagation Letters*, Vol. 14, 2015, pp. 787–790.

[14] Ittipiboon, A., D. Roscoe, R. Mongia, and M. Cuhaci, "A Circularly Polarized Dielectric Guide Antenna with a Single Slot Feed," *IEEE Symposium on Antenna Technology and Applied Electromagnetics*, 1994, pp. 427–430.

[15] Huang, C. Y., J. Y. Wu, and K. W. Wong, "Cross Slot Coupled Microstrip Antenna and Dielectric Resonator Antenna for Circular Polarization," *IEEE Transactions on Antennas and Propagation*, Vol. 47, No. 4, 1999, pp. 605–609.

[16] Zou, M., J. Pan, L. Zuo, and Z. P. Nie, "Investigation of a Cross-Slot-Coupled Dual-Band Circularly Polarized Hybrid Dielectric Resonator Antenna," *Progress in Electromagnetics Research C*, Vol. 53, 2014, pp. 187–195.

[17] Chauthaiwale, P., R. K. Chaudhary, and K. V. Srivastava, "Circularly Polarized Bowtie-shaped Dielectric Resonator Antenna Excited with Asymmetric Cross Slot," *Microwave and Optical Technology Letters*, Vol. 57, 2015, pp. 1732–1727.

[18] Almpanis, G., C. Fumeaux, and R. üdiger Vahldieck, "Offset Cross-Slot-Coupled Dielectric Resonator Antenna for Circular Polarization," *IEEE Microwave and Wireless Components Letters*, Vol. 16, No. 8, 2006, pp. 461–463.

[19] Zhang, M., B. Li, and X. Lv, "Cross-Slot-Coupled Wide Dual-Band Circularly Polarized Rectangular Dielectric Resonator Antenna," *IEEE Antennas and Wireless Propagation Letters*, Vol. 13, 2014, pp. 532–535.

[20] Yang, M., Y. Pan, and W. Yang, "A Singly Fed Wideband Circularly Polarized Dielectric Resonator Antenna," *IEEE Antennas and Wireless Propagation Letters*, Vol. 17, No. 8, 2018, pp. 515–1518.

[21] Zou, M., and J. Pan, "Wide Dual-Band Circularly Polarized Stacked Rectangular Dielectric Resonator Antenna," *IEEE Antennas and Wireless Propagation Letters*, Vol. 15, 2016, pp. 1140–1143.

[22] Abdulmajid, A. A., Y. Khalil, and S. Khamas, "Higher-Order-Mode Circularly Polarized Two-Layer Rectangular Dielectric Resonator Antenna," *IEEE Antennas and Wireless Propagation Letters*, Vol. 17, No. 6, 2018, pp. 1114–1117.

[23] Akbari, M., S. Gupta, M. Farahani, A. R. Sebak, and T. A. Denidni, "Gain Enhancement of Circularly Polarized Dielectric Resonator Antenna Based on FSS Superstrate for MMW Applications," *IEEE Transactions on Antennas and Propagation*, Vol. 64, No. 12, 2016, pp. 5542–5546.

[24] Kumar, R., and R. K. Chaudhary, "Circularly Polarized Rectangular DRA Coupled Through Orthogonal Slot Excited with Microstrip Circular Ring Feeding Structure for Wi-MAX Applications," *International Journal of Microwave and Millimeter-Wave Computer-Aided Engineering*, Vol. 28, No. 1, 2018, pp. 1–9.

[25] Wang, X. C., L. Sun, X. L. Lu, S. Liang, and W. Z. Lu, "Single-Feed Dual-Band Circularly Polarized Dielectric Resonator Antenna for CNSS Applications," *IEEE Transactions on Antennas and Propagation*, Vol. 65, No. 8, 2017, pp. 4283–4287.

[26] Leung, K. W., W. C. Wong, K. M. Luk, and E. K. N. Yung, "Annular Slot-Coupled Dielectric Resonator Antenna," *IET Electronics Letters*, Vol. 34, No. 13, 1998, pp. 1275–1277.

[27] Leung, K. W., and S. K. Mok, "Circularly Polarised Dielectric Resonator Antenna Excited By Perturbed Annular Slot with Backing Cavity," *IET Electronics Letters*, Vol. 37, No. 15, 2001, pp. 934–936.

[28] So, K. K., K. W. Leung, and H. K. Ng, "Frequency Design of Circularly Polarized Dielectric Resonator Antenna," *IEEE Antennas and Propagation Society Symposium*, 2004.

[29] Huang, C. Y., and J. S. Kuo, "Frequency-Adjustable Circularly Polarized Dielectric Resonator Antenna," *Microwave and Optical Technology Letters*, Vol. 34, No. 3, 2002, pp. 211–213.

[30] Lin, C. C., and J. S. Sun, "Circularly Polarized Dielectric Resonator Antenna Fed by Off-Centered Microstrip Line for 2.4-GHz ISM Band Applications," *IEEE Antennas and Wireless Propagation Letters*, Vol. 14, 2015, pp. 947–949.

[31] Altaf, A., Y. Yang, K.-Y. Lee, and K. C. Hwang, "Circularly Polarized Spidron Fractal Dielectric Resonator Antenna," *IEEE Antennas and Wireless Propagation Letters*, Vol. 14, 2015, pp. 1806–1809.

[32] Varshney, G., V. S. Pandey, R. S. Yaduvanshi, and L. Kumar, "Wide Band Circularly Polarized Dielectric Resonator Antenna with Stair-Shaped Slot Excitation," *IEEE Transactions on Antennas and Propagation*, Vol. 65, No. 3, 2017, pp. 1380–1383.

[33] Zhou, Y. D., Y. C. Jiao, Z. B. Weng, and T. Ni, "A Novel Single-Fed Wide Dual-Band Circularly Polarized Dielectric Resonator Antenna," *IEEE Antennas and Wireless Propagation Letters*, Vol. 15, 2016, pp. 930–933.

[34] Kumar, R., and R. K. Chaudhary, "Investigation of Higher Order Modes Excitation through F-Shaped Slot in Rectangular Dielectric Resonator Antenna for Wideband

Circular Polarization with Broadside Radiation Characteristics," *International Journal of RF and Microwave Computer-Aided Engineering,* Vol. 28, No. 6, 2018.

[35] Kumar, R., S. R. Thummaluru, and R. K. Chaudhary, "Improvements in Wi-MAX Reception: A New Dual-mode Wideband Circularly Polarized Dielectric Resonator Antenna," *IEEE Antennas and Propagation Magazine,* Vol. 61, No. 1, 2019, pp. 41–49.

[36] Yang, N., K. W. Leung, K. Lu, and N. Wu, "Omnidirectional Circularly Polarized Dielectric Resonator Antenna with Logarithmic Spiral Slots In The Ground," *IEEE Transactions on Antennas and Propagation,* Vol. 65, No. 2, 2016, pp. 839–844.

[37] H. N. Chen, J. M. Song, and J. D. Park, "A Compact Circularly Polarized MIMO Dielectric Resonator Antenna over Electromagnetic Band-Gap Surface for 5G Applications," *IEEE Access,* Vol. 7, 2019, pp. 40889–140898.

[38] Das, G., A. Sharma, and R. K. Gangwar, "Dielectric Resonator Based Circularly Polarized MIMO Antenna with Polarization Diversity," M*icrowave and Optical Technology Letters,* Vol. 60, No. 3, 2018, pp. 685–693.

[39] Gupta, S., A. Sharma, G. Das, R. K. Gangwar, and M. Khalily, "Wideband Circularly Polarized Dielectric Resonator Antenna Array with Polarization Diversity," *IEEE Access,* Vol. 7, 2019, pp. 49069–49076.

[40] Bahreini, B., H. Oraizi, N. Noori, and S. Fakhte, "Design of a Circularly Polarized Parasitic Array with Slot-Coupled DRA with Improved Gain for the 5G Mobile System," *IEEE Antennas and Wireless Propagation Letters,* Vol. 17, No. 10, 2018, pp. 1802–1806.

[41] Sahoo, A., A.Kumar, R. D. Gupta, and M. S. Parihar, "Circularly Polarised Filtering Dielectric Resonator Antenna for X-Band Applications," *IET Microwaves, Antennas and Propagation,* Vol. 12, No. 9, 2018, pp. 1514–1518.

7

Coaxial Probe-Fed Dielectric Resonator Antennas for Circular Polarization

This chapter scrutinizes various dielectric resonator designs for circularly polarized antennas using a simple coaxial probe feeding technique. Initially, an overview to this feeding method is presented. Several dielectric resonator-based antenna designs will be discussed along with a comparison between them in each method. A brief review of the probe fed technique in different applications of antenna, such as dual-band, array, and MIMO antenna, is also included in this chapter.

7.1 Introduction to the Coaxial Probe-Feeding Technique

In the previous chapter, the most popular aperture coupling method was investigated for generation of circularly polarized radiation from DRAs. However, this chapter examines another simple way to couple the energy to a DRA by using the coaxial probe-feeding method. This coaxial probe approach is also the oldest feeding scheme to be implemented for a single-feed circularly polarized planar DRA array, which will be discussed later in this chapter [1]. The coaxial probe is simply comprised of a center metallic pin (known as a core) covered by a dielectric coating that is the middle layer followed by an upper conductor layer called cladding. This is a simple coaxial transmission line whose inner conductor, which is called a core, is extended through a small hole in the ground plane as shown in Figure 7.1.

As perceived from the feeding arrangement, this method of coupling does not allow any modification with its feed arrangement (a noteworthy exception

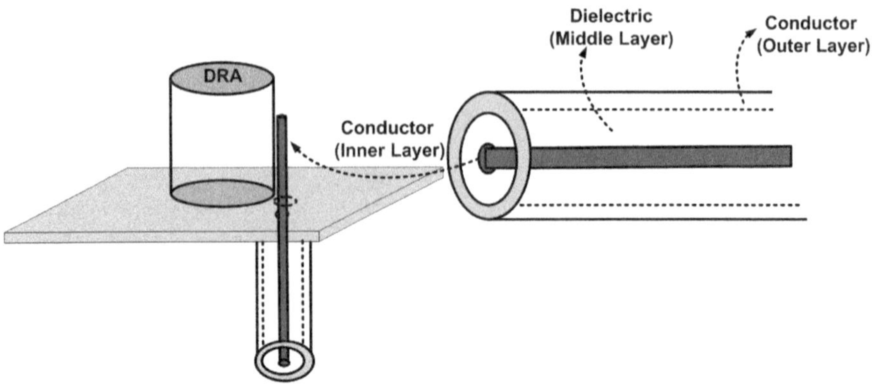

Figure 7.1 General construction of coaxial probe feeding in a dielectric resonator antenna.

is the *L*-shape probe feeding presented in [2]). Therefore, the investigation of circularly polarized antennas based on this feeding type mostly depends on the geometry and the relative location of dielectric resonators that will be surveyed in this chapter. However, the center pin can also be soldered to a metallic microstrip line network in some cases. With this feeding, there are only two feed parameters available to the antenna designer for getting optimized results: probe location and probe height. The probe location is responsible for exciting different modes in a DRA. For example, when the coaxial probe is located at the center of a cylindrical DRA, it excites the $TM_{01\delta}$ mode, whereas when the same probe is positioned near the outer surface of the same cylindrical DRA it excites the fundamental $HE_{11\delta}$ mode. The corresponding electric field distribution for different locations of the probe coupled to the DRA is schematically shown in Figure 7.2. Also, in the case of rectangular DRAs, its fundamental $TE_{11\delta}$ mode can be easily excited by locating the coaxial probe adjacent to its surface boundary.

The other parameter is the probe height that basically controls the coupling and matching of modes and by properly choosing probe height, desired matching can be achieved [3, 4]. The effect of probe height on input reflection coefficient is depicted in Figure 7.3 for a hemispherical DRA excited in its TM_{101} mode, where it has been observed that the variation in probe height controls the coupling and matching performance of the mode.

The coaxial probe behaves as a vertical equivalent electric current source and therefore, in order to excite any particular mode in a DRA using this coupling method, the center pin should be placed in a location of maximum electric field intensity to ensure strong coupling. Thus, knowlegde of internal field configuration (particularly electric field distribution) of the mode in DRA is prerequistic criteria. The electric and magnetic field is sketched in Figure 7.4 inside a DRA to display their field distribution around the probe.

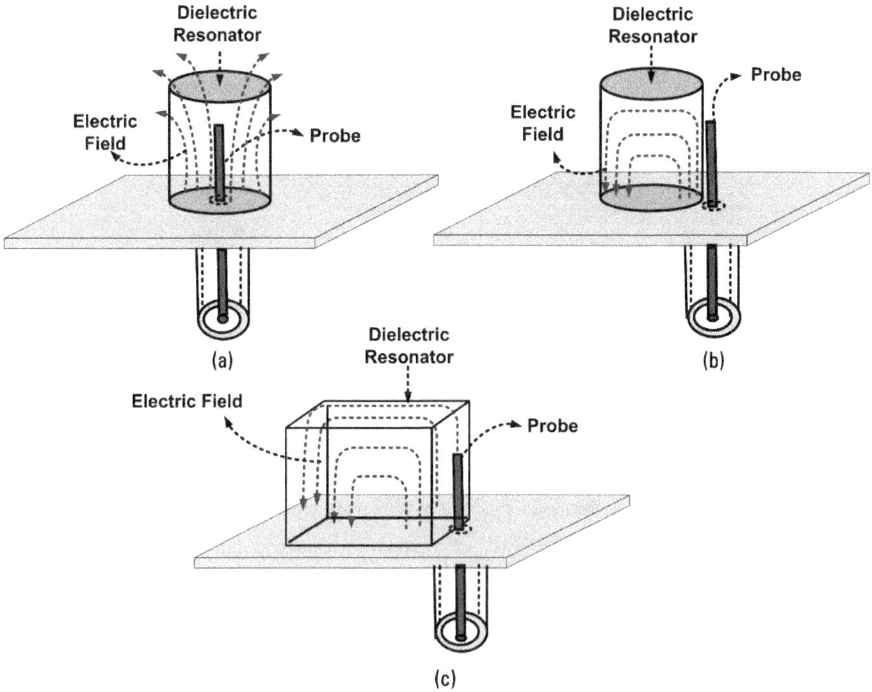

Figure 7.2 Excitation of different modes in a cylindrical DRA by placing probes at different locations: (a) $TM_{01\delta}$ mode in cylindrical DRA, (b) $HE_{11\delta}$ mode in a cylindrical DRA, and (c) $TE_{11\delta}$ mode in rectangular DRA.

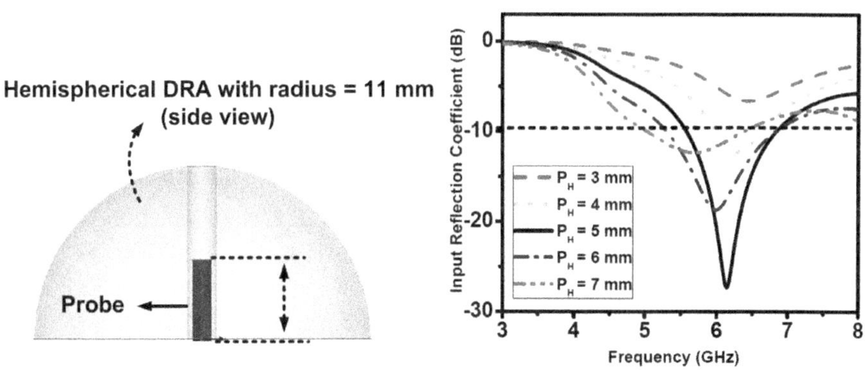

Figure 7.3 Example of a probe-fed hemispherical DRA showing the effect of probe height on input reflection coefficient plot.

Some of the key features of coaxial probe excitation are that it involves direct coupling to a 50Ω antenna system without the requirement of any bulky matching network, and these types of excitations are suitable at lower frequencies where aperture coupling could create practical problems such as excessive

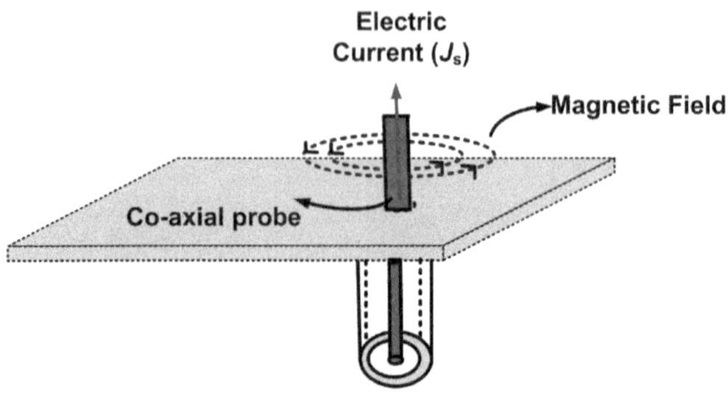

Figure 7.4 Probe acting as a vertical electric current source and the corresponding magnetic field surrounding it.

radiation due to large aperture size at lower frequenices. Also, to avoid radiation from the coaxial probe, its height is kept smaller than the height of the DRA. There is only one area of flexibility in terms of locating the probe: adjacent to the DRA or embeded within the DRA). There is no simple equation available for a probe-feeding scheme for the given DRA parameters; however, the following guidelines may help ease the design process:

1. Identify the mode to be excited in the DRA and recognize its corresponding electric field arrangement. Locate the probe feed at a location of strong electric field.
2. The probe height can be adjusted to achieve the desired matching. In practical cases, start with probe height taller than DRA and then trim its height until the desired matching is achieved [5].
3. To gain more flexibility in design, a metallic strip network of the desired shape can be printed on the surface of DRA where the inner conductor of probe can be soldered to the one end of micrsotrip network.
4. To realize a CP signal in a conventional DRA, the probe feed is usually placed assymmetric to the DRA dimensions.

7.2 CP Generation in Conventional DRAs

This section inspects various designs reported using coaxial probe feeding techniques starting from conventional DRAs. It is worth mentioning here that when using a probe feed for CP generation, usually the DRA shape must be modified (to get better axial ratio bandwidth), because there is no such flexibility possible in probe feed geometry, unlike other feeding methods where geometry of the feeding network can be altered to get CP. Thus, CP generation using this feeding scheme generally depends on the shape and relative placement of DRAs, although some antenna designs are reported for CP generation in conventional DRAs in previous years using a dual-feed and single-feed approach, which is surveyed next.

7.2.1 Dual-Feed Technique

The easiest way to achieve CP using the dual-feed technique was first generated by exciting two orthogonally arranged linearly polarized components, as was reported in [6]. Here, the authors proposed a direct method to obtain CP radiation for the first time using the dual-feeding technique. Circularly polarized radiation was obtained by exciting fundamental $HE_{11\delta}$ modes of a cylindrical ring DRA by using two vertical probe conductors placed at orthogonal locations. Likewise, orthogonal probes using a 3-dB quadrature coupler assembly may be used with a rectangular ring DRA to realize circular polarization, as shown in Figure 7.5.

The required phase difference between the orthogonal modes was then obtained by using a 3-dB quadrature coupler assembly. Another CP design using a dual-probe feeding technique was reported in [7] with some minor modifications from the above example. In this case, the orthogonal modes were excited again by using two orthogonally placed coaxial probes that were soldered on a microstrip line network (printed on the substrate) in the design. Furthermore, unlike a 3-dB quadrature coupler (used in the previous example), the required phase difference condition was obtained by using a $\lambda/4$ transmission line in the shape of a semicircle in the design, which provided a 90° phase shift. The possible antenna configuration could be realized by the modification in the feed network and rectangular DRA shape that is shown in Figure 7.6, which has been adapted from [7].

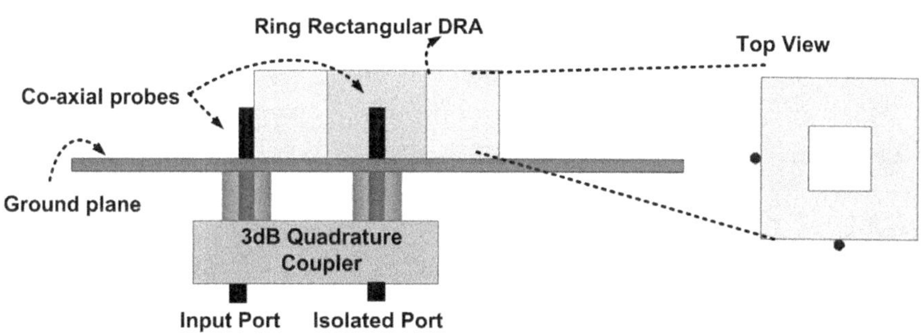

Figure 7.5 Example of a dual-probe-fed DRA for CP in a rectangular ring DRA. (Modified after [6].)

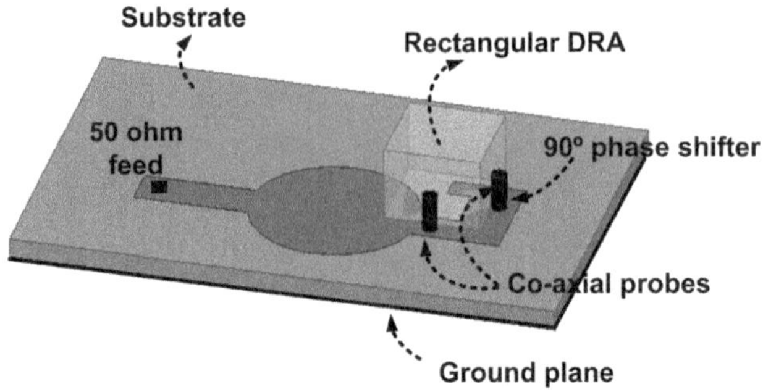

Figure 7.6 Example of a dual-probe-fed rectangular DRA connected on a microstrip line for CP generation. (Modified after [7].)

7.2.2 Single-Feed Technique

In the last two examples of a double-point feed, additional circuitry was needed for achieving phase difference conditions, which increase the design complexity. In the next example, CP designs based on a single-feed technique is discussed, which does not achieve the same ARBW but simplifies the design construction. The general procedure for the development of CP radiation using a single-probe feed requires the detuning of orthogonal degenerate modes in the DRA, such that the resonance frequency of degenerate modes can be separated in a way that net phase response from an individual mode is ±90°. One of the easiest approaches using these ideas was presented in [8]. In this reported design [8], a circularly polarized DRA was excited by a single coaxial probe feeding and the feed was located at one of the corners of the rectangular DRA. The whole design was rested over a circular ground plane. In this case, the fundamental TE^x_{111} and TE^y_{111} degenerate modes were excited at slightly different frequencies by proper selection of the rectangular DRA dimensions (and also feed location) when the following (7.1) is satisfied:

$$\frac{f_1}{Q_1}+\frac{f_2}{Q_2}=f_2-f_1 \tag{7.1}$$

where f_1 and f_2 are the resonance frequencies of lower and upper mode, respectively. Q_1 and Q_2 denote the unloaded radiation quality factor of the two modes.

The probable antenna design is shown in Figure 7.7, which was adapted from [8]. The rectangular DRA will radiate a CP signal when the lower and upper resonance frequencies are driven at their upper and lower 3-dB points providing 45° phase shifts separately, resulting in a 90° net phase difference between the modes. This idea of CP generation has been borrowed from

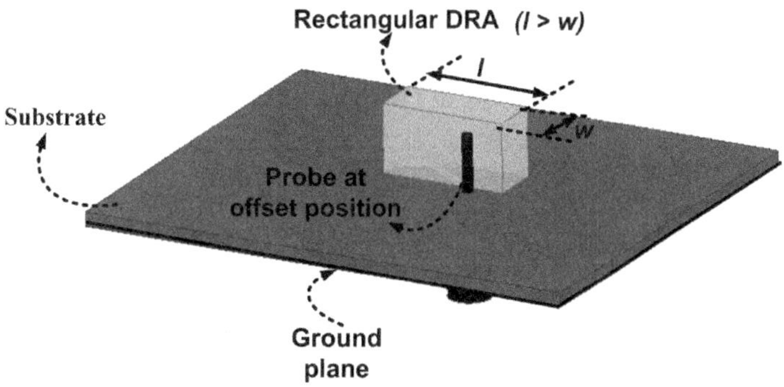

Figure 7.7 A simple quasi-rectangular DRA fed by coaxial probe at the corner for the generation of CP waves. (Modified after [8].)

microstrip antennas [9, 10]. Another example of a single-probe-fed CP design was reported in [11], in which a hexagonal-shaped DRA was coupled through a coaxial probe at one of the corners along the path to the opposite corner of the design. This arrangement has reported a wide axial ratio bandwidth of 14.84% (3–3.475 GHz) due to the excitation of orthogonal $HE^{x}_{11\delta}$ and $HE^{y}_{11\delta}$ modes. Figure 7.8 shows a decagonal (polygon with 10 sides) DRA excited by a coaxial feeding along its diagonal, which is adapted from the design idea of [11].

The next example uses a square spiral-shaped metallic strip attached to a rectangular DRA surface for excitation of CP radiation designed in [12]. This spiral shaped metallic strip is then fed by a coaxial probe at one of its ends. Unlike in previous examples, the proposed technique in this work relies on the geometry of metallic strips, which provide some degree of flexibility for CP generation. Also in this case, a wide axial ratio bandwidth of 14% has been reported. Similar results may be obtained from the antenna structure as shown in Figure 7.9 where a rectangular DR is coupled with an F-shaped metallic.

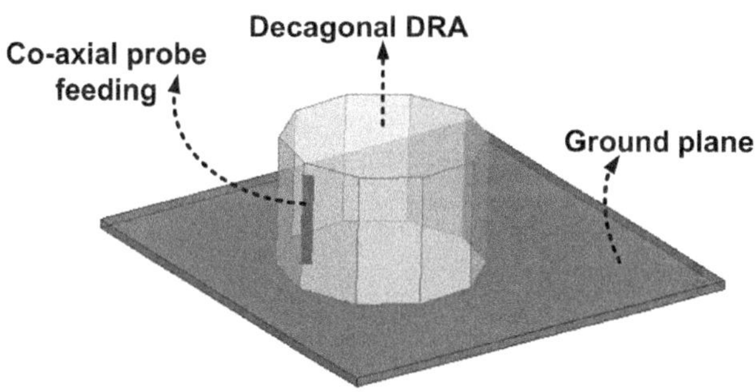

Figure 7.8 A simple decagonal-shaped DRA fed by coaxial probe feeding for generation of CP waves. (Modified after [11].)

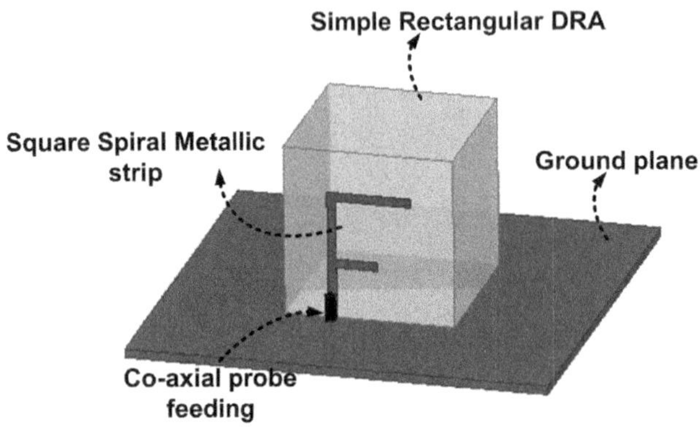

Figure 7.9 Rectangular DRA with F-shaped metallic strip excited by a coaxial probe. (Modified after [12].)

To conclude, a probe feed can be used in dual-feed or single-feed arrangement as per the requirements. In the case of dual feeding, the feed itself is placed orthogonally and an extra circuit is required for establishing the phase quadrature, whereas in the case of single feeding, the placement of a single probe with respect to the quasi-dimensions of the DRA is essential to generate CP by detuning the degenerate modes in the dielectric resonator. A summary of various designs reported in conventional DRAs using a simple coaxial probe coupling is given in Table 7.1.

7.2.3 Design Example for CP Generation in a Conventional DRA

Next, a design example is discussed for CP generation in a simple rectangular DRA without modifying its geometry using a simple coaxial probe feed. The

Table 7.1
Brief Comparison of Various CP Designs Using Probe Feeding Techniques in Conventional DRAs

Ref	DRA Shape (ε_r)	Feed Arrangement	IMBW (%)	ARBW (%)	f_c (GHz)	Modes Excited
[6]	Cylindrical ring (36.2)	Dual-probe feed with 3-dB quadrature coupler	NM	~10.9	4.55	$HEM_{11\delta}$
[7]	Cylindrical (37)	Dual-probe feeding with 90° phase shifter	3.9	NM	NM	$HEM_{11\delta}$
[11]	Hexagonal (59)	Single-probe feeding	14.84	14.84	3.2	$HEM_{11\delta}$-like
[12]	Rectangular (9.3)	Probe feeding with spiral metallic strip	11	14	4.26	TE_{112}

IMBW = fractional −10-dB impedance bandwidth, ARBW = fractional 3-dB axial ratio bandwidth, NM = not mentioned, ε_r = dielectric constant of DRA, f_c = center frequency with regard to axial ratio plot.

considered antenna configuration for the design example is shown in Figure 7.10. A rectangular DRA having a dielectric constant of 9.8 (alumina) is used, which has unequal lateral dimensions of l= 15 mm, w= 6 mm, and h= 8 mm. These dimensions are optimized where maximum axial ratio bandwidth is obtained. The knowledge of relative strength of electric and magnetic fields of a particular mode ($TE_{11\delta}$ in this case) as a function of location within the DRA is important to determine the position where the placement of the feed will efficiently excite the DRA modes.

In this design example, the probe is placed at an offset position near the corners of the DRA in order to excite both TE^x_{111} and TE^y_{111} modes simultaneously. In a rectangular DRA, TE^x, TE^y, and TE^z are supported, which radiates like a short horizontal magnetic dipole along the x-, y-, and z-directions. These different TE mode configurations have an orthogonal electric field arrangement. It is known that the resonance frequency of each mode is a function of its dimensions in a rectangular DRA [13]. Thus, by choosing appropriate dimensions, any two TE mode configurations can be simultaneously excited in the DRA. Depending on the feed position, phase difference from both modes can also be obtained, which will eventually realize a circular polarized wave. A detailed discussion of the selection of the proper dimensions of an RDRA with a simulated input reflection coefficient and axial ratio plot is discussed next.

Figure 7.11(a) and (b) show the input reflection coefficient and axial ratio plot for the rectangular DRA. The circular polarization mechanism using DRA geometry is adapted from single-feed microstrip antennas. With reference to Figure 7.11(a), at lower frequency (6 GHz), the field is x-directed as verified from Figure 7.12(a). At this resonance point the corresponding phase shift from the mode is 0°. Now at higher frequency (7 GHz), the field is y-directed as verified from Figure 7.12(b). At this resonance frequency, the corresponding phase shift is also 0°. At a lower frequency, the mode has only an x-component and no y-component, and therefore it is linearly polarized at this frequency. Similarly,

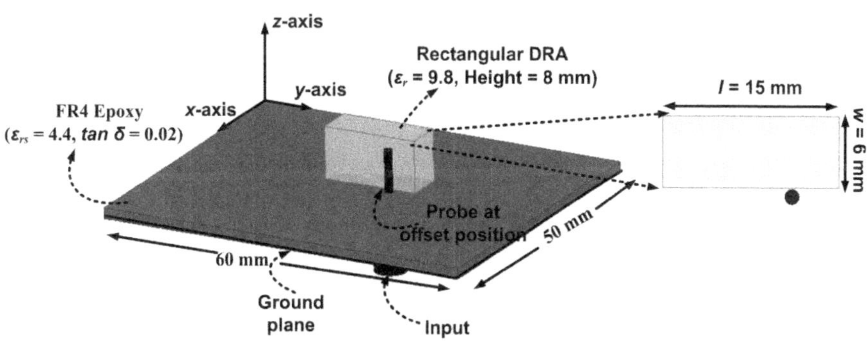

Figure 7.10 A design example showing rectangular DRA fed with a probe for circular polarization generation.

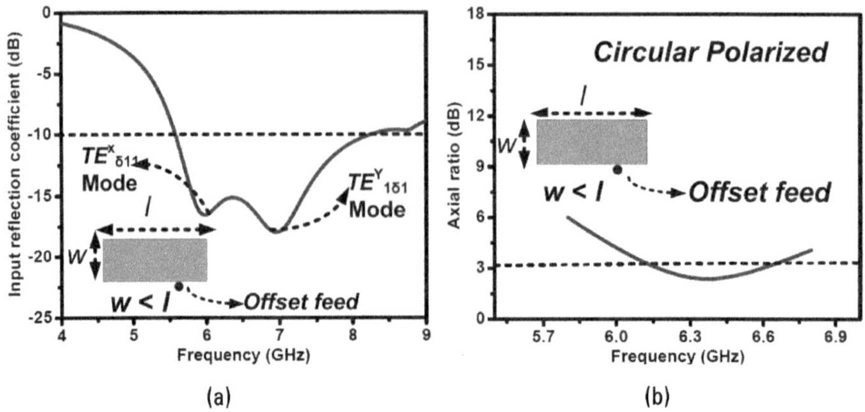

Figure 7.11 Simulated (a) input reflection coefficient and (b) axial ratio for conventional RDRA with unequal dimensions along x- and y-directions (w < l) excited with probe feeding.

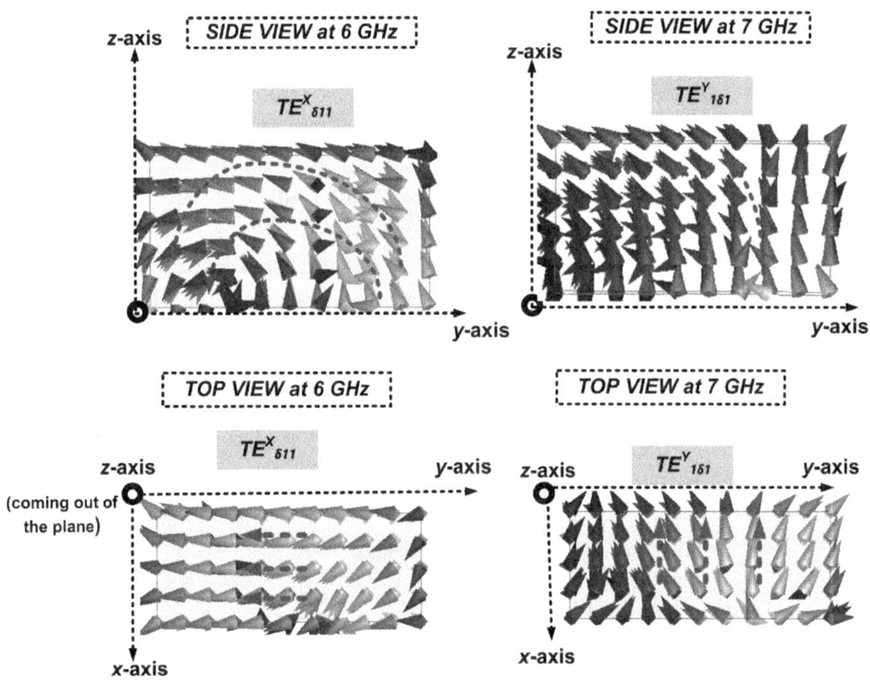

Figure 7.12 Simulated electric field distribution in the proposed rectangular DRA with (a) $TE^x_{\delta11}$ mode, and (b) $TE^y_{1\delta1}$ mode.

at a higher frequency, the mode only has a *y*-component and no *x*-component, and therefore it is also linearly polarized at this frequency. In a frequency in between these two resonances, there will be some frequency range in which a rectangular DRA will have both *x*- and *y*- components of the electric field. As well, if these two modes can be separated by a distance so that phase difference between them is 90°, a rectangular DRA will produce a circular polarized signal.

This is the basic operation of CP in a conventional rectangular DRA by using its dimensions as a design parameter. Figure 7.12 shows the simulated electric field distribution inside a rectangular DRA at a lower frequency (6 GHz) and a higher frequency (7 GHz) to confirm the excitation of orthogonal $TE^x_{\delta 11}$ and $TE^y_{1\delta 1}$ modes. It has been observed that the $TE^x_{\delta 11}$ mode is excited at a lower frequency as compared to the $TE^y_{1\delta 1}$ mode, which is excited at a slightly higher frequency. These two modes have orthogonal electric field distribution and the asymmetry in the structure with respect to the feed point offers the requisite conditions for radiating CP waves in a narrow band. The simulated results for some special values of length (l) and width (w) are discussed next. Figure 7.13 shows the input reflection coefficient and axial ratio plot when the lateral dimensions are swapped (i.e., $l = 6$ mm and $w = 15$ mm). The result obtained is the same, but in this case, the $TE^y_{1\delta 1}$ mode is excited at a lower frequency as compared to the $TE^x_{\delta 11}$ mode, which is excited at a slightly higher frequency.

Figure 7.14 shows simulated input reflection coefficients and the axial ratio for the same rectangular DRA under consideration with equal dimensions (15 mm in this case) now along the *x*- and *y*-axis, which excites the fundamental $TE^y_{1\delta 1}$ mode at 4.54 GHz with linear polarization. When the same DRA is excited at different feed positions as shown in Figure 7.15, they excite the fundamental $TE^x_{\delta 11}$ mode at the same 4.54 GHz with linear polarization.

Thus, both $TE^x_{\delta 11}$ and $TE^y_{1\delta 1}$ modes have the same resonant frequencies when dimensions are equal, but when any one dimension is larger as compared to other dimensions, it excites both modes at slightly different frequencies.

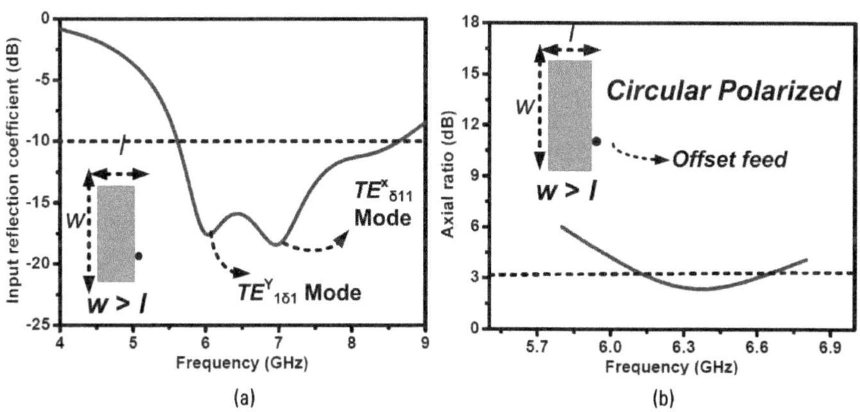

Figure 7.13 Simulated (a) input reflection coefficient and (b) axial ratio for a conventional RDRA with unequal dimensions along x- and y-directions (w > l) excited with a probe feeding.

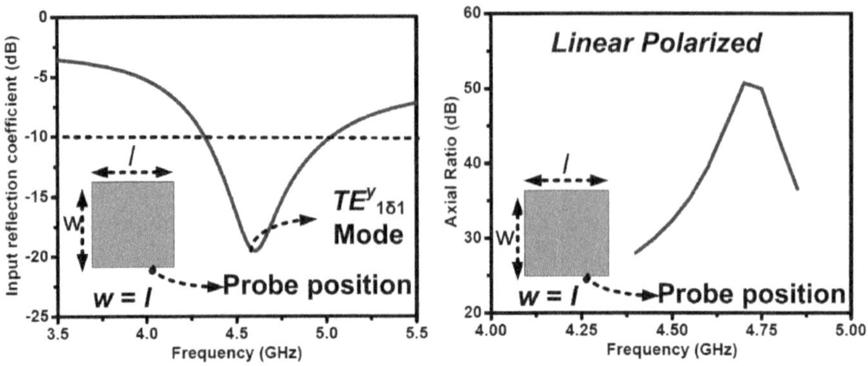

(a) (b)

Figure 7.14 Simulated (a) input reflection coefficient and (b) axial ratio for a conventional RDRA with equal dimensions along x- and y-directions excited with a probe feeding.

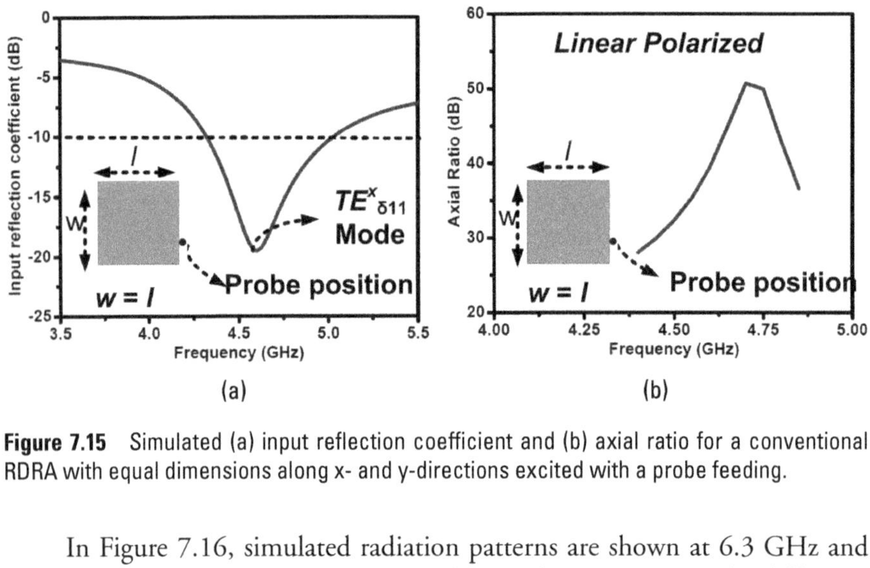

(a) (b)

Figure 7.15 Simulated (a) input reflection coefficient and (b) axial ratio for a conventional RDRA with equal dimensions along x- and y-directions excited with a probe feeding.

In Figure 7.16, simulated radiation patterns are shown at 6.3 GHz and 6.4 GHz, respectively. From the normalized radiation pattern, the difference between the RHCP and LHCP fields are –23 dB and –24 dB on the *xz*-plane. Thus, proposed antennas show RHCP behavior as it is more dominant then LHCP fields in broadside direction. In Figure 7.17, the electric field distribution is initially demonstrated inside rectangular DRA with equal dimensions for different phase angles (0°, 90°, 180°, and 270°) at 4.54 GHz.

It has been observed that the orientation of the net electric field is confined in a single direction and varying in a linear polarized manner. Next, E-field distribution is shown for a rectangular DRA with unequal dimensions (design example under consideration) at 6 and 6.4 GHz with reference to Figure

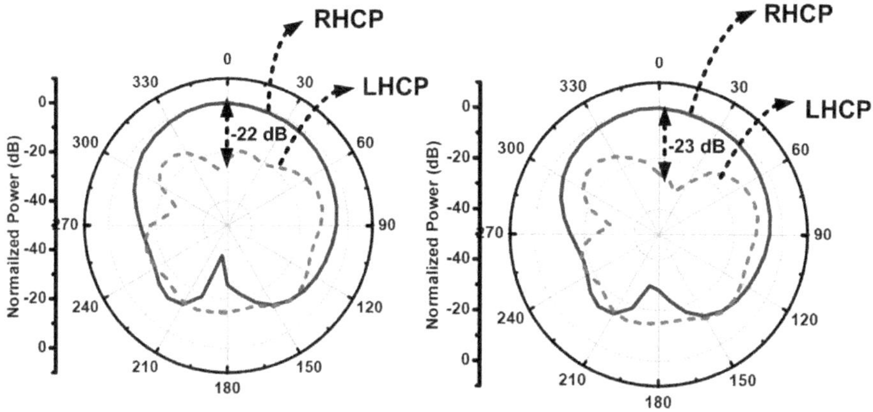

(a) xz Plane: 6.3 GHz (b) xz Plane: 6.4 GHz

Figure 7.16 Example of a conventional rectangular DRA with unequal dimensions coupled by an offset coaxial probe feeding for circular polarization generation.

7.11. In this case, the net electric field is found to be varying in a clockwise circular manner, which reveals its circularly polarized behavior. As well, the counterclockwise motion of the E-field confirms the RHCP mode of circular polarization.

7.3 CP Generation in Nonconventional DRAs

In the previous section, CP was achieved by proper optimization of DRA dimensions using a single probe, which is only possible when the geometry of the DRA supports the degenerate modes that can be detuned by optimizing their dimensions only (without altering the DRA geometry). This type of approach results in a narrow axial ratio bandwidth. This section presents the next method for CP generation in DRAs by altering their geometries. Generation of circularly polarized waves by modifying the dielectric resonator geometry is one of the most popular techniques used by antenna designers using coaxial probe-feeding schemes. Although this method involves fabrication complexity as compared to previous examples, it provides better CP performance and more design flexibility, which will be discussed in this section. For example, a rectangular DRA was modified with four inclined slits on the sidewall in [14]. It was then excited by a simple coaxial probe feeding at the center and by using the perturbation in DRA geometry, a circularly polarized omnidirectional radiation was obtained and a narrow ARBW of 3.8% was reported. A similar situation is portrayed in Figure 7.18 where a cylindrical DRA geometry is transformed in

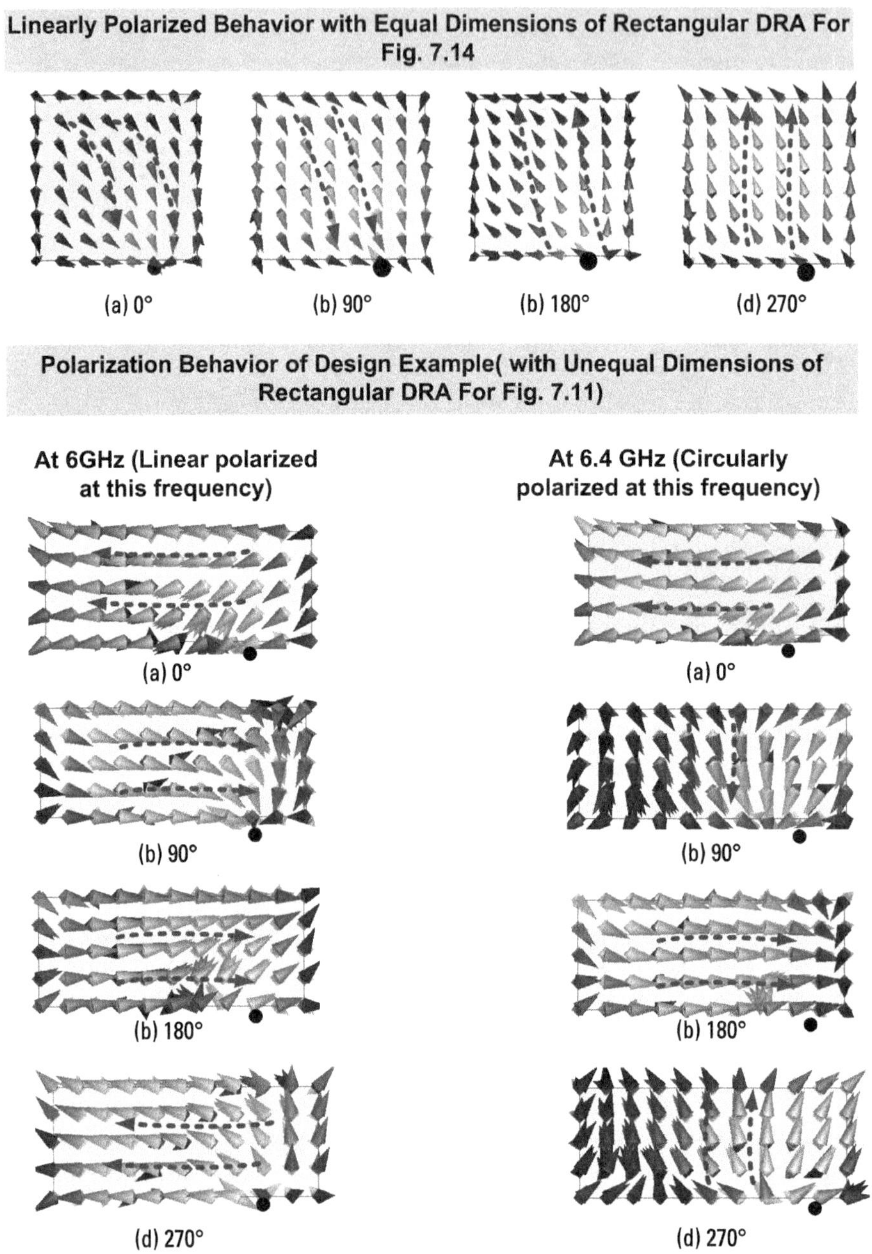

Figure 7.17 Electric field distribution inside rectangular DRA (for equal and unequal dimensions) with a probe feed for different phase angles showing linearly polarized behavior at 6 GHz and circularly polarized behavior at 6.4 GHz.

a similar manner and better CP performance could be obtained. The antenna geometry shown in Figure 7.18 is modified after [14].

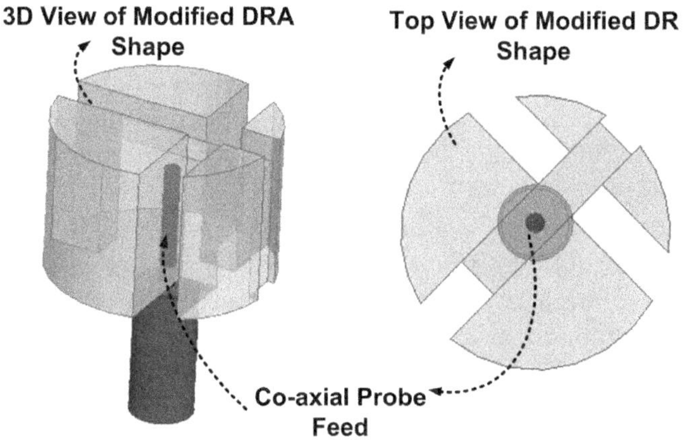

Figure 7.18 Modified cylindrical DRA fed by a coaxial probe for CP. (Modified after [14].)

Likewise, in [15] the authors modified an elliptical DRA in a semieccentric annular form. Also in this case, the modified geometry resulted in the decomposition of the electric field in orthogonal directions. Then, the aspect ratio and the feeding position were optimized to get CP radiation from the structure. This design offered an impedance and axial ratio bandwidth of 29.14%, and 5.71%, respectively. Figure 7.19 shows a similarly modified annular sectored DRA adapted after [15].

Another interesting modification was reported in [16] in which a rectangular DRA was strategically modified and excited by a probe feeding at the center to generate CP waves. In this design, the vertical longitudinal electric field was perturbed into orthogonal components by introducing several diagonal slots on a rectangular DRA. Also, the slotted dielectric behaved effectively as an anisotropic dielectric, and hence orthogonal components traveled at different

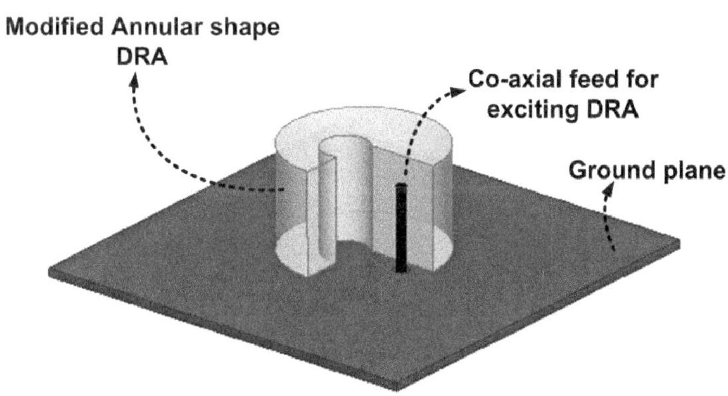

Figure 7.19 Modified annular semicircular DRA fed by a coaxial probe. (Modified after [15].)

velocities, leading to a phase difference between them. By correct selection of slot dimensions, phase quadrature requirement was met, leading to the excitation of CP waves. This modification approach in the design provided a wide ARBW of 22%. A similar tactic as shown in [16] may be applied to the cylindrical DRA shown in Figure 7.20 to excite CP radiation and improved results could be achieved.

Subsequently, there was another design reported in [17] in which a comb-shaped cylindrical DRA with metallic strips was suggested. In this configuration, the slots in the DRA were strategically oriented with respect to the feed to produce two orthonormal components of the electric field with a quadrature phase shift to get CP radiation. This DRA modification approach offered an IBW and ARBW of 9% and 4%, respectively. The idea in [17] could be tried with rectangular DRA as depicted in Figure 7.21 to get CP radiation with improved antenna parameters.

In [18], the corner of a rectangular DRA was chamfered, and a simple metallic strip was attached to it. This configuration was also coupled through a probe feeding and provided an IBW of 59.8% and an ARBW of 10.6% and 13.5%. The compactness and wide ARBW in this design were attributed to the modified geometry and added metallic strip. A circularly polarized DRA design is shown in Figure 7.22 (adapted from the design in [18]) and could provide comparable results where a rectangular DRA (resting on a grounded substrate) fed through a probe is chamfered along its corner in a curved form and a metallic patch is pasted on it.

In [19], the authors reported two CP antennas with wide AR bandwidths by inserting two different types of square-shaped slots in a rectangular DRA in the same antenna configuration. Primarily, the CP was obtained by the procedure of perturbation in the form of four and two square slots along the diagonal

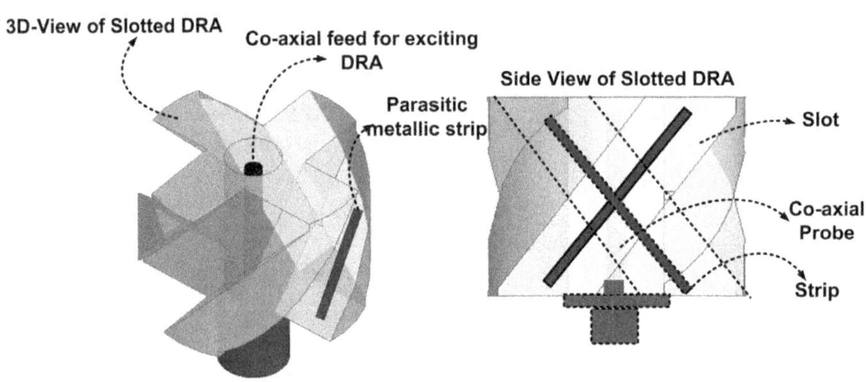

Figure 7.20 Configuration of modified cylindrical DRA for wideband omnidirectional circular polarization. (Modified after [16].)

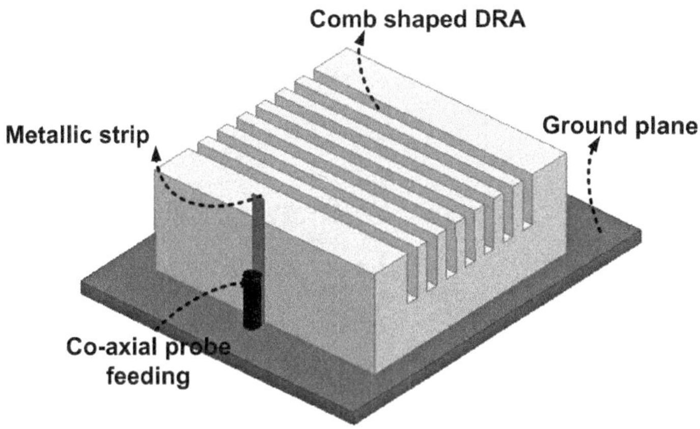

Figure 7.21 Configuration of a modified rectangular DRA in comb shape for circular polarization generation. (Modified after [17].)

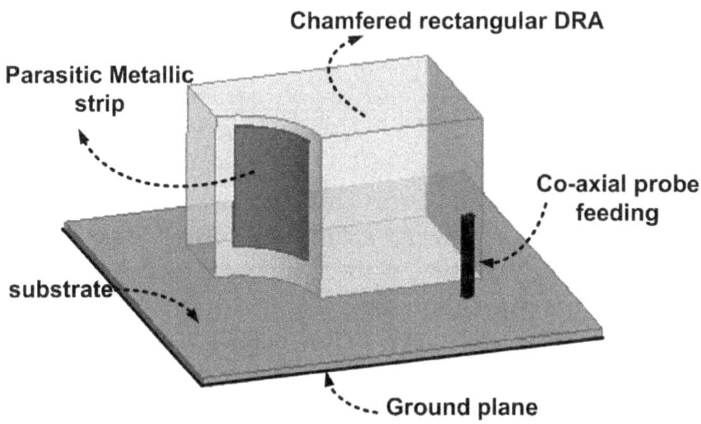

Figure 7.22 Modified rectangular DRA for multiband circular polarization. (Modified after [18].)

of a rectangular DR. Therefore, this antenna configuration was able to generate two different modes, TE_{111} and TE_{113}, and these modes were merged to enhance the impedance bandwidth up to 48%. For achieving wider CP bandwidth, two orthogonal pair modes, TE_{111} (TE^x_{111} and TE^y_{111}) and TE_{113} (TE^x_{113} and TE^y_{113}), were merged and enhanced the AR bandwidth from 11.5% to 24%. Similar results could be obtained by realizing the antenna geometry as shown in Figure 7.23, in which a cylindrical DRA can be formed with circular slots and proper positioning of feed locations to get CP waves.

Another example [20] is based on the asymmetric cross-shaped rectangular DRA coupled with a coaxial probe for wideband circular polarization

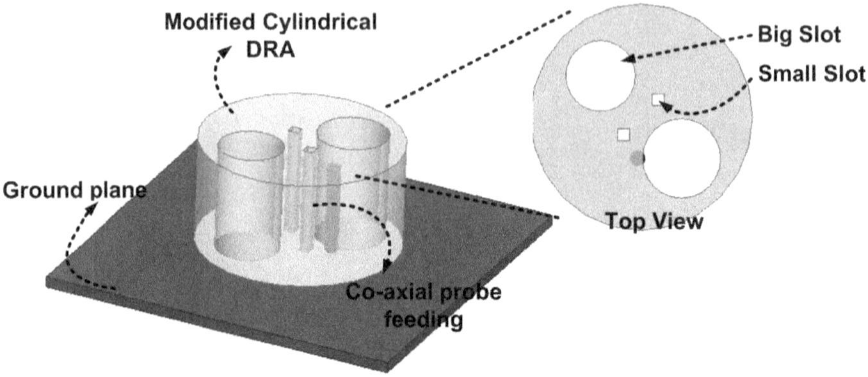

Figure 7.23 Cylindrical DRA with four circular slots fed by a coaxial probe feeding for CP generation. (Modified after [19].)

applications as shown in Figure 7.24. By using a metal strip in the corner of the asymmetric cross-shaped dielectric resonator, a pair of orthogonal $TE^x_{\delta 11}$ and $TE^y_{1\delta 1}$ modes are excited and these modes are responsible for circularly polarized fields. This design operates in the frequency band of 4.06–6.34 GHz, while it can also use 5.04–5.64 GHz for circular polarization applications; therefore it is very much suitable for the WLAN/Wi-MAX band applications. This antenna offered an overlapping bandwidth of 11.23% (centered at 5.25 GHz), where both the impedance bandwidth and ARBW fall in the same passband.

To summarize, in the previous examples, geometry of the DRA has been reformed in order to perturb the electric field into its orthogonal counterparts resulting in CP radiation. A summary of various designs reported in nonconventional DRAs using simple coaxial probe coupling is given in Table 7.2.

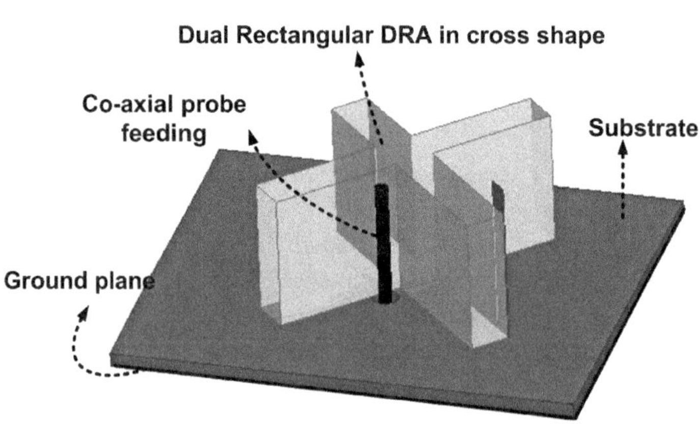

Figure 7.24 Example of a cross-shaped DRA coupled by a probe and metallic strip for CP generation [20].

Table 7.2
Brief Comparison of Various CP Designs Using Probe Feeding Techniques in Nonconventional DRAs

Ref	DRA Shape (ε_r)	Feed Arrangement	IMBW (%)	ARBW (%)	f_c (GHz)	Modes Excited
[14]	Modified rectangular DRA(10)	Single-probe feeding	4.5	~4	5.25	TM Mode
[15]	Semieccentric annular DRA (10)	Single-probe feeding	29.14	5.71	10.67	NM
[16]	Slotted rectangular DRA with parasitic metallic strips (15)	Single-probe feeding	22	25.4	3.62	NM
[17]	Comb-shaped cylindrical DRA (10)	Single-probe feeding	4	6.4	4.55	NM
[18]	Chamfered rectangular DRA (10)	Single-probe feeding (dual-band CP)	59.8	10.6, 13.5	8.77, 10.92	NM
[19]	Rectangular DRA with square shaped slots (10.2)	Single-probe feeding (two designs proposed)	48 and 41	11.5 and 24	6.05, 5.35	TE^x_{111} and TE^y_{111}, TE_{113}
[20]	Asymmetric cross shaped DRA (20)	Single-probe feeding	43.84	11.23	5.34	TE^x_{111} and TE^y_{111}

IMBW = fractional –10-dB impedance bandwidth, ARBW = fractional 3-dB axial ratio bandwidth, NM = not mentioned, ε_r = dielectric constant of DRA, f_c = center frequency with regard to axial ratio plot.

7.4 CP Generation in Multielement DRAs

The next few designs make use of two dielectric resonators in different arrangements for getting circularly polarized fields. The first design is shown in Figure 7.25 where circular polarization performance is dependent on the relative arrangement of the two rectangular dielectric resonators of unequal dimensions and the position of a coaxial feed with respect to the radiators [21]. By utilizing two different rectangular DRs, fundamental orthogonal TE^x_{111} and TE^y_{111} modes have been generated in individual DRs to get a wide axial ratio bandwidth of 26.85%.

The next example is based on a dual C-shaped rectangular DRA to obtain wideband circularly polarized radiation reported in [22]. A metallic strip is placed at the surface of a dual C-shaped rectangular DR to achieve the orthogonal modes as shown in Figure 7.26. By creating an optimal gap between the two C-shaped DRs, the designed antenna is able to generate wideband circularly polarized waves to cover the entire Wi-MAX (3.3–3.7 GHz) band. This multielement configuration using nonconventional DRAs offered an overlapping axial ratio bandwidth 14.81% (centred frequency at 3.51 GHz).

In [23], a wideband CP bird-nest DRA fed with a coaxial probe is reported. This configuration consists of a feeding probe and parasitic dielectric

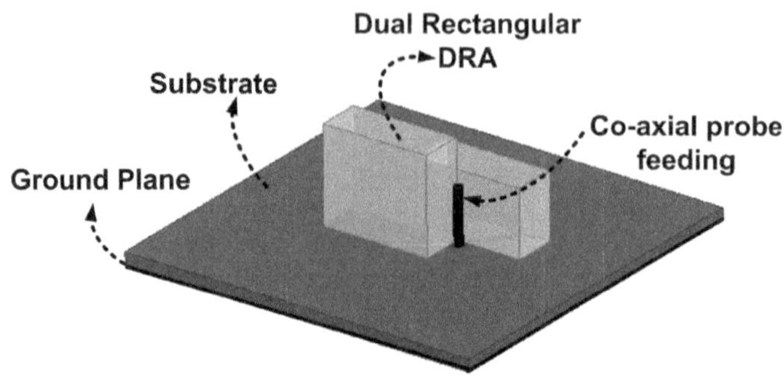

Figure 7.25 Example of a dual rectangular DRA with unequal dimensions fed by a simple probe for CP [21].

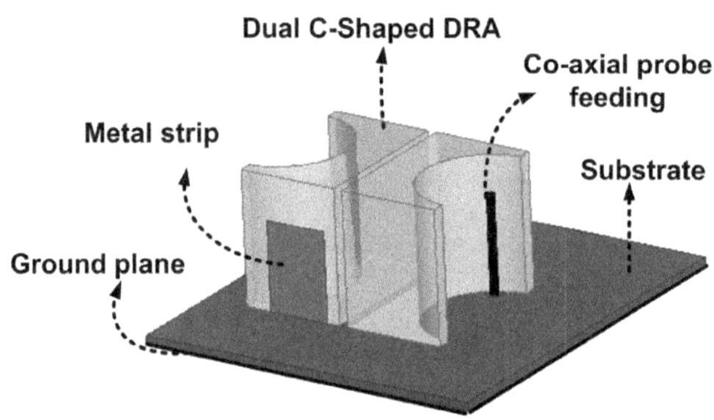

Figure 7.26 Example of a dual C-shaped DRA with a parasitic metallic excited by a probe for CP generation [22].

parallelepiped elements that surround the probe. The dielectric parallelepipeds work as a wave polarizer that converts omnidirectional LP fields into omnidirectional CP fields. This antenna offers an axial ratio bandwidth of 54.9% in the direction of $\theta = 30°$ and $\Phi = 0°$, and shows omnidirectional radiation patterns. Improved antenna results may be obtained with the design configuration as portrayed in Figure 7.27, where it increases the number of elements and the angle of inclination in the configuration of [23].

A summary of various designs reported in multielement DRAs using simple coaxial probe coupling is given in Table 7.3.

7.5 Probe-Fed Circularly Polarized DRAs for Other Applications

There are some additional applications where probe feeding has been used in DRA technology with respect to circular polarization. For example, a circularly

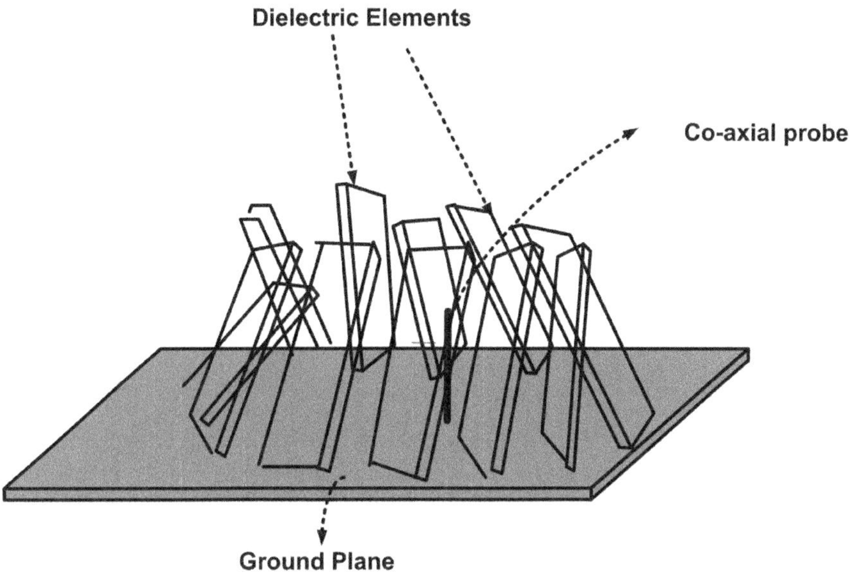

Figure 7.27 Configuration of the omnidirectional CP birdhouse antenna. (Modified after [23].)

Table 7.3
Brief Comparison of Various CP Designs Using Probe Feeding Techniques in Multielement DRAs

Ref	DRA Shape (ε_r)	Feed Arrangement	IMBW (%)	ARBW (%)	f_c (GHz)	Modes Excited
[21]	Simple rectangular DRA, 2 (9.8)	Single-probe feeding	26.07	26.85	6.14	TE^x_{111} and TE^y_{111}
[22]	C-shaped DRA, 2 (9.8)	Single-probe feeding	30.07	14.81	3.51	TE^x_{111} and TE^y_{111}
[23]	Parasitic dielectric parallelepipeds, 8, (79)	Single-probe feeding	42.5	57.7	NM	NM

IMBW = fractional –10-dB impedance bandwidth, ARBW = fractional 3-dB axial ratio bandwidth, NM = not mentioned, ε_r = dielectric constant of DRA, f_c = center frequency with regard to axial ratio plot.

polarized MIMO design was reported in [24], where two *L*-shaped self-complementary dielectric resonators were excited by a 50Ω coaxial probe. It has been observed that a probe can be strategically placed around the corners of a DRA to generate a pair of orthogonal modes. Moreover, a defected ground structure (DGS) was implemented to provide low interport correlation. An analogous MIMO design configuration with improved isolation and antenna parameters is depicted in Figure 7.28 after substantial modifications were made in the DR orientation and DGS of [24].

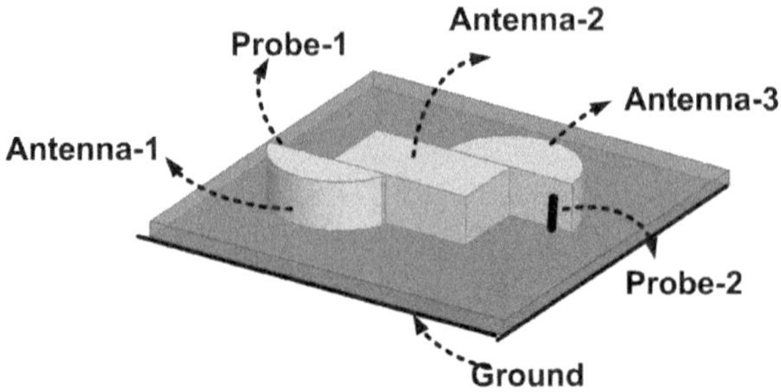

Figure 7.28 L-shaped dielectric resonator-based MIMO antenna coupled using probe feeding for CP generation. (Modified after [24].)

A second example of circularly polarized MIMO design is presented in [25]. In this design, fundamental $TE^x_{\delta 11}$ and $TE^y_{1\delta 1}$ modes are excited by using the probe to excite three rectangular dielectric resonators arranged in stepped structure form. The reported axial ratio bandwidth is 13.23% with pattern diversity in this design. Better results with high isolation and antenna parameters may be claimed with MIMO design configuration as depicted in Figure 7.29 after significant modification of DR shapes and orientation in [25].

The probe-feeding method has also been utilized in dielectric resonator based array antennas for realizing CP fields [1, 26]. The first reported example of CP in a DRA using a probe feeding used chamfered rectangular DRA [1] to generate CP radiation. Later, the axial ratio bandwidth was enhanced by arranging the same DRA elements using the sequential rotation technique [27]. In one more illustration, the same DRA arrangement was analyzed numerically in [26] where different arrangements and excitations for the DRA elements (chamfered rectangular DRA) were used to study their effect on the axial ratio bandwidth to find the best arrangement. The DRA elements that were arranged in the sequential rotation technique provided the best axial ratio bandwidth.

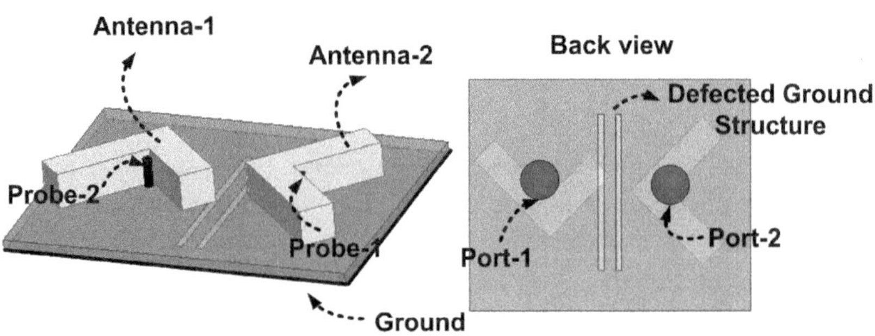

Figure 7.29 A DRA configuration based on MIMO antenna with pattern diversity using a simple probe feed. (Modified after [25].)

In [28], a differential system (to eliminate the common mode interference) was designed using a probe-fed circularly polarized ring cylindrical DRA for dual-band applications. In the design, the probe is attached to an arc-shaped metallic strip and its orientation controls the polarization state. The relative size of the two DRAs and the shape of the strip independently controls the two bands. The fundamental $HE_{11\delta}$ mode of cylindrical DRA constitutes the lower band, whereas the arc shaped strips generate the upper band. A differential CP design configuration has been attempted by implementing the same line in a rectangular DRA as portrayed in Figure 7.30, which is adapted from the configuration of [28].

As a final example, the next two designs [29, 30] use a cylindrical DRA loaded with a modified Alford loop patch to get omnidirectional CP antenna, which is a challenging task. In [29], two designs were proposed in which the first design is an omnidirectional antenna top-loaded by a modified Alford loop and the antenna is fed by an axial feeding probe at its center excited in its fundamental mode. The technique of combining equivalent electric and magnetic monopoles was successfully implemented in the design. The idea of generating CP fields with omnidirectional radiation is to tune the dimensions of top-loaded Alford loop and ground plane. Due to this, the amplitude of the radiation fields from the electric-magnetic monopole arrangement becomes equal but different in phase by 90. This omnidirectional CP-DRA design is then converted into a three-port full CP pattern diversity antenna. Also in [30], a CP-DRA was designed based on the same structure as [29] but for dual-band applications, which is again centrally fed by a coaxial probe feed for omnidirectional patterns with different senses of circular polarization in two bands.

In the design, the loaded patch is used to convert LP radiation into CP radiation and also provides another resonance for dual-band antenna designing. The lower and upper bands are controlled by the patch and dielectric resonator. A design configuration has been implemented and offers acceptable results, as

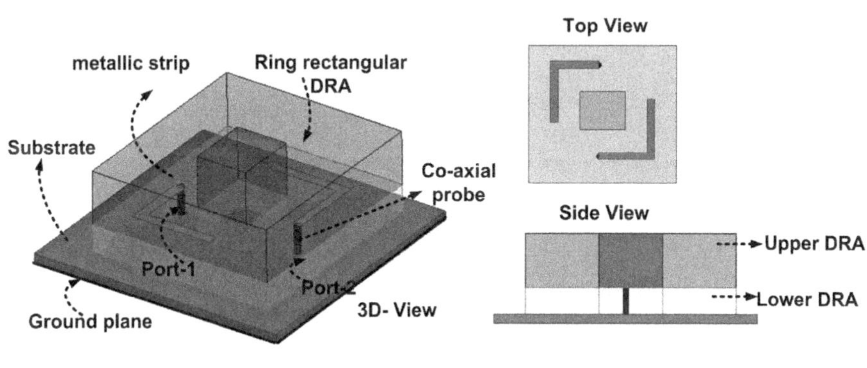

Figure 7.30 Differentially inserted probe-fed dual-band CP ring CDRA. (Modified after [28].)

shown in Figure 7.31, which is adapted from [29, 30], where a square metallic strip is loaded instead of an Alford loop and a square-shaped microstrip feeding. A brief summary of this section is given in Table 7.4.

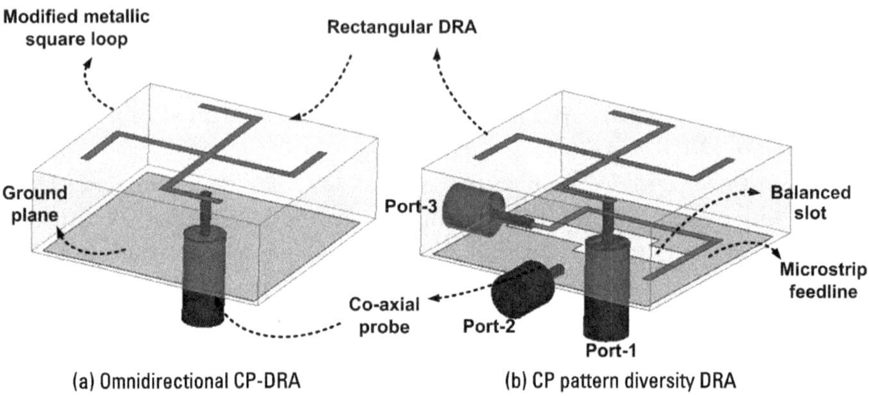

Figure 7.31 Square loop loaded rectangular DRA with c-axial probe feeding. (Modified after [29] and 30].)

Table 7.4.
Brief Comparison of Various CP Designs Using Miscellaneous Slot-Shape Coupling Techniques

Ref	DRA Shape (ε_r)	Feed Arrangement	IMBW (%)	ARBW (%)	f_c (GHz)	Modes Excited
[24]	Two *L*-shaped DRA (9.8)	Probe feeding with DGS (MIMO)	15.6	7.05	5.39	$TE^x_{\delta 11}$ and $TE^y_{1\delta 1}$
[25]	Three stepped rectangular DRA (9.8)	Only probe feeding (MIMO)	21.51	13.23	5.51	$TE^x_{\delta 11}$ and $TE^y_{1\delta 1}$
[1]	Chamfered square DRA	Probe feed (sequential rotation method)		17		
[28]	Two ring cylindrical DRA of unequal height	Arc-shaped strips connected with probe-fed differential system	30.4 and 18.5	8.06 and 6.38	3.47 and 4.85	$HE_{11\delta}$
[29]	Cylindrical DRA with small ground plane (10)	Probe feeding with modified Alford loop (omnidirectional CP)	7	10.5	2.38	$TM_{01\delta}$
[30]	Cylindrical DRA (6.85)	Probe feeding with modified Alford loop (omnidirectional dual-band dual-sense CP)	2.64, 18.03	3.16, 5.06	1.9, 2.37	TM_{01} and TM_{011}

IMBW = fractional –10-dB impedance bandwidth, ARBW = fractional 3-dB axial ratio bandwidth, NM = not mentioned, ε_r = dielectric constant of DRA, f_c = center frequency with regard to axial ratio plot.

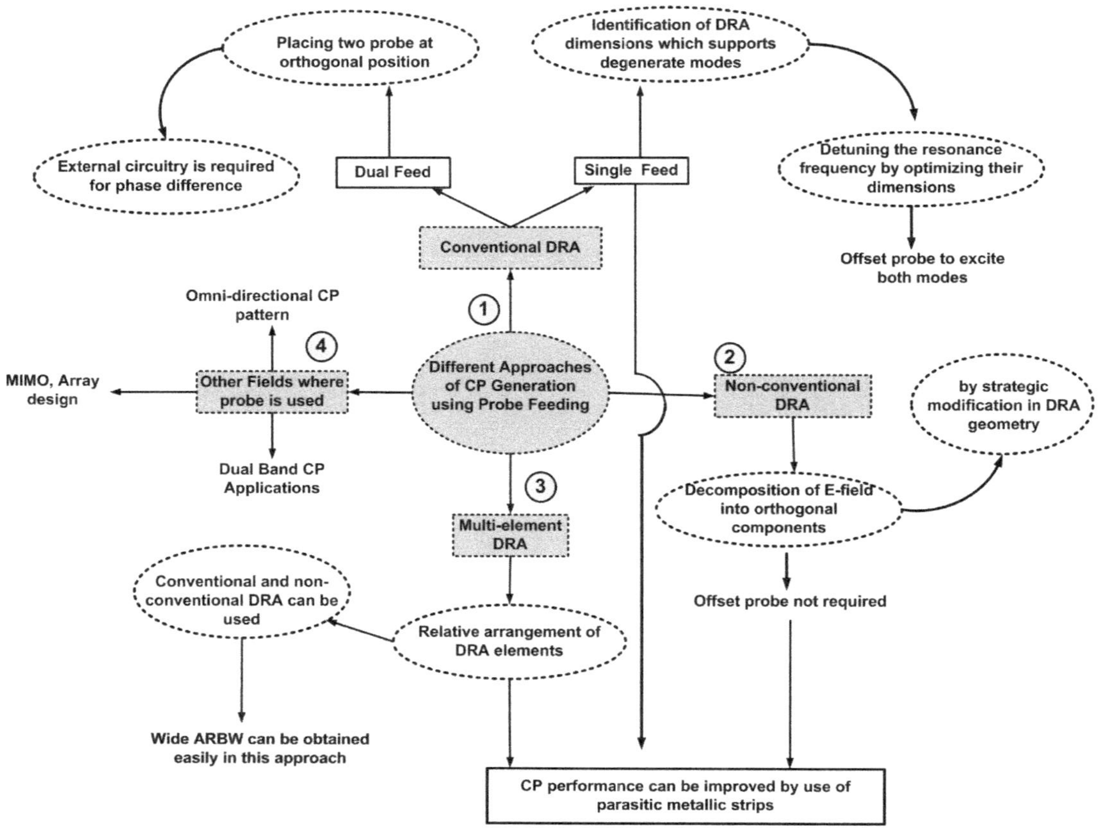

Figure 7.32 An overview of probe coupling approaches in DRA for CP generation.

7.6 Summary

A comprehensive investigation of the circular polarization performance in DRAs using probe-feed techniques was presented with modified and new antenna geometries in the preceding sections. This chapter provided insight into the different probe-feeding schemes and their properties, such as their construction, advantages, disadvantages, and key features. Figure 7.32 summarizes the different approaches being used for designing circularly polarized antenna in DRA technology. Different possible approaches for CP generation are possible through conventional, nonconventional, and multiple DRAs. In the case of conventional and multiple DRAs, circularly polarized fields can be obtained by setting appropriate locations of feeds and the relative position of DRAs with slightly unequal dimensions.

In contrast, for the nonconventional DRAs, the 3D geometry of the dielectric resonator is used as a design parameter and the detuning of degenerate modes approach is adopted. The shape of the DRA can be adapted into various forms to decompose the electric field into its orthogonal components. In the last section, the use of probe coupling also extended to MIMO, array, differential form, omnidirectional CP antenna, and so on.

References

[1] Haneishi, M., and H. Takazawa, "Broadband Circularly Polarised Planar Array Composed of a Pair of Dielectric Resonator Antennas," *Electronics Letters*, Vol. 21, No. 10, 1985, pp. 437–438.

[2] Kishk, A. A., R. Chair, and K. F. Lee, "Broadband Dielectric Resonator Antennas Excited by L-Shaped Probe," *IEEE Transactions on Antennas and Propagation*, Vol. 54, No. 8, 2006, pp. 2182–2189.

[3] Leung, K. W., K. M. Luk, K. Y. A. Lai, and D. Lin, "Theory and Experiment of a Coaxial Probe Fed Hemispherical Dielectric Resonator Antenna," *IEEE Transactions on Antennas and Propagation*, Vol. 41, No. 10, 1993, pp. 1390–1397.

[4] Kishk, A. A., "A Triangular Dielectric Resonator Antenna Excited by a Coaxial Probe," *Microwave and Optical Technology Letters*, Vol. 30, No. 5, 2001, pp. 340–341.

[5] Petosa, A., *Dielectric Resonator Antenna Handbook*, Norwood, MA: Artech House, 2007.

[6] Mongia, R. K., A. Ittipiboon, M. Cuhaci, and D. Roscoe, "Circularly Polarised Dielectric Resonator Antenna," *IET Electronics Letters*, Vol. 30, No. 17, 1994, pp. 1361–1362.

[7] Drossos, G., Z. Wu, and L.E. Davis, "Circular Polarised Cylindrical Dielectric Resonator Antenna," *IET Electronics Letters*, Vol. 32, No. 4, 1996, pp. 281–282.

[8] Oliver, M. B., R. K. Mongia, and Y. M. M. Antar, "A New Broadband Circularly Polarized Dielectric Resonator Antenna," *IEEE Antennas and Propagation Society International Symposium Digest*, Vol. 1, 1995, pp. 738–741.

[9] Lo, Y. Y., and W. F. Richards, "Perturbation Approach to Design of Circularly Polarised Microstrip Antennas," *Electronics Letters*, Vol. 17, No. 11, 1981, pp. 383–385.

[10] Sharma, P. C., and K. C. Gupta, "Analysis and Optimized Design of Single Feed Circularly Polarized Microstrip Antennas," *IEEE Transactions on Antennas and Propagation,* Vol. 31, No. 6, 1983, pp. 949–955.

[11] Hamsakutty, V., A. V. P. Kumar, J. Yohannan, and K. T. Mathew, "Coaxial Fed Hexagonal Dielectric Resonator Antenna for Circular Polarization," *Microwave and Optical Technology Letters*, Vol. 48, No. 3, 2006, pp. 581–582.

[12] Sulaiman, M. I., and S. K. Khamas, "A Singly Fed Rectangular Dielectric Resonator Antenna with a Wideband Circular Polarization," *IEEE Antennas and Wireless Propagation Letters*, Vol. 9, 2010, pp. 615–618.

[13] Mongia, R. K., A. Ittipiboon, "Theoretical and Experimental Investigations on Rectangular Dielectric Resonator Antennas," *IEEE Transactions on Antennas and Propagation*, Vol. 45, No. 9, 1997, pp. 1348–1356.

[14] Khalily, M., M. R. Kamarudin, M. Mokayef, and M. H. Jamaluddin, "Omnidirectional Circularly Polarized Dielectric Resonator Antenna for 5.2-GHz WLAN Applications," *IEEE Antennas and Wireless Propagation Letters*, Vol. 13, 2014, pp. 443–446.

[15] Lee,J. M., S.-J. Kim, G. Kwon, C. M. Song, Y. Yang, K. Y. Lee, and K. C. Hwang, "Circularly Polarized Semi-Eccentric Annular Dielectric Resonator Antenna for X-Band Applications," *IEEE Antennas and Wireless Propagation Letters*, Vol. 14, 2015, pp. 1810–1813.

[16] Pan, Y. M., and K. W. Leung, "Wideband Omnidirectional Circularly Polarized Dielectric Resonator Antenna with Parasitic Strips," *IEEE Transactions on Antennas and Propagation*, Vol. 60, No. 6, 2012, pp. 2992–2997.

[17] Chu, L. C. Y., D. Guha, and Y. M. M. Antar, "Comb-shaped Circularly Polarised Dielectric Resonator Antenna," *IET Electronics Letters*, Vol. 42, No. 14, 2006.

[18] Abedian, M., S. K. A. Rahim, S. Danesh, M. H. Jamaluddin, and M. T. Islam, "Compact Wideband Circularly Polarized Dielectric Resonator Antenna," *IET Electronics Letters*, Vol. 53, No. 1, 2017, pp. 5–6.

[19] Patel, P., B. Mukherjee, and J. Mukherjee, "Wideband Circularly Polarized Rectangular Dielectric Resonator Antennas Using Square-Shaped Slots," *IEEE Antennas and Wireless Propagation Letters*, Vol. 15, 2016, pp. 1309–1312.

[20] Kumar, R., and R. K. Chaudhary, "Compact Asymmetric Cross-Shaped Rectangular Dielectric Resonator Antenna for Wideband Circular Polarization," *Microwave and Optical Technology Letters*, Vol. 61, No. 7, 2019, pp. 1863–1867.

[21] Chowdhury, R., R. Kumar, and R. K. Chaudhary, "A Coaxial Probe Fed Wideband Circularly Polarized Antenna Using Unequal and Adjacent-Slided Rectangular Dielectric Resonators for WLAN Applications," *International Journal of RF and Microwave Computer-Aided Engineering*, Vol. 28, No. 4, 2018.

[22] Kumar, R., N. Nasimuddin, and R. K. Chaudhary, "A New Dual C-Shaped Rectangular Dielectric Resonator Based Antenna for Wideband Circularly Polarized Radiation," *International Journal of RF and Microwave Computer-Aided Engineering*, Vol. 29, No. 6, 2019.

[23] Pan, Y. M., and K. W. Leung, "Wideband Circularly Polarized Dielectric Bird-nest Antenna with Conical Radiation Pattern," *IEEE Transactions on Antennas and Propagation,* Vol 61, No. 2, 2012, pp. 563–570.

[24] Sahu, N. K., G. Das, and R. K. Gangwar, "L-shaped Dielectric Resonator based Circularly Polarized Multi-input-multi-output (MIMO) Antenna for Wireless Local Area Network (WLAN) Applications," *International Journal of RF and Microwave Computer-Aided Engineering,* Vol. 28, No. 9, 2018.

[25] Sahu, N. K., G. Das, and R. K. Gangwar, "Dielectric Resonator-based Wide Band Circularly Polarized MIMO Antenna with Pattern Diversity for WLAN Applications," *Microwave and Optical Technology Letters,* Vol. 60, No. 12, 2018, pp. 2855–2862.

[26] Kishk, A. A., "Performance of Planar Four-element Array of Single-fed Circularly Polarized Dielectric Resonator Antenna," *Microwave and Optical Technology Letters,* Vol. 38, No. 5, 2003, pp. 381–384.

[27] Yang, S. S., R. Chair, A. A. Kishk, K. F. Lee, and K. M. Luk, "Study on Sequential Feeding Networks for Subarrays of Circularly Polarized Elliptical Dielectric Resonator Antenna," *IEEE Transactions on Antennas and Propagation,* Vol. 55, No. 2, 2007, pp. 321–333.

[28] Tong, C., H. Tang, W. Qin, W. Yang, X. Zhang, and J. Chen, "Differentially Inserted-Fed Compact Dual-Band Circularly Polarized Dielectric Resonator Antenna," *IEEE Antennas and Wireless Propagation Letters,* Vol. 18, No. 12, 2019, pp. 2498–2502.

[29] Li, W. W., and K. W. Leung, "Omnidirectional Circularly Polarized Dielectric Resonator Antenna with Top-Loaded Alford Loop for Pattern Diversity Design," *IEEE Transactions on Antennas and Propagation,* Vol. 61, No. 8, 2013, pp. 4246–4256.

[30] Pan, Y. M., S. Y. Zheng, and W. Li, "Dual-Band and Dual-Sense Omnidirectional Circularly Polarized Antenna," *IEEE Antennas and Wireless Propagation Letters,* Vol. 13, 2014, pp. 706–709.

8

Sectorization Approach in Dielectric Resonator Antennas for Circular Polarization

This chapter investigates the sectorization technique in dielectric resonator antennas for circular polarization. To develop an elementary foundation of the sectorization technique in DRAs, a brief overview is presented followed by a discussion on the advantages and disadvantages associated with this technique. Thereafter, analysis of circular polarization in cylindrical and hemispherical shaped DRAs is presented with a design example. A new approach for CP generation is then introduced in the last section followed by discussion of CP-MIMO antennas and future directions.

8.1 Introduction to the Sectorization Technique

In the previous chapters, different types of feeding techniques have been discussed for CP analysis that were mainly grounded on the geometry and relative dimensions of the feeding networks (with an exception in Chapter 7, where the DRA geometries were modified, since the probe feeding does not allow any modification in its geometry). This chapter introduces the concept of the sectorization method (one type of uniform perturbation technique) in DRAs that is potentially advantageous for CP applications and offers less fabrication complexity.

Sectorization [1, 2] in DRAs implies removing an equal volume of dielectric blocks (wedge shaped) from a conventional DRA geometry. After removing each block, one modified geometry is obtained and is called one form of sectored DRA. Each of the removed blocks form the original DRA when assembled, as depicted in Figure 8.1.

This approach of sectorization is applicable to DRA geometries with body of revolution (a structure obtained by rotation of a plane about its axis) property like cylindrical DRA (CDRA) and hemispherical DRA (HDRA). A new tunable parameter is obtained in this process called as sector angle (denoted by β), which is the angular separation between the two newly formed flat faces due to sectorization. This parameter offers extra design flexibility to optimize the ARBW for better CP performance. A pictorial description of the above statements is shown in Figure 8.1 for CDRA and HDRA separately.

The sectorization method [1] has various advantages over conventional geometries. such as a new adjusting parameter and sector angle (β), which is obtained and can be tuned for better antenna performance. It is also useful for predicting the resonance frequency in modified geometry and controlling the polarization purity of the radiated wave. The second advantage of this approach is its simplicity in fabrication as compared to available modification techniques that are generally complex and arbitrarily modified [3, 4], which increases the fabrication difficulty. This approach also provides compactness and less weight (due to volume reduction), among other advantages.

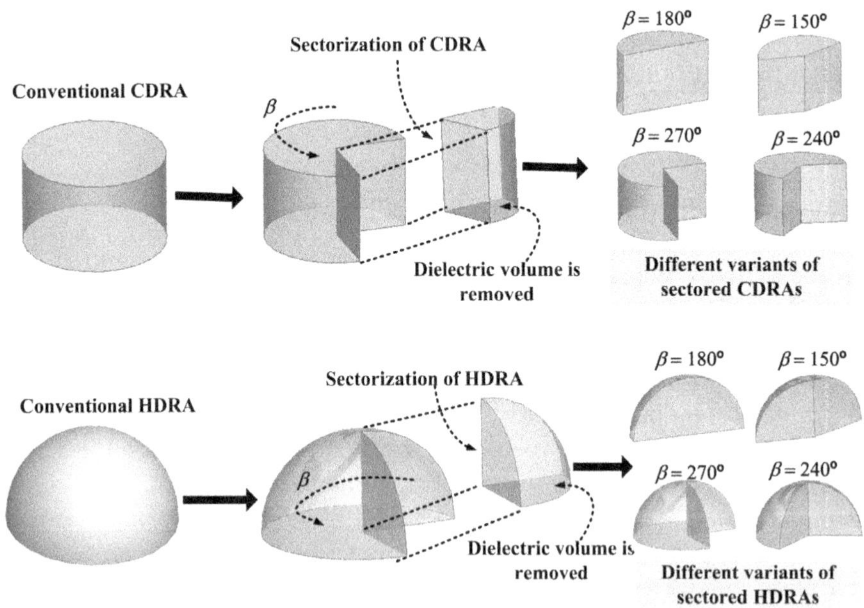

Figure 8.1 Pictorial description of the sectorization method in (a) cylindrical DRA, and (b) hemispherical DRA.

One of the most important uses of sectored DRAs would be in defense or spaceborne applications, where there are situations when a very large number of antenna elements are used in array form, such as in massive MIMO applications. In that case, sectored DRAs will be beneficial because sectored DRAs could provide the same or better performance as compared to conventional DRAs with reduced volume and weight. Thus, required volume is minimized for installing antennas as more numbers of antenna elements can be implemented for different applications simultaneously. Furthermore, more design flexibility can be incorporated in sectored DRAs by introducing multisegment and multipermittivity approaches [5] as shown in Figure 8.2.

There are also certain disadvantanges to this technique. Due to the removal of dielectric material, the frequency of the mode shifts to a higher range. One solution to this problem is investigated in [6] by proper utilization of metallic plates at its flat faces (a similar situation is shown in Figure 8.3). Another limitation associated with this approach is that DRA geometry becomes irregular after modification and the boundary conditions are changed. As a consequence, it becomes difficult to predict and calculate the behavior of the electromagnetic mode inside the DRA.

To conclude from the above discussion, sectored forms of DRAs offer the same performance as conventional DRAs with some drawbacks and can be considered a new candidate in the DRA community. This chapter focuses on the sectorization approach in the context of circular polarization only. The concept of circular polarization generation in sectored DRA is based on the principle

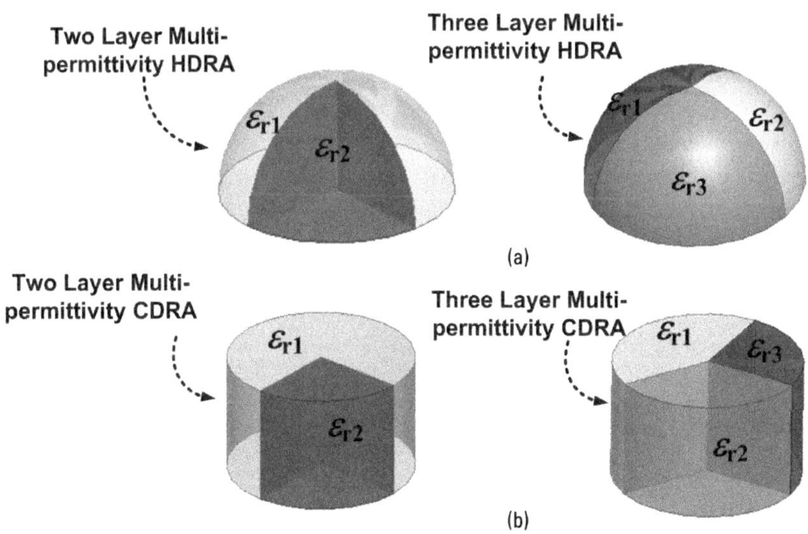

Figure 8.2 Possible alteration in sectored (a) HDRAs and (b) CDRAs by employing multisegment and multipermittivity to further enhance their performance.

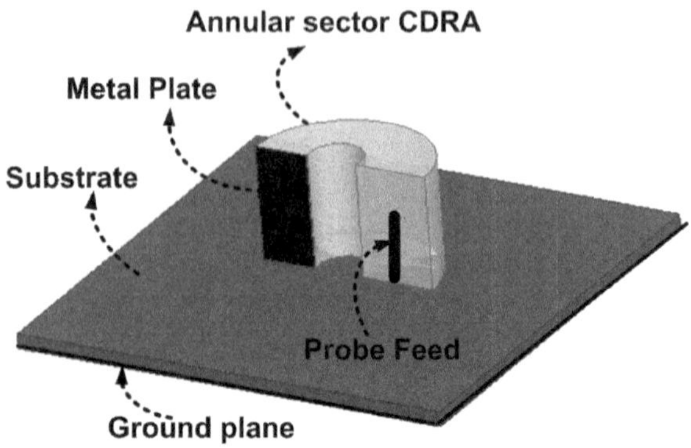

Figure 8.3 A similar example of annular sector DRA with a metal sheet for compact applications. (Modified after [9].)

of detuning and optimizing the degenerate orthogonal modes in the structure [7]. It is just a simple perturbation technique that can be used to detune the orthogonal degenerate modes in the DRA to get CP radiation. The ARBW can also be improved by taking into account the sector angle with any design parameter of the DRA. In view of this, the sectorization technique is presented in the next two sections in cylindrical and hemispherical DRA separately.

8.2 The Sectorization Technique in Cylindrical DRAs

This section introduces investigations on the sectorization technique reported in CDRA in order to allow an understanding of circular polarization analysis discussed in subsequent sections. To compare the geometries of a conventional CDRA and sectored CDRA, the former has one degree of freedom that is the aspect ratio. Different values of aspect ratio (dimeter to height ratio) may provide different ranges of bandwidth, directivity, and so on, for same resonance frequency. In the case of sectored CDRA, another tunable parameter is obtained, which is sector angle (β). Thus, more flexibility is available to the antenna designers for optimizing the performance of sectored DRAs. Before the discussion of CP, a quick glimpse of modes nomenclature in sectored CDRAs is now presented.

8.2.1 Mode Nomenclature in Sectored CDRAs

A circular cylindrical DRA mainly supports three type of electromagnetic modes [8]:

- Transverse electric (*TE*) modes;

- Transverse magnetic (*TM*) modes;
- Hybrid (*HE* or *EH*) modes.

TE and *TM* modes have no azimuthal dependence, whereas hybrid modes have azimuthal dependency. Thus, hybrid modes have their degenerate counterparts. One such mode is the fundamental $HE_{11\delta}$ mode, which radiates like a horizontal magnetic dipole with broadside radiation. In the case of sectored cylindrical DRAs, the cavity resonator model has been used for analysis of approximate field behavior and predicts their resonance frequency. This model (that assumes perfect magnetic conductor boundary conditions) delivers an accurate prediction for resonant frequency but does not guess more complicated field behavior such as hybrid modes. Thus in the literature, *TM* and *HE* have been simply grouped as *TM* mode [6]. This is because for *HE* mode, the *H*-field component is negligible in propagation direction and also the *H*-field component for a *TM* mode is zero in propagation direction. Similarly, *TE* and *EH* can be grouped as *TE* mode. The use of *TM* mode nomenclature is popular in sectored DRAs and hence the same mode nomenclature has been used in this chapter.

8.2.2 Analysis of Circular Polarization in Sectored Cylindrical DRAs

The first reported examination of a sectored DRA was reported by Tam and Murch in 1999 for compact applications [9]. However, the major focus was on the compactness and theoretical study of different forms of circular sectored and annular sectored DRAs in their study. In this work [9], circular polarization was not investigated in all forms of sector and annular sector DRAs. The discussion of various operating modes was also investigated for compact sector DRAs but in the context of volume minimization. The same situation can be realized by the design shown in Figure 8.3 for CP generation adapted from [9].

In the next example, the same authors generated CP for the first time in 2000, in a semicylindrical DRA [10]. In this design [10], the probe feed location and radius to height ratio was used to generate CP radiation by exciting two resonant modes in spatially orthogonal arrangement and phase quadrature. However, the role of sector angle was missing in this design and the analysis was demonstrated in a single form of a sector DRA. Hence, compared to [9], the investigation was performed on different forms of sector DRAs without CP, whereas compared to [10], CP was generated but the effect of ARBW for different sector angle was not analyzed. Since, then very few studies have been devoted to sector CDRAs [11, 12]. In [11] also, one sector form of DRA was presented with narrow axial ratio bandwidth. In the final example of circularly polarized sectored CDRA, a complete investigation on CP with detailed analysis of sector angle was presented in [13]. The sectored CDRA geometry is shown in Figure 8.4. The sector angle (β) and the aspect ratio (α) of CDRA was utilized to improve the ARBW based on (8.1) [9].

$$f_{vpm} = \frac{c}{2\pi r\sqrt{\varepsilon_r}}\sqrt{X_{vp}^2 + \left[\frac{(2m+1)\pi r}{2H}\right]^2} \tag{8.1}$$

$$f_{vpm} = K_1\sqrt{\frac{K_3}{\beta^2} + K_2\alpha^2} \tag{8.2}$$

Equation (8.2) is found by reformulation of (8.1) and the dependence of resonant frequency on α and β has been established.

In this design, both parameters, β (varied from 270° to 120°) and α (varied from 1.2 to 2.2 mm) have been varied simultaneously and their CP performance was analyzed for the best performance. In the first case, β varied from 120° to 270° with a 30° increment and further, these sectored geometries were studied at different values of aspect ratios. The radius of the sectored DRA was fixed (at 11 mm) and their heights were varied to get different aspect ratios. The result of fractional ARBW obtained is shown in Figure 8.5(a). Where the graph shows first an increase in bandwidth up to a maximum value then decreases,

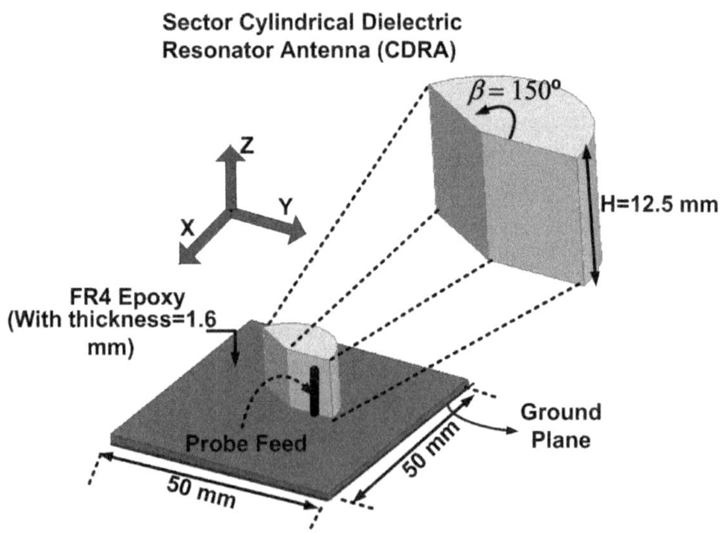

Figure 8.4 Geometry of circular sector cylindrical dielectric resonator antenna with $\alpha = 1.6$ and $\beta = 150°$ [13].

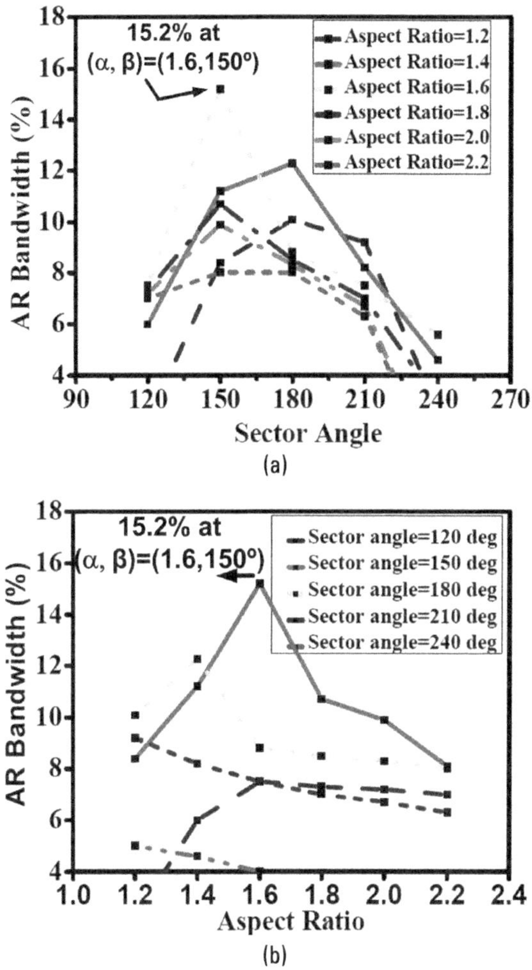

Figure 8.5 Simulated axial ratio bandwdith with (a) varying β at different values of α and (b) varying α at different β. (From [13].)

and the best results (i.e., an ARBW of 15.2% (5.47–6.37 GHz)) have been reported for $(\alpha, \beta) = (1.6, 150^o)$.

In the second case, a similar method was repeated and α varied from 1.2 to 2.2 at a fixed value of β and further studied using different values of β as shown in Figure 8.5(b). A summary of various designs reported for CP antenna in sectored CDRAs is collectively briefed in Table 8.1. It has been claimed that sector angle β is responsible for detuning the degenerate modes and the parameter α is responsible for altering the spacing between the two modes.

Next, a design example will be discussed for CP generation in a sectored CDRA emphasizing the importance of a sector angle and the considered

Table 8.1
Brief Comparison of Various Sectored CDRAs for Circular and Linear Polarization

Ref	Sector CDRA Feature (ε_r)	Feeding	Feature	IMBW (%)	ARBW (%)	f_c (GHz)	Modes Excited
[9]	Sector CDRA (β = 45°to 330°), 12	Probe coupling	Compact antenna	9.5	NA	NA	$TM_{\upsilon 11}$ (υ is dependent on β)
[10]	Sector CDRA (β = 180°), 12	Probe coupling	CP antenna	31.5	10%	2.675	$TM_{11\delta}$ and $TM_{21\delta}$
[11]	Sector CDRA (β = 90°), 13	Probe coupling	CP antenna	15.3	3.3	2.26	$TM_{01\delta}$ and $TM_{41\delta}$
[12]	Sector CDRA (β = 180°), 12	Microstrip coupling	Wideband antenna	36.7	NA	NA	$TM_{21\delta}$
[13]	Sector CDRA (β = 120° to 240°), 9.8	Probe coupling	CP antenna	48.3	15.4	5.88	$TM^x_{11\delta}$ and $TM^y_{11\delta}$

IMBW = fractional –10-dB impedance bandwidth, ARBW = fractional 3-dB axial ratio bandwidth, ε_r = dielectric constant of DRA, f_c = center frequency with regard to axial ratio plot, NA = not applicable.

example is adapted from our work presented in [13]. The general sectored geometries considered in this example are shown in Figure 8.6. The antenna consists of a sectored cylindrical DRA having sector angle β varying from 120° to 270° (with fixed radius r = 10 mm) and height H. The alumina dielectric resonator (with ε_r = 9.8 and $tan\,\delta$ = 0.001) is placed over a 50 × 50 mm^2 FR4 epoxy glass substrate (with ε_r = 4.4 and $tan\,\delta$ = 0.02) below which there is a copper layer serving as a ground plane. Different variations of sectored geometries are obtained by removing a wedge-shaped block (at center angle of 30°) from the conventional cylindrical DRA and each sectored geometry obtained has been investigated for circular polarization at a different β. To excite a degenerate $TM_{11\delta}$ mode, a coaxial probe feed is placed near the edge of the sectored CDRA as shown in Figure 8.6.

The manner of getting circular polarization radiation from a sectored dielectric resonator antenna using a single-feed technique generally requires the identification, excitation, and detuning of degenerate orthogonal modes in the structure. Then, the proper frequency of excitation and feed location is chosen to detune the center frequency (f_0) in between the resonance frequency of orthogonal modes and there by affecting the magnitude and phase difference between them. The sectored cylindrical DRA supports TM^x and TM^y modes (field configuration similar to HE^x and HE^y) as already stated before. These modes have orthogonal electric field configurations and their resonance frequency is the same when no sectorization is applied. Now, these TM modes have mode degeneracy property. The aspect ratio is fixed at 1.6 and sector angle is varied from 270° to 120°. Upon varying the sector angle at fixed aspect

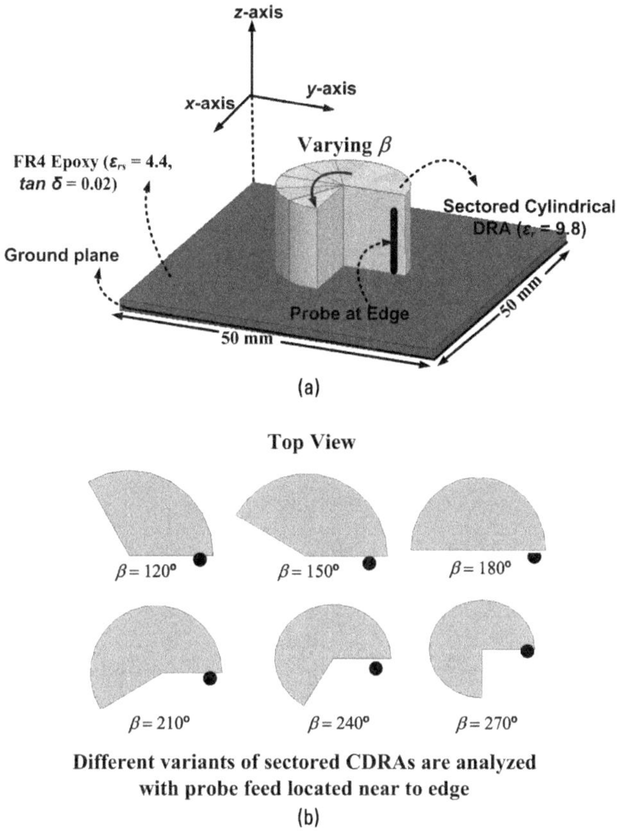

Figure 8.6 (a) A design example showing a sectored cylindrical DRA fed with a coaxial probe feed for circular polarization generation, and (b) the top view of various sectored cylindrical DRA that are formed.

ratios, the corresponding input reflection coefficient and axial ratio plot for different variants of sectored cylindrical DRAs formed are plotted in Figures 8.7 and 8.8, respectively. From observation, it is revealed that the resonance frequency of degenerate modes for $\beta = 240°$ and $210°$ has been slightly separated (partially for $\beta = 180°$). For the case of $\beta = 150°$ and $120°$, mode degeneracy has been disturbed, but both modes are not properly coupled at this feed position. Feed position will be later optimized to couple both modes properly in this section. But first, generation of CP mechanism is discussed in detail. The axial ratio plot is below 3 dB in a frequency band in between the two orthogonal resonances for $\beta = 240°$, $210°$ as seen in Figure 8.8. In order to explain the consequence of varying β on CP, one particular sectored form of CDRA is selected in this design example, which is $\beta = 210°$ and which is examined in detail.

Figure 8.9 shows the simulated input reflection coefficient and axial ratio plot for $\beta = 210°$ only. It can be observed that a wide -10-dB impedance

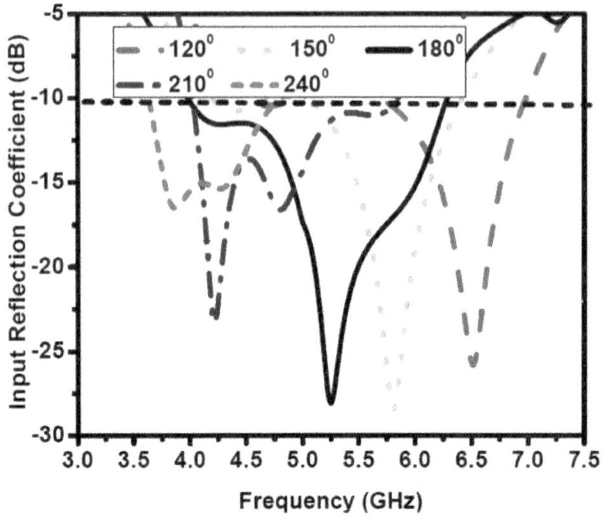

Figure 8.7 Simulated input reflection coefficient for various sectored cylindrical DRA with a fixed aspect ratio (1.6) and feed position neat to edge.

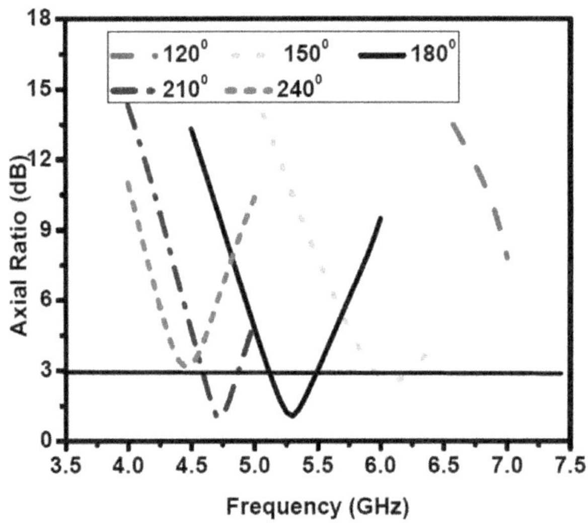

Figure 8.8 Simulated axial ratio plot for various sectored cylindrical DRA with a fixed aspect ratio (1.6) and feed position neat to edge.

bandwidth of 48.2% (3.9–6.39 GHz) and a narrow 3-dB axial ratio bandwidth of 6.3% (4.58–4.88 GHz) have been obtained, respectively, for $\beta = 210°$. It is worth mentioning here that the purpose of using this design example is to show the CP generation by sectorization only. Thus, ARBW can be further improved

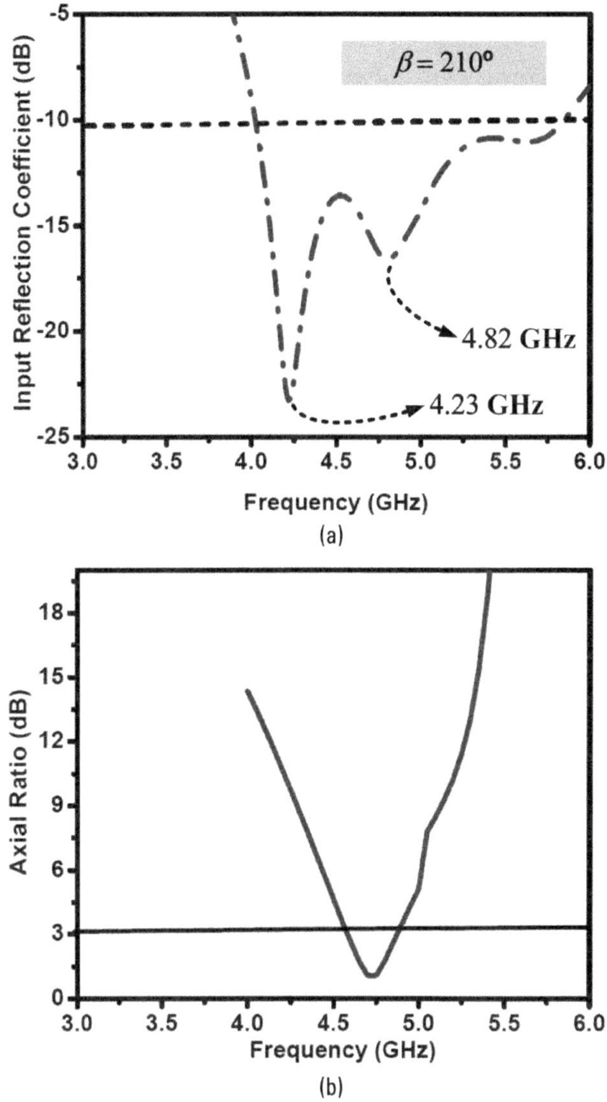

Figure 8.9 Simulated (a) input reflection coefficient, and (b) an axial ratio plot for β = 210° only at aspect ratio =1.6.

by optimizing other parameters, such as feed location, aspect ratio, and sector angle, which is not attempted in this example. Next, the lower mode frequency is obtained at 4.23 GHz, whereas higher mode frequency is obtained at 4.82 GHz. The primary cause for the detuning of these degenerate modes due to sectorization is now discussed. Sectored CDRA supports degenerate orthogonal $TM^x_{11\delta}$ and $TM^y_{11\delta}$ modes that have same resonance frequency and orthogonal electric field configuration but vary as sin(ϕ) and cos(ϕ) with equal magnitude in azimuthal direction.

It is also known that the sectorization technique perturbs the conventional CDRA geometry along azimuthal direction, and therefore the resonant frequency and quality factor of these modes ($TM^x_{11\delta}$ and $TM^y_{11\delta}$) will be altered to some extent and as a result their resonance frequency will be separated. In a conventional CDRA, these two modes have the same resonance frequency but due to sectorization their resonance frequencies are slightly detuned. To verify the orthogonality at these separated frequencies, electric field distribution is plotted at these two frequencies depicted in Figure 8.10, where the direction of the electric field can be seen in an orthogonal arrangement. From field distribution, it is identified that $TM^y_{11\delta}$ is excited at a lower 4.23 GHz, whereas $TM^x_{11\delta}$ is excited at a higher 4.82 GHz. In order to validate the change in the field variation along azimuthal direction, Figure 8.11 shows the electric field magnitude at a lower frequency (4.23 GHz) and a higher frequency (4.82 GHz) at different phase angles. From observation it can be verified that there is a difference in electric field concentration at these two frequencies. Also note that electric field concentration is more for as compared to the mode, and therefore it () is excited at a lower frequency.

Here in this geometry, the CP mechanism is also the same as that discussed in the design example of Chapter 7. The difference is that in the present case, sectorization is responsible for separating the resonance frequencies of degenerate orthogonal modes resulting in CP radiation from the structure. With reference to Figure 8.10(a), at a lower frequency (4.23 GHz), the field component is *y*-directed. At this resonance phase shift is also 0°; at a higher frequency (4.82 GHz) the field is *x*-directed where the resonance phase shift is also 0°. At a lower frequency, the mode has only a *y*-component and no *x*-component and therefore it is linearly polarized at this frequency. Similarly, at a higher frequency, the mode has only an *x*-component and no *y*-component and therefore

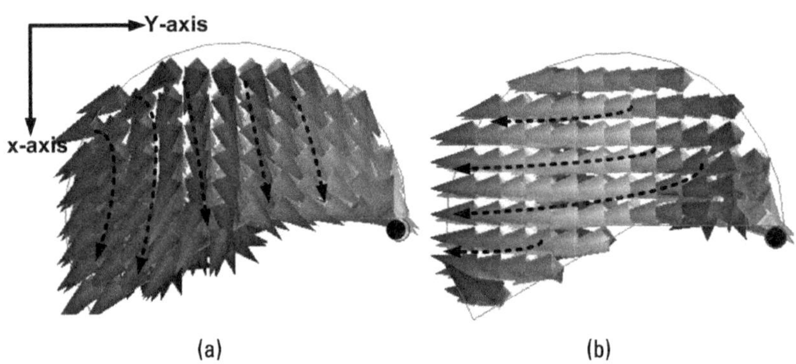

Figure 8.10 Simulated electric field distribution at (a) 4.23 GHz, and (b) 4.82 GHz for $\beta = 210°$.

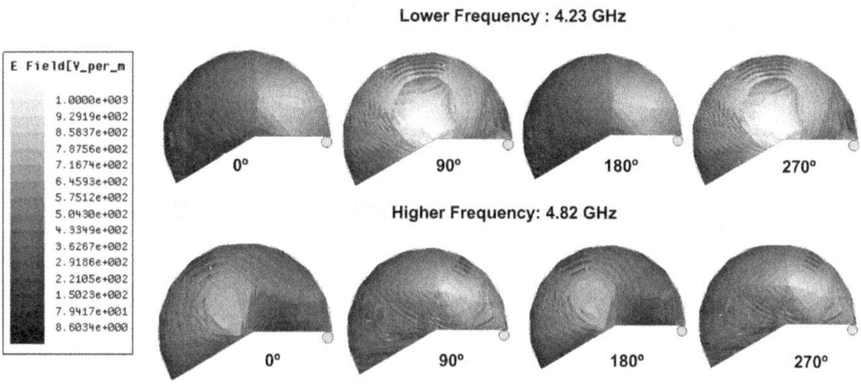

Figure 8.11 Simulated electric field magnitude distribution at different phase angles for (a) 4.23 GHz, and (b) 4.82 GHz ($\beta = 210$) with feed position near to edge.

it is also linearly polarized at this frequency. In a frequency in between these two resonances, there will be some frequency range in which sectored CDRA will have both an *x*- and *y*- component of electric field. Also, if these two modes can be separated by a distance so that phase difference between them is 90°, it will create a circular polarized signal as verified from Figure 8.9(b).

Selection of feed position with respect to a dielectric resonator is another important factor in determining the CP performance and improving the ARBW. It makes sure the excitation of both degenerate counterparts simultaneously. Following in the same manner, analysis performed in previous sections are repeated with different probe feed locations as shown in Figures 8.12(a) and 8.13(a). The corresponding input reflection coefficients are shown in Figures 8.12(b) and 8.13(b). With reference to these figures, no mode degeneracy is observed for the case when a feed is positioned at the center. This is because, at this location, the second resonant mode $TM_{21\delta}$ of a sectored cylindrical DRA is strongly coupled and it is not a degenerate mode.

8.3 The Sectorization Technique in Hemispherical DRAs

This section discusses the sectorization method in HDRAs in order to allow an understanding of circular polarization analysis discussed in subsequent sections. To compare the geometries of a conventional HDRA and a sectored HDRA, the former has zero degrees of freedom that makes them less attractive for antenna applications due to less design flexibility [14, 15]. In sectored HDRAs, another tunable parameter is obtained, which is sector angle (β). Thus, more flexibility is available to the antenna designers for optimizing the performance of sectored DRAs using β, although, little research is available that claims circular polarization in HDRA geometries use sectorization or any other approach.

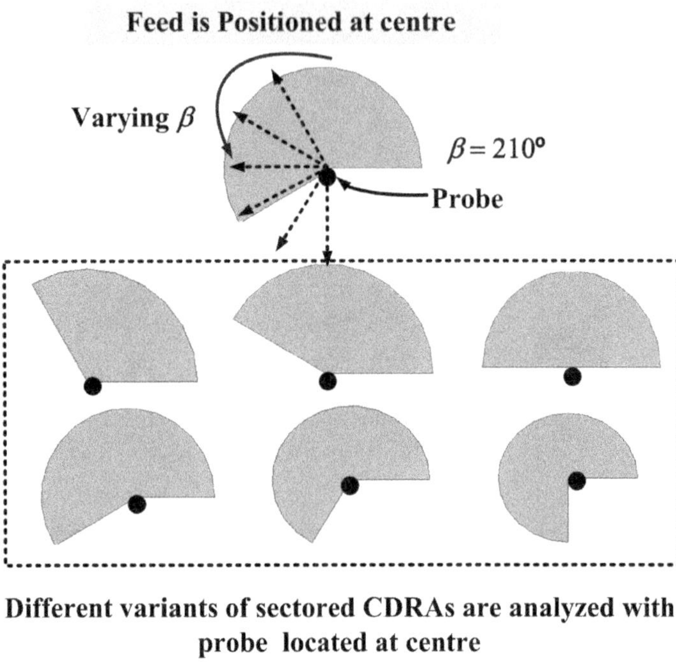

(a)

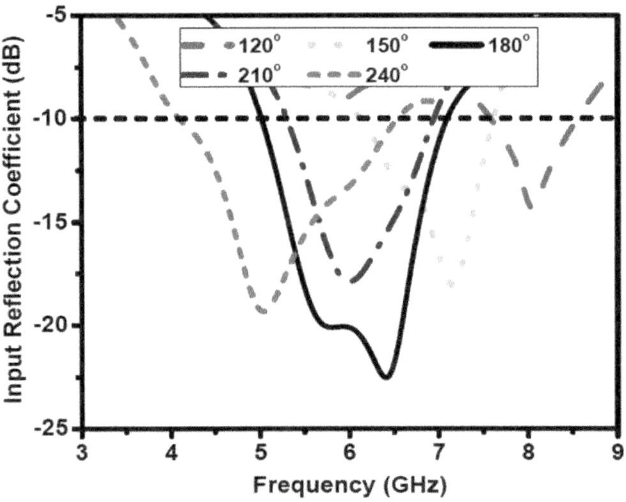

(b)

Figure 8.12 (a) Geometry of various sectored CDRA with a fixed aspect ratio and feed position near to center, and (b) the corresponding simulated input reflection coefficient.

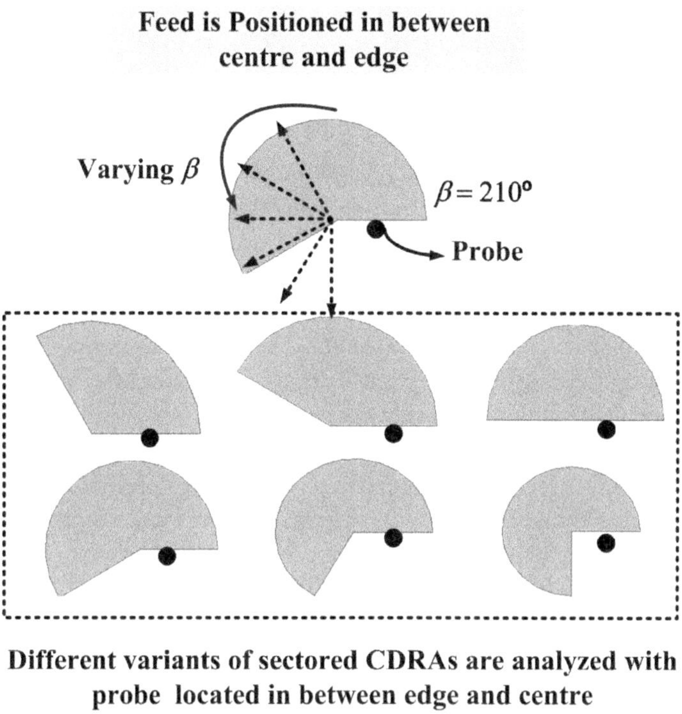

(a)

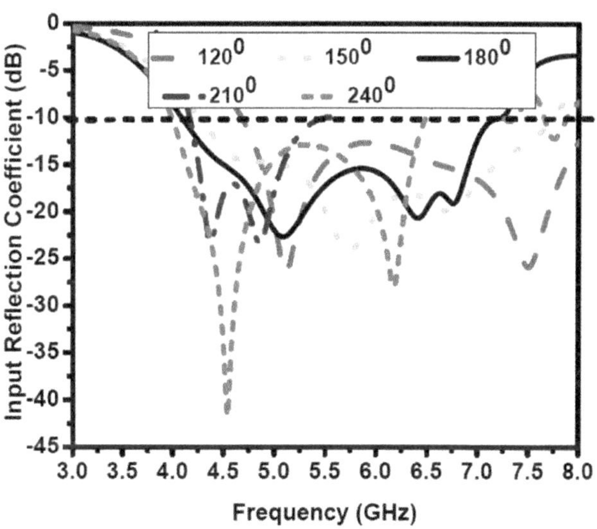

(b)

Figure 8.13 (a) Geometry of various sectored CDRA with a fixed aspect ratio and feed position in between the center and edge positions, and (b) the corresponding simulated input reflection coefficient.

Before the discussion of CP mechanism in sectored HDRAs, a quick glimpse of its history is presented.

8.3.1 Past Research on Sectorization Techniques in Hemispherical DRAs

Early research on HDRAs were mostly grounded in numerical studies for calculating the various parameters. However, several techniques have been reported to improve their performance of HDRAs by tailoring its conventional shape, such as a two-layer nonhomogeneous HDRA that was reported in [16] for LP, and a maximum 31.9% impedance bandwidth was achieved. This technique increased the system complexity and is not feasible for practical applications. In another example, an impedance bandwidth of 83% was achieved in [17] using a conventional HDRA in an inverted arrangement, but stability of this structure is not preferable for practical use and this design was also linearly polarized. In a similar way, various linearly polarized designs were presented [18, 19] that did not include the sector angle in the analysis. In the research of HDRAs, there are only three papers that claim circular polarization in a HDRA as per the author's knowledge.

In [20], circularly polarized fields were generated with the help of a parasitic patch whose location and width controlled the CP performance. In this design, analysis of CP was based on numerical and theoretical approaches only. In the next example, a converging fractal geometry was designed for C-band applications where perturbations were introduced in the HDRA by applying a fractal approach to enhance the impedance bandwidth [21]. In this design, the CP generation was dependent on the fractal geometry, the alignment of the layers, and number of iterations. A similar design illustration for [20, 21] is shown in Figures 8.14 and 8.15, respectively. The sectorization approach for getting CP radiation from DRAs has been reported in [22] for the first time in which the generation of CP is dependent on the sector angle and feed location. Since the focus of this chapter is the sectorization method, a detailed analysis of this design [22] is investigated in the next section. A summary of various HDRA designs reported for LP and CP applications is given in Table 8.2.

8.3.2 Analysis of Circular Polarization in Sectored HDRAs

In this section, a complete investigation on circular polarization is discussed in detail for sectored hemispherical DRAs based on the work presented in [22]. The general geometry considered is shown in Figure 8.16 in which the new geometries of nonconventional HDRAs formed after sectorization have a radius r. A spherical wedge-shaped volume has been removed repeatedly from reference a conventional HDRA to obtain different sectored geometries formed by setting $90 \leq \beta < 360°$. The new design parameter obtained in this case is also the sector angle, which is again symbolized by β. The process of getting circular polariza-

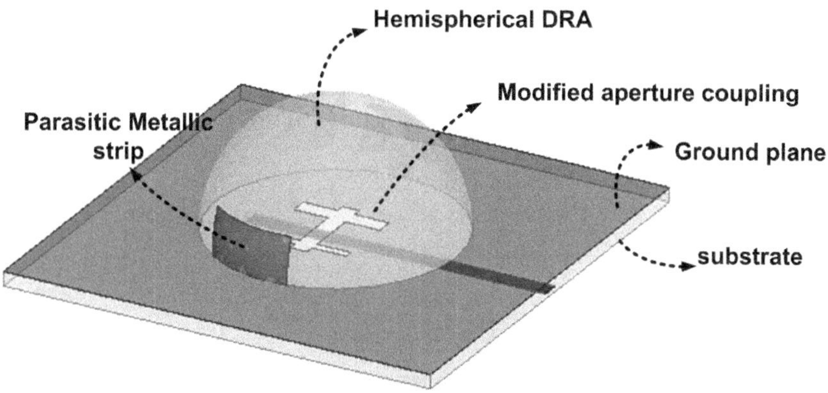

Figure 8.14 Example of a circularly polarized HDRA with parasitic patch element excited by a slot coupling. (Modified after [20].)

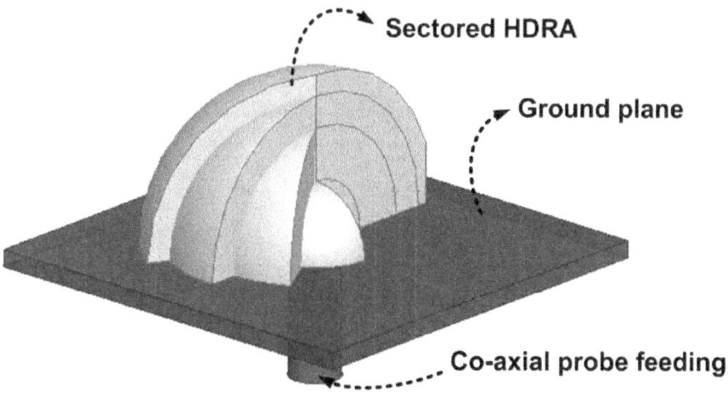

Figure 8.15 Example of a circularly polarized HDRA with fractal approach. (Modified after [21].)

tion radiation from a sectored HDRA using a single-feed technique is similar to the sectored CDRA discussed in the previous section. In a probe-feed HDRA, the *z*-directed probe current is resolved into the θ-directed and *r*-directed current components.

$$J_z\left(\vec{r'}\right)\hat{z} = J_r\left(\vec{r'}\right)\hat{r} + J_\theta\left(\vec{r'}\right)\hat{\theta}$$

where

$$J_r\left(\vec{r'}\right) = J_z\left(\vec{r'}\right)\cos\theta'$$
$$J_\theta\left(\vec{r'}\right) = -J_z\left(\vec{r'}\right)\sin\theta'$$

Table 8.2
Summary of Various Sectored HDRAs Available for Linear and Circular Polarization in the Literature

Ref	DRA Shape (ε_r) Sector Angle	Feeding Used	Polarization	IMBW (%)	ARBW (%)	f_c (GHz)	Modes Excited
[16]	Two-layer HDRA (6.85, 2.5), 360°	Probe coupling	Linear	31.9%	NA	NA	TM_{101}
[17]	Inverted HDRA geometry (9.2), 360°	Probe coupling	Linear	83%	NA	NA	TM_{101}-like and $TM_{01\delta}$-like
[18]	Two element sector HDRA, (10), 180°	Probe coupling	Linear	35%	NA	NA	TM_{101}-like and $HEM_{11\delta}$-like
[19]	Four element sector HDRA, (10), 90°	Probe coupling	Linear	26%	NA	NA	TM_{101} and $HEM_{11\delta}$-like
[20]	HDRA with parasitic patch (9.5), 360°	Aperture coupling	Circular	22%	3.4%	3.51	TE_{111}
[21]	Fractal HDRA (4.3), *NA*	Probe coupling	Circular	57%	~23%	4.75	TE_{123}
[22]	Sector HDRA (9.8), 360° to 90°	Probe coupling	Circular	47.4%	8%	5.34	TE_{111}

IMBW = fractional −10-dB impedance bandwidth, ARBW = fractional 3-dB axial ratio bandwidth, ε_r = dielectric constant of DRA, f_c = center frequency with regard to axial ratio plot, NA = not applicable.

where, r refers to the source point, J_r is the r-directed current, and J_θ is the θ-directed current. When a feed is placed at the center, probe current is dominated by J_r only because in this case the elevation angle is zero degrees and therefore it excites only the *TM* mode in HDRA (where the e-fields are in radial direction only), but as soon as the probe is displaced from the center the elevation angle is no more than zero degrees. Now in this case, both J_r and J_θ exist, which excites both the *TM* and *TE* modes. Thus in sectored HDRAs, the probe offset will always excite the θ-directed current component, which in turn affects the angular fields in the HDRA that are necessary to detune the degenerate modes. Figure 8.17 shows the pictorial description of this explanation.

The sectored hemispherical DRA supports the TE^x and TE^y modes that have orthogonal electric field configuration and their resonance frequency is the same when no sectorization is applied. The pictorial field distribution for fundamental TE modes in HDRA is shown in Figure 8.18.

Next, these *TE* modes have mode degeneracy property. Sectored HDRA support degenerate orthogonal $TE^x_{11\delta}$ and $TE^y_{11\delta}$ modes that have mode degeneracy along azimuthal direction. Both modes have the same resonance frequency and orthogonal electric field configuration but vary as sin(ϕ) and cos(ϕ) with equal magnitude. Also, their resonance frequencies can be slightly separated after sectorization is performed. Then, the probe is fed near the edge to excite the degenerate modes in the HDRA and sectorization is performed. The spectrum

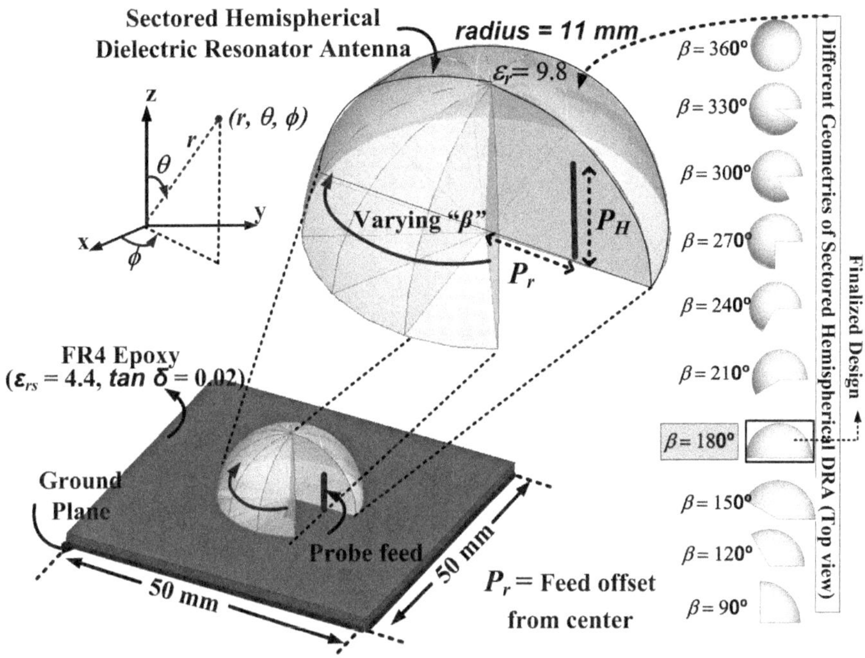

Figure 8.16 General geometry of a sectored hemispherical DRA. Sector angle varies from $\beta = 360°$ to 90° with a step size of 30° [22].

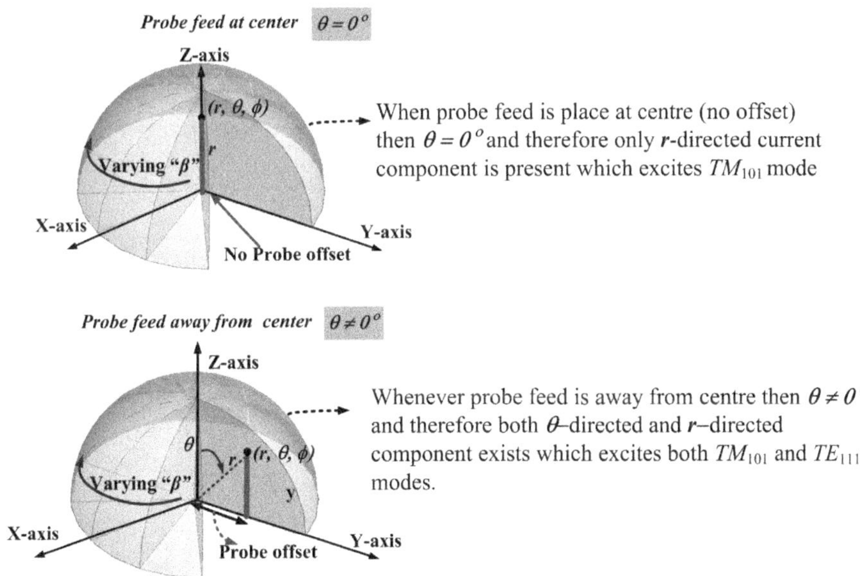

Figure 8.17 Picture showing (a) a probe feed at the center, and (b) a probe feed away from the center (with probe offset).

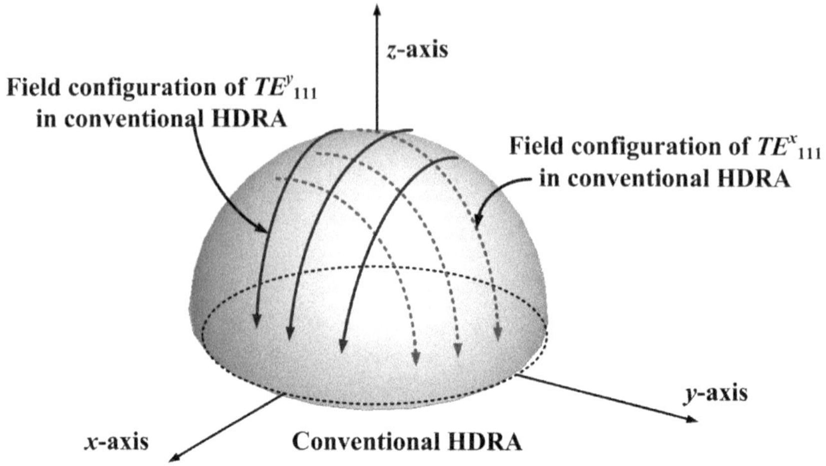

Figure 8.18 Schematic diagram of field distribution for TE modes in a conventional HDRA.

of different modes excited in sectored HDRAs when a probe is fed near the edge is shown in Figure 8.19. It is revealed here that a fundamental TE_{111} mode is excited, but as β decreases, a second resonant mode appears in close proximity to the TE_{111} mode from $210° \leq \beta$.

This new excited mode has been recognized as a degenerate counterpart of the TE_{111} mode. In conventional HDRA, both the TE^x_{111} and TE^y_{111} modes have the same resonant frequency and propagation constant. But due to sectorization, the TE_{111} mode has been detuned into its orthogonal degenerate counterparts, which are identified as the TE^x_{111} and TE^y_{111} modes. When sectorization is employed according to perturbation theory, it slightly separates the resonance frequency of both modes and this change in resonant frequency is approximated by change in their electric and magnetic field energies confined in the removed volume.

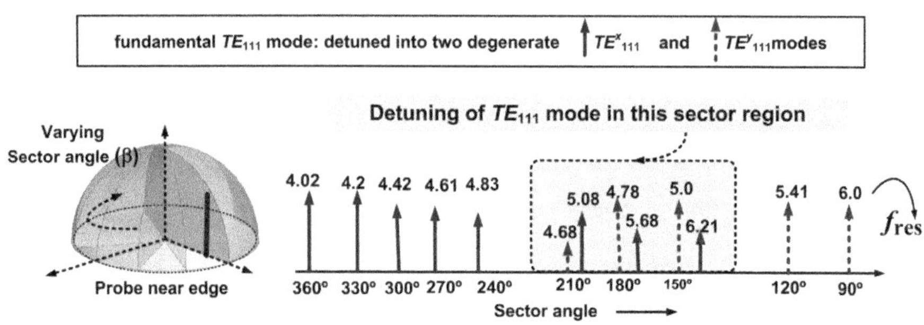

Figure 8.19 Spectrum of different modes excited in a sectored HDRA for Pr = 7 mm with varying sector angle β. (From [22].)

Next, in order to realize the CP generation, one form of sector DRA is considered, which is $\beta = 180°$ in this case. Figure 8.20 shows the input reflection coefficient and axial ratio plot for $\beta = 180°$. It is revealed that a –10-dB impedance bandwidth of 26% (4.5–7 GHz) and a narrow 3-dB axial ratio bandwidth of 6.9% (5.18–5.56 GHz) have been obtained for $\beta = 180°$. It is worth mentioning here that the purpose of taking this design example is to show the CP generation by sectorization only. Thus, ARBW can be further improved by optimizing the feed location. From electromagnetic simulations, it is revealed that the TE^{y}_{111} mode is excited at 4.78 GHz and the TE^{x}_{111} mode is excited at 5.68 GHz. The ratio of two orthogonal degenerate modes should be in between 1.01 to 1.10 and may be extended up to 1.2 [7] when an individual bandwidth of mode is large. Here the ratio is 1.18 and satisfies the criteria. When frequency of excitation is properly selected at a center frequency (f_o) in between these two orthogonal resonances, the amplitude of both modes will be equal and phase shift from both will be 45° resulting in phase quadrature relationship between them. Figure 8.21 shows the corresponding electric field distribution in a sectored HDRA at orthogonal resonant frequencies for $\beta = 180°$. For a conventional HDRA, there is no path difference between E_ϕ and E_θ components due to a geometrical symmetrically form of HDRA and thus phase difference is not possible. However, when sectorization is introduced, a HDRA becomes asymmetric in azimuthal direction and consequently path length for E_ϕ is altered according to β values, whereas path length for E_θ almost remains fixed.

This is accountable for path difference and consequently phase difference in the structures. Figure 8.22 shows the plot of the magnitude and phase angle of individual electric field components responsible for CP generation in the structure for $\beta = 180°$. It has been realized that the magnitude is approximately

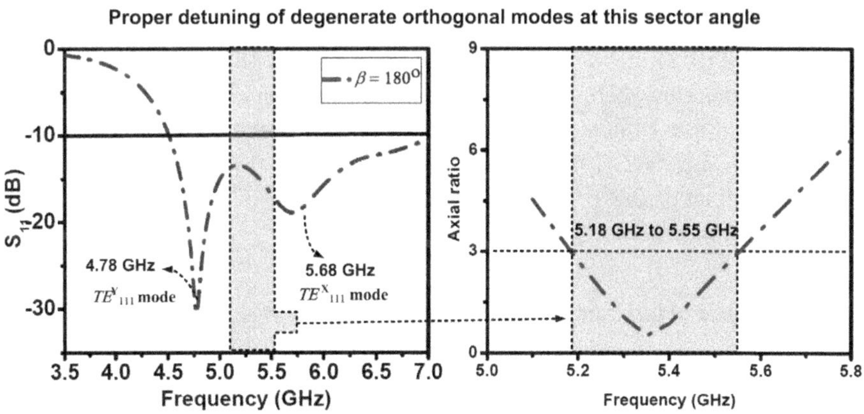

Figure 8.20 Simulated (a) input reflection coefficient, and (b) axial ratio plot for $\beta = 180°$ only. (From [22].)

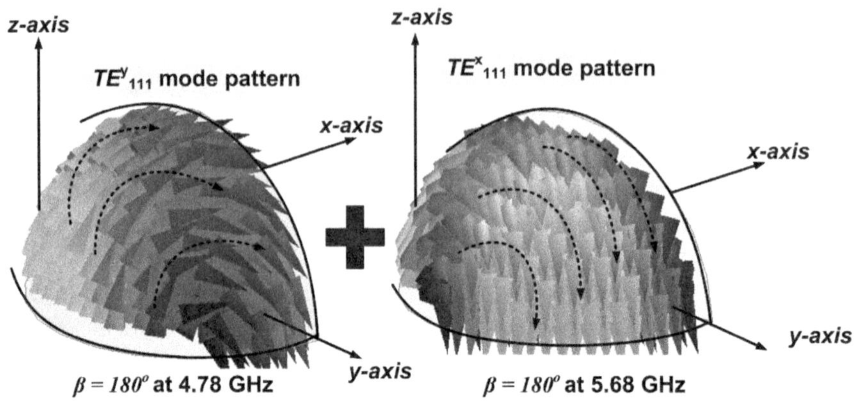

Figure 8.21 Simulated electric field distribution at (a) 4.78 GHz, and (b) 5.68 GHz for $\beta = 180°$ in sectored HDRA. (From [22].)

equal across the operating band with nearly a 90° phase difference between them. However, there is amplitude variation in which a minimum axial ratio is obtained around 3.6 GHz where both magnitudes of both field components are equal as can be perceived from Figure 8.20(b). The probe feed location can be used to change the polarization state of the antenna and to verify this, simulations have been performed and the obtained radiation pattern is shown in Figure 8.23. It is observed that that polarization state of CP fields is changed from LHCP to RHCP mode.

8.4 A New CP Generation Approach in DRAs Using Sectorization

In the previous two sections, an analysis of CP was conducted based on the same principle for the detuning of the degenerate orthogonal modes using the sectorization method (in the cases of CDRAs and HDRAs). In this section, a novel modification technique is discussed for CP generation using a sectorization technique while considering the PMC boundary approximation at the dielectric-air interface. This method does not involve any mode degeneracy requirement. The design considered in this section is also based on our work on the sectorization method presented in [23].

8.4.1 Analysis of Circular Polarization in Modified Geometry

A novel modified sectorization method has been reported in [23] for circular polarization generation. The antenna design is shown in Figure 8.24. In [23], the geometrical shape of a simple cylindrical dielectric resonator has been deformed periodically with the same angular factor ($\beta = 45°$) along the azimuthal

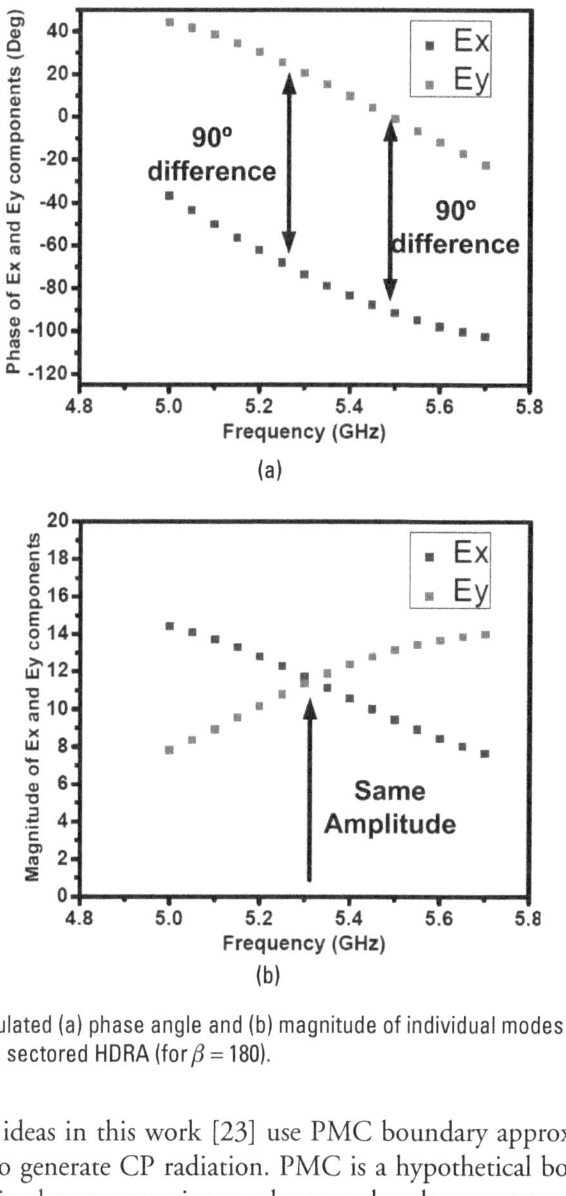

Figure 8.22 Simulated (a) phase angle and (b) magnitude of individual modes in an orthogonal mode pair in a sectored HDRA (for $\beta = 180$).

direction. The ideas in this work [23] use PMC boundary approximation for the first time to generate CP radiation. PMC is a hypothetical boundary that is readily used in electromagnetics to solve complex electromagnetic problems. The mathematical description of electromagnetic fields inside a dielectric resonator is very complex. To solve the field equations and analyze its behavior, the PMC approximation has been widely used in the literature [24]. Under this scenario, the dielectric medium having a higher permittivity value shares an open circuit boundary condition with air. The higher the permittivity ratio at the boundary, the more accurate the behaviors of fields at the interface will be as per PMC approximation.

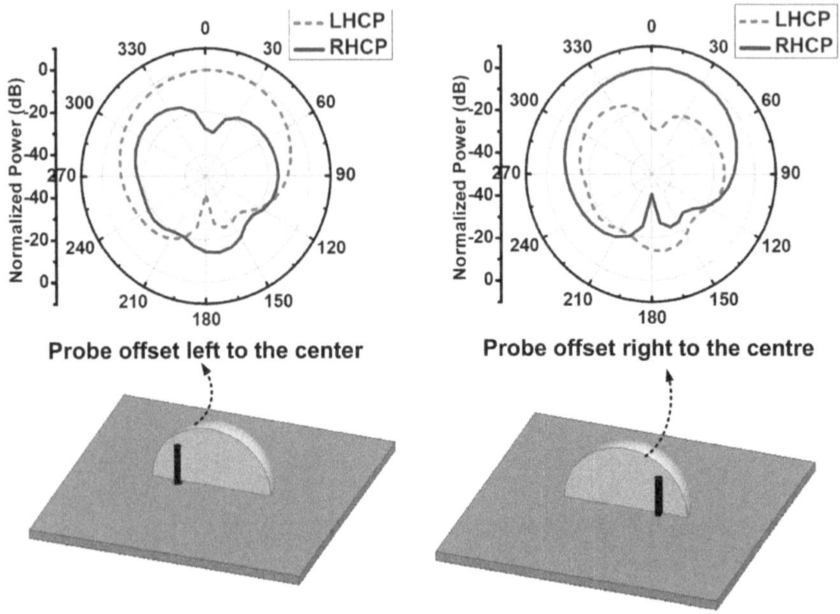

Figure 8.23 Simulated radiation pattern for $\beta = 180°$ at different feed positions (extreme left and extreme right) to show the polarization state.

At this condition, the tangential magnetic field and normal component of an electric field vanishes, whereas a tangential component of an electric field and normal magnetic field component is present at the air-dielectric interface. In order to generate CP in a DR, the same concept has been applied in [24] assuming that DRs form a leaky boundary approximately satisfying magnetic boundary conditions at the air-dielectric interface. The idea implemented was to change the air-dielectric interface geometrical shape so that the *E*-fields also reorient their direction by decomposing their components at different angles in accordance with the applied modification.

Next, a design example will be discussed for CP generation using the above-mentioned approach in a sectored CDRA, the considered example is adapted from work presented in [23]. The antenna geometry considered in this design example is partially modified from [23] and shown in Figure 8.25. The antenna design consists of a cylindrical DR made of alumina material with $\varepsilon_r =$ 9.8 ($r = 11$ mm, $H = 16$ mm).

The design is periodically modified with the same angular factor β (= 45°) in spiral form along azimuthal direction. The antenna structure is settled on an FR4 epoxy substrate over a 50 mm × 50 mm finite-sized ground plane. A conformal microstrip line feeding is used to excite the dielectric resonator in this design example. Figure 8.26 displays the DRA geometry in forward antenna stages with corresponding electric field variations.

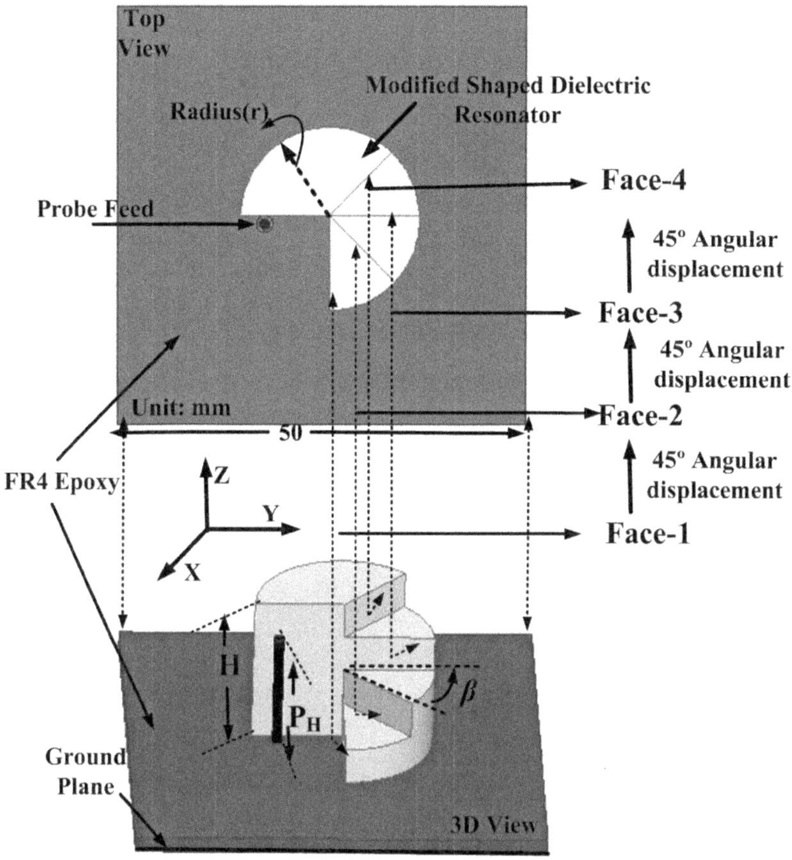

Figure 8.24 Diagram of modified dielectric resonator antenna using sectorization for CP generation [23].

For example, in Antenna 1, shown in Figure 8.26(a), only Face 1 is present and all the fields are tangential to that Face 1 to satisfy the PMC boundary conditions. This field component is named as $E_{\text{face-1.}}$ When Antenna 2 is formed by removing a dielectric block with a center angle 45° in a counterclockwise direction from Antenna 1, due to PMC approximation, $E_{\text{face-1}}$ is then decomposed into $E_{\text{face-1}}$ and $E_{\text{face-2}}$. Then, it can be seen from Figure 8.26(b) that these two electric field components are separated in space by 45°. When Antenna 3 is formed (shown in Figure 8.26(c), $E_{\text{face-2}}$ is further decomposed into $E_{\text{face-3}}$ following the PMC approximation. Next, due to these modifications in DR structure, $E_{\text{face-3}}$ becomes 90° separated from $E_{\text{face-1}}$ in space, whereas it becomes 45° with $E_{\text{face-2}}$. Now $E_{\text{face-1 and}}$ $E_{\text{face-3}}$ satisfy the orthogonality condition of CP generation. Furthermore, following the same procedure when Antenna 4 is formed, $E_{\text{face-3}}$ is decomposed into $E_{\text{face-4}}$ as shown in Figure 8.26(d). Next,

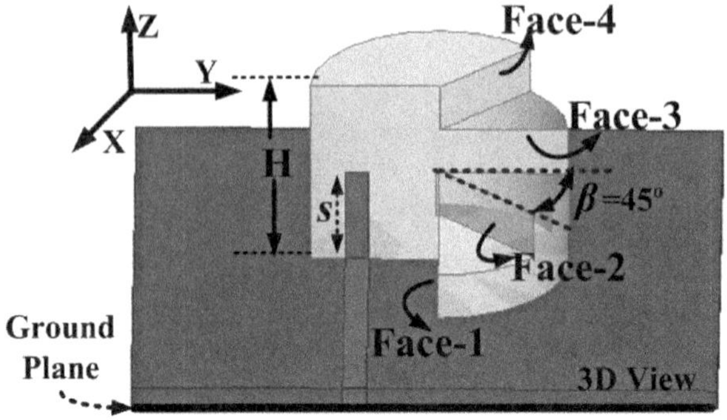

Figure 8.25 A design example showing a modified CDRA fed with a conformal microstrip line for circular polarization generation.

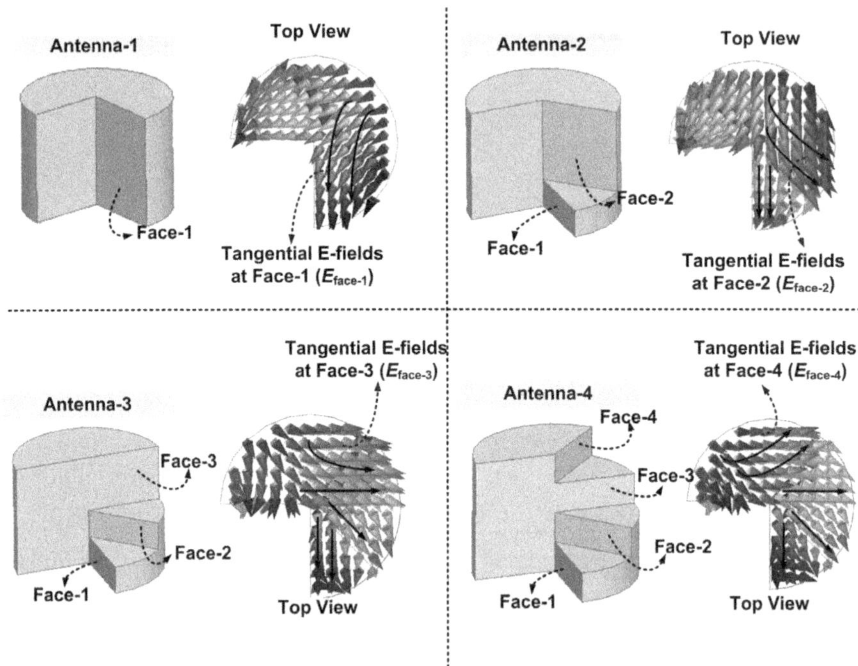

Figure 8.26 Simulated electric field distribution at different faces of sectored antenna visualizing the mode distribution at 4.3 GHz for the presented design example. (From [23].)

Antenna 4 is the final modification stage where $E_{face\text{-}4}$ becomes 90° separated from $E_{face\text{-}2}$ and also $E_{face\text{-}3}$ becomes 90° separated from $E_{face\text{-}1}$.

In this approach, the fields on several adjacent sectors are responsible for creating the orthogonality condition in the dielectric resonator. Note that the nonorthogonal field components between adjacent faces also contribute to the far-field and thus degrade the axial ratio bandwidth in this structure. The explanation for phase difference is related to the different amount of distance that is covered by each field component within the same dielectric medium. Since the dielectric medium and frequency of operation is the same whereas the distance traveled by each component is different, this creates a path difference between the orthogonal components that is accountable for required phase difference. The corresponding simulated input reflection coefficient and axial ratio plots are shown in Figure 8.27 for the design example under consideration. It is observed that a –10-dB impedance bandwidth of 30.96% and a 3-dB axial ratio bandwidth of 9.08% has been obtained, respectively.

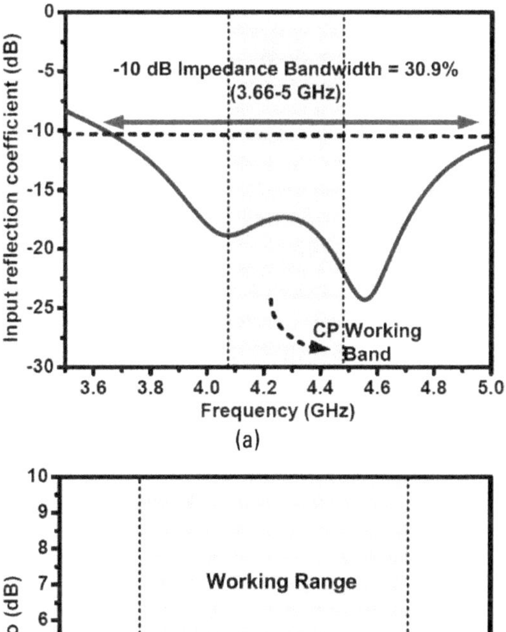

(a)

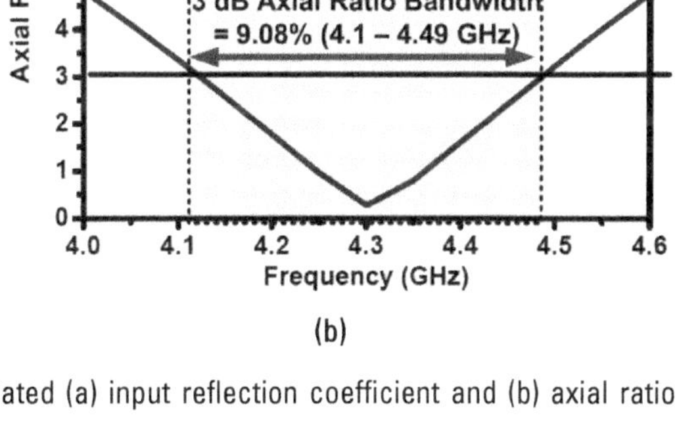

(b)

Figure 8.27 Simulated (a) input reflection coefficient and (b) axial ratio plot of presented design example.

8.5 The Sectorization Technique in CP-MIMO Antennas

In the introduction section, it was mentioned that sectored DRAs can be very useful for multiantenna systems because of compactness. Recently, one such MIMO design has been reported using this technique in [25]. A wideband circularly polarized MIMO antenna system based on quarter cylindrical DRA was designed in [25]. The design consists of two adjacently facing quarter (β = 90°) CDRAs coupled through a microstrip feeding network. The dimensions of the resonator are reduced by incorporating electrical and magnetic walls of orthogonal arrangement in the design. Due to this antenna arrangement, it excites orthogonal HEM_{11} & HEM_{12} modes in the structure that are responsible for CP. With two identical quarter CDRAs (ε_r = 10 each, radius = 10 mm and height = 6.35 mm), there is a wide ARBW of 24.4% (4.5–5.75) at the center frequency of 5.12 GHz. The same situation can be realized by the design shown in Figure 8.28 for CP generation adapted after [25].

8.6 Summary

An investigation of the sectorization technique was presented for circular polarization in CDRAs and HDRAs followed by research surveys in this chapter. In the case of cylindrical and hemispherical dielectric resonators, mode degeneracy is along their azimuthal direction. The degenerate orthogonal modes are detuned with the help of a sectorization technique to generate CP, which has been established in this chapter. For cylindrical DRAs, a sector angle β and aspect ratio α constitute a new optimization framework to analyze CP parameters, whereas in the case of hemispherical DRAs, the only parameter sector angle β provides CP conditions. A design example was also presented for β = 210° (in

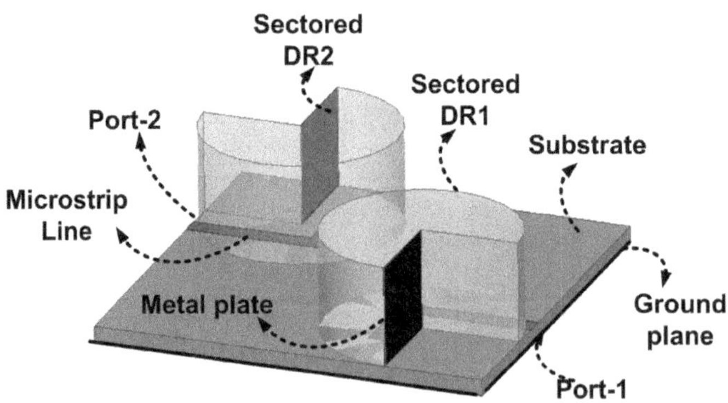

Figure 8.28 A sectored dielectric resonator antenna-based MIMO antenna for wideband circular polarization. (Modified after [25].)

the case of CDRAs) and $\beta = 180°$ (in the case of HDRAs) to show the generation of CP radiation due to sectorization. A novel geometry modification approach was also discussed with a design example that can be employed in any shape of DRA for CP generation by using the sectorization technique and behavior of electric fields at the air-dielectric interface (PMC approximation). In the last example, the applications of a sectored DRA were extended to MIMO antennas as in the last section.

Sectored CDRAs could be used for situations where compact and wide axial ratio bandwidth is required. This is because the ARBW can be tuned by the aspect ratio and sector angle in the case of CDRAs. For HDRAs, although axial ratio is limited, sectored HDRAs provide curved surfaces in both azimuthal and elevation directions and thus could be useful for the applications for implementing microstrip structures on the surface of the DRAs. Hemispherical structures can be used to model parts of structures such as vehicles, airplanes, tracking systems, airborne radars, and rockets where the antenna is mounted on the surface.

References

[1] Tam, M. T. K., and R. D. Murch, "Compact Cylindrical Sector Dielectric Resonator Antennas," *IEEE Antennas and Propagation Society International Symposium*, 1998.

[2] Kumar, A. V. P., J. Yohannan, A. Lonappan, G. Bindu, T. Vinu, V. Hamsakkutty, and K. T. Mathew, "Microstripline fed Circular Sector Dielectric Resonator Antenna," *IEEE Antennas and Propagation Society International Symposium*, Vol. 2, 2005, pp. 192–195.

[3] Pan, Y. M., and K. W. Leung, "Wideband Omnidirectional Circularly Polarized Dielectric Resonator Antenna With Parasitic Strips," *IEEE Transactions on Antennas and Propagation*, Vol. 60, No. 6, 2012, pp. 2992–2997.

[4] Chu, L. C. Y., D. Guha, and Y. M. M. Antar, "Comb-shaped Circularly Polarised Dielectric Resonator Antenna," IET Electronics Letters, Vol. 42, No. 14, 2006.

[5] Chaudhary,R. K., K. V. Srivastava, and A. Biswas, "Three-Element Multi-Layer Multi-Permittivity Cylindrical Dielectric Resonator Antenna for Wideband Applications with Omnidirectional Radiation Pattern and Low Cross-Polarization," *Microwave and Optical Technology Letters (MOTL)*, Wiley Journals, Vol. 54, No. 9, 2012, pp. 2011–2016.

[6] Luk, K. M., and K. W. Leung, *Dielectric Resonator Antennas*, Baldock, England: Research Studies Press Ltd, 2003.

[7] Sharma, P. C., and K. C. Gupta, "Analysis and Optimized Design of Single Feed Circularly Polarized Microstrip Antennas," *IEEE Transactions on Antennas and Propagation*, Vol. 31, No. 6, 1983, pp. 949–955.

[8] Balanis, C. A., *Antenna Theory: Analysis and Design*, Hoboken, NJ: John Wiley & Sons, 2005.

[9] Tam, M. T. K., and R. D. Murch, "Compact Circular Sector and Annular Sector Dielectric Resonator Antennas," *IEEE Transactions on Antennas and Propagation*, Vol. 47, No. 5, 1999, pp. 837–842.

[10] Tam, M. T. K., and R. D. Murch, "Circularly Polarized Circular Sector Dielectric Resonator Antennas," IEEE Transactions on Antennas and Propagation, Vol. 48, No. 1, 2000, pp. 126–128.

[11] Bezerra, J. W. O., et al., "Circularly Polarized Quarter-Cylinder Shaped Dielectric Resonator Antenna using a Single Probe Feed," Microwave and Optical Technology Letters, Vol. 57, No. 3, 2015, pp. 722–726.

[12] Gopakumar, C., J. Yohannan, U. G. Kalapurra, and K. T. Matthew, "An Investigation on Wideband Circular Cylindrical Sector Dielectric Resonator Antenna with Microstrip Feed Line," Microwave and Optical Technology Letters, Vol. 51, No. 12, 2009, pp. 2861–2865.

[13] Chowdhury, R., and R. K. Chaudhary, "Investigation on Different Forms of Circular Sectored-Dielectric Resonator Antenna for Improvement in Circular Polarization Performance," IEEE Transactions on Antennas and Propagation, Vol. 66, No. 10, 2018, pp. 5596–5601.

[14] Leung, K. W., K. M. Luk, K. Y. A. Lai, and D. Lin, "Theory and Experiment of a Coaxial Probe Fed Hemispherical Dielectric Resonator Antenna," IEEE Transactions on Antennas and Propagation, Vol. 41, No. 10, 1993, pp. 1390–1397.

[15] Luk., K. M., K. W. Leung, K. Y. A. Lai, and D. Lin, "Analysis of Hemispherical Dielectric Resonator Antenna," Radio Science, Vol. 28, No. 6, 1993, pp. 1211–1218.

[16] Fang, X. S., and K. W. Leung, "Design of Wideband Omnidirectional Two-Layer Transparent Hemispherical Dielectric Resonator Antenna," IEEE Transactions on Antennas and Propagation, Vol. 62, No. 10, 2014, pp. 5353–5357.

[17] Mukherjee, B., P. Patel, and J. Mukherjee, "A Novel Cup-Shaped Inverted Hemispherical Dielectric Resonator Antenna for Wideband Applications," IEEE Antennas and Wireless Propagation Letters, Vol. 12, 2013, pp. 1240–1243.

[18] Guha, D., and Y. M. M. Antar, "New Half- Hemispherical Dielectric Resonator Antenna for Broadband Monopole-Type Radiation," IEEE Transactions on Antennas and Propagation, Vol. 54, No. 12, 2006, pp. 3621–3628.

[19] Guha, D., B. Gupta, C. Kumar, and Y. M. M. Antar, "Segmented Hemispherical DRA: New Geometry Characterized and Investigated in Multi-Element Composite Forms for Wideband Antenna Applications," IEEE Transaction on Antenna and Propagation, Vol. 60, No. 3, 2012, pp. 1605–1610.

[20] Leung, K. W., and H. K. Nge, "The Slot-Coupled Hemispherical Dielectric Resonator Antenna with a Parasitic Patch: Applications to the Circularly Polarized Antenna and Wideband Antenna," IEEE Transactions on Antennas and Propagation, Vol. 53, No. 5, 2005, pp. 1762–1769.

[21] Gupta, S., V. Killamsetty, M. Chauhan, and B. Mukherjee, "Compact and Circularly Polarized Hemispherical DRA for C-Band Applications," Frequenz Journal of RF-Engineering and Telecommunications, Vol. 73, No. 7–8, 2019, pp. 227–234.

[22] Chowdhury, R., and R. K. Chaudhary, "Investigation of New Sectored Hemispherical Dielectric Resonator Antennas Operating at $TM_{101}$and TE_{111} Mode for Circular Polarization," Progress In Electromagnetics Research (PIER), Vol. 67, 2020, pp. 95–109.

[23] Chowdhury, R., and R. K. Chaudhary, "An Approach to Generate Circular Polarization in a Modified Shaped Cylindrical Dielectric Resonator Antenna Using PMC Boundary Approximation," IEEE Antennas and Wireless Propagation Letters, Vol. 17, No. 9, 2018, pp. 1727–1731.

[24] Kajfez, D., and P. Guillon, Dielectric Resonators, Norwood, MA: Artech House, 1986.

[25] Belazzoug, M., I. Messaoudene, S. Aidel, T. A. Denidni, and A. A. Kishk, "Quarter Cylindrical Dielectric Resonator Antenna with Circular Polarisation for Wideband MIMO Systems," Electronics Letters, Vol. 56, No. 4, 2019, pp. 174–176.

9

The State of the Art in Circularly Polarized Dielectric Resonator Antennas

This chapter provides a comprehensive survey on the current state of research being carried out in circularly polarized DRA technology. This chapter discusses substrate integrated waveguide (SIW) technology in circularly polarized DRA arrays for millimeter-wave applications. This chapter also includes a discussion of the liquid DRA concept. Miscellaneous research fields explored in circularly polarized DRAs are also detailed. A summary concludes the chapter.

9.1 Circularly Polarized DRAs in Substrate Integrated Waveguide Technology

The current state of the art in DRAs is more focused on millimeter-wave technology where the DRAs offer more efficient performance than metallic antennas. This is due to the fact that a major portion of DRA systems are made of dielectrics that help in maintaining high efficiency even at higher frequencies. However, this has certain drawbacks: mm-wave DRAs become extremely small in dimensions and therefore practical realization becomes challenging. Other drawbacks concern the positioning of DRA arrays at high frequencies. The various feeding techniques discussed in previous chapters (i.e., slot coupling [1], probe coupling [2], and microstrip coupling [3]) are not so efficient for realizing circular polarization in DRA at mm-wave range and for array applications. A promising candidate for solving this issue is SIW technology [4–7], which is shown in Figure 9.1. It is designed by a rectangular waveguide formed in a dielectric substrate with top and bottom metallic plates. An array of metallized

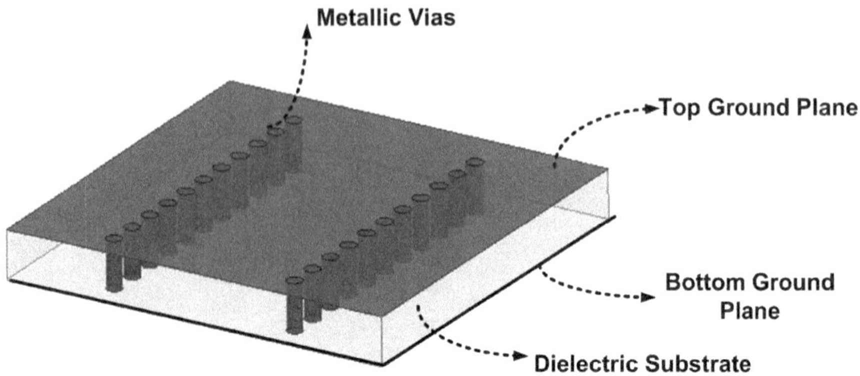

Figure 9.1 Pictorial description of an SIW.

posts via holes are embedded inside the dielectric substrate in some regular arrangements and connected to the top and bottom ground plane.

Another challenging task in mm-wave circularly polarized DRAs is in the designing of arrays in mm-wave. This is due to the introduction of conventional feeding networks discussed in previous chapters because array structures become complex, array efficiencies decrease, and array gain enhancements are generally limited. This issue is also solved by various solutions in SIW technology. Recently, a few solutions have been proposed for implementing SIW technology in mm-wave circularly polarized DRAs for array applications, which will be discussed in the next examples.

In [8], a circularly polarized DRA array was designed using a cylindrical DRA operating at 60 GHz by exciting the fundamental $HE_{11\delta}$ mode. The antenna setup was comprised of two different substrates in which the cylindrical DRA in the upper substrate is isolated from the lower substrate using concentric arrays of metallic vias and hole. To obtain CP waves, a circular-shaped aperture with a pair of 45 inclined slits was etched on the radiating side of the DRA. The DRA was excited through a rectangular slot in the ground plane positioned below the DRA. An axial ratio bandwidth of 15.9% was achieved with an 11.43-dBic peak gain in the 2 × 2 arrays of proposed antenna.

The same situation can be realized with a SIW setup with minor modification in [8] and is shown in Figure 9.2. This design uses a rectangular DRA instead of a cylindrical DRA.

In [9], a novel concept was designed for DRA arrays in which the DRA elements and their feeding network are shaped together and fabricated using the same dielectric medium, unlike in the previous example [8]. For the purpose of CP generation, a pair of open-ended U-shaped striplines symmetrically etched on different sides of the DRA were utilized and connected to the top and bottom ground plane. This pair of striplines is responsible for orthogonal TE_{111}

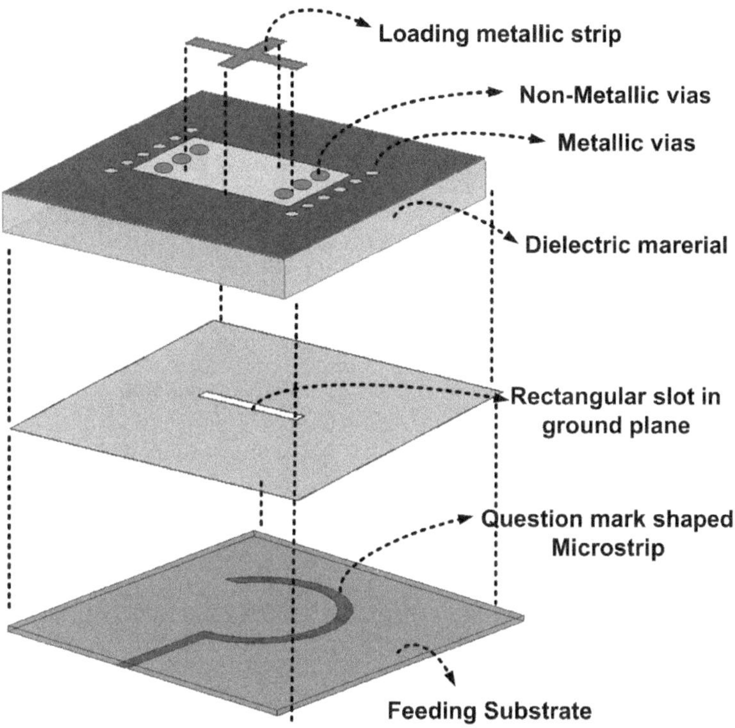

Figure 9.2 Pictorial description of CP substrate-integrated rectangular DRA. (Modified after [8].)

modes in the design. This approach offered 400 MHz of axial ratio bandwidth at 30 GHz with maximum boresight gain of 7.6 dBic. There is another example of SIW DRA for mm-wave application reported in [10], in which the antenna arrangement consisted of an inner cylindrical dielectric resonator surrounded by an outer ring DRA.

Due to this embedded arrangement, the inner DRA is excited in its fundamental $HE_{11\delta}$ mode, whereas the outer ring DRA is excited in its higher-order $HE_{12\delta+1}$ mode simultaneously at adjacent frequencies. The excitation of these modes was accomplished by using a nonuniform cross slot technique that serves a dual purpose (i.e., to generate CP fields and enhance the bandwidth). This reported design provided a wide impedance bandwidth of 34.6% (20.1–28.5 GHz) and an axial ratio bandwidth of 26.3% (20.3–26.45 GHz). In addition, the gain of the antenna is also enhanced due to the surrounding SIW cavity environment, with the maximum gain of 8.15 dBic. The same situation can be realized by a SIW setup with minor modification (we have used a rectangular DRA instead of a cylindrical DRA) in [10] and is shown in Figure 9.3.

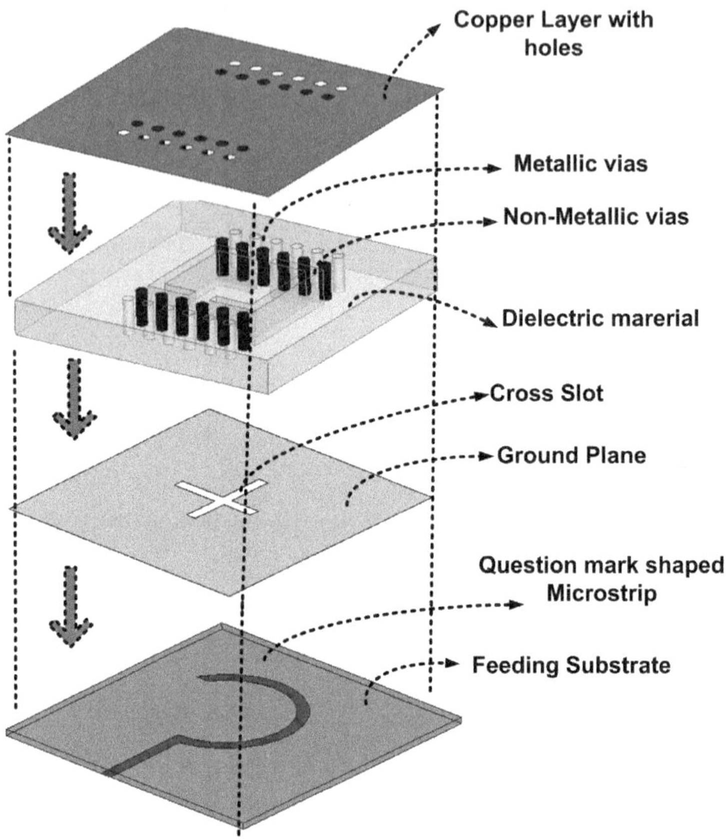

Figure 9.3 Pictorial description of wideband CP substrate-integrated DRA. (Modified after [10].)

In the next example of a mm-wave DRA array [11], the shape of a 4 × 4 element rectangular DRA is truncated around its corner in a diagonal direction to generate CP radiation. This antenna setup is cavity backed by slotted SIW technology without any extra feeding networks or baluns. In this design, fundamental orthogonal TE_{111} modes were excited in order to get CP fields by asymmetric placement of truncated DRA over a rectangular slot embedded in the SIW. In this antenna, an axial ratio bandwidth of 240 MHz and a 16 dBi peak gain is achieved. A summary of various designs reported in circularly polarized DRA arrays using SIW technology is given in Table 9.1.

9.2 Techniques Using Circularly Polarized Liquid DRAs

This section presents the most recent techniques using circularly polarized liquid DRAs with broadband and reconfigurable ability. The idea of circular

Table 9.1
Brief Summary of Various Circularly Polarized DRA Arrays Based on SIW Technology

Ref	DRA Shape	AR Frequency Range (GHz)	Gain (dBic)	ARBW (%)	Modes Excited
[8]	Cylindrical shape	55–64.5 GHz	11.43	15.9	Orthogonal $HE_{11\delta}$ modes
[9]	Rectangular shape	29.7–30.05 GHz	12.7	400 MHz	Orthogonal TE_{111} modes
[10]	Inner cylindrical and outer ring cylindrical DRA	20.3–26.45 GHz	8.15	26.3	$HE_{11\delta}$ and $HE_{12\delta+1}$ mode
[11]	Truncated rectangular	9.88–10.12 GHz	16	240 MHz	Orthogonal TE_{111} modes

AR = 3 dB axial ratio, ARBW = fractional 3-dB axial ratio bandwidth, DRA = dielectric resonator antenna.

polarization in a liquid DRA is new and it offers several advantages to realize small, reconfigurable, flexible, and transparent antennas as compared to solid material-based DRAs. Moreover, any complex feeding design can be inserted inside DRA when an antenna feed is required to be inserted into it, thus avoiding fabrication difficulties.

Recently, authors in [12] proposed a reconfigurable circularly polarized liquid DRA. In this design, the orthogonal fields were first generated by using the principle of detuning the degenerate TE_{111} modes in the rectangular DRA. The idea of using liquid solvent in the design is as follows: the liquid solvent was an ethyl acetate solvent and reconfigurability was achieved by controlling the flow of the liquid solvent across the left and right zones. Using a liquid injection method for reconfigurability is very useful because the conventional technique of using PIN diode is tough to apply to the DRA. The additional

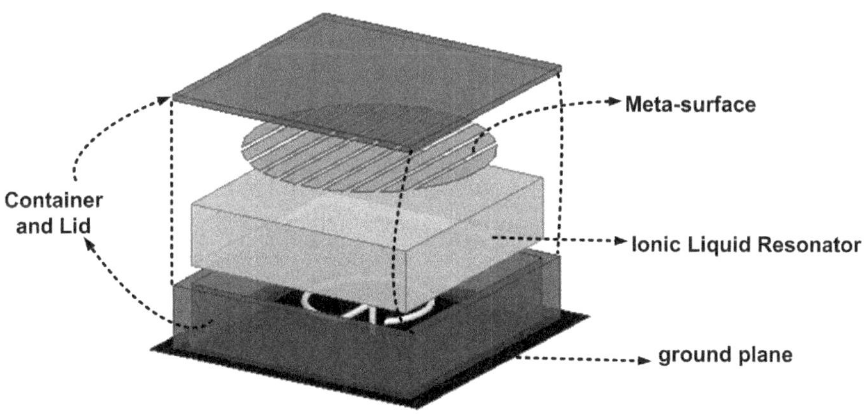

Figure 9.4 Example of a liquid DRA with a metasurface. (Modified after [13].)

PIN diodes on the DRA will cause unwanted radiation and reduce the radiation efficiency. The proposed design was integrated with a 3D printed container providing access for the liquid solution to be pumped in and out appropriately.

In another example, a wideband circularly polarized DRA was designed using a new organic ionic liquid material with attractive material properties in [13]. Also in this case, the DRA was excited by a modified probe feeding to excite the DRA to get CP radiation. To further improve the CP characteristics, a metasurface is utilized in the proposed design. The reported liquid CP antenna in [13] can be realized by a similar design configuration (with slight modification) as shown in Figure 9.4.

9.3 Miscellaneous Research Areas in Circularly Polarized DRAs

Antennas are also important in electromagnetic energy harvesting systems. Energy harvesting refers to collecting spatial electromagnetic waves and converting them into electric currents for further processing. For this purpose, an antenna should be capable of receiving maximum electromagnetic waves of any arbitrary polarizations (linear or circular). Therefore, a CP antenna is usually preferred. Other requirements of an energy harvesting antenna are to have a wide axial ratio bandwidth and its conversion efficiency should be high. There is one such design that was reported using DRAs in [14], in which a rectangular DRA was excited by a series feeding structure consisting of four crossing slots and one microstrip line. This feeding structure acts as a traveling wave structure, and hence ensures a broad bandwidth in the antenna design. The proposed antenna was designed to receive not only CP waves but also LP waves. The reported liquid CP antenna in [14] is shown in Figure 9.5 with slight modifications in the feeding network and DRA shape. There are some other miscellaneous applications, such as microwave image sensing [15, 16] and wearable antennas [17], where circularly polarized DRA can find applications for future use.

9.4 Concluding Remarks

As modern wireless technology becomes more integral to our lives, antenna polarization becomes an essential consideration when designing antennas for effective communication. The importance of successfully retrieving electromagnetic signals cannot be underestimated in today's wireless communication technology, which is one of the most important requirements in any wireless system for effective information transfer.

This book has highlighted the advancement of dielectric resonator antennas from a circular polarization point of view. The intention of this book has been to provide a path for the advancement of the DRA community with

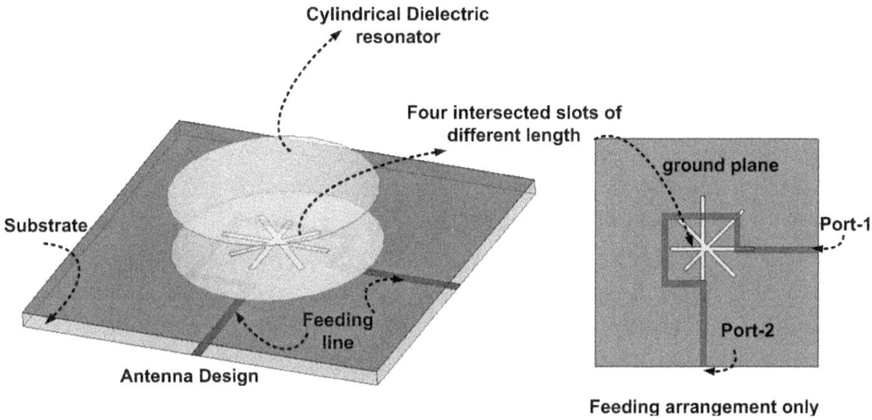

Figure 9.5 Example of a DRA for energy harvesting applications. (Modified after [14].)

regard to the proposed application-oriented DRA models, as well as to encourage researchers to further enhance the performance of the existing structures.

Comprehensive technical studies were carried out on various aspects of circularly polarized DRAs and we have highlighted the generation of circular polarization with design examples in each chapter. This book also presented some of the new design concepts by using five different feeding techniques, in which different shapes of DRs and ground plane arrangements were utilized for obtaining wideband circular polarization radiation. All the antennas we have presented are suitable for a variety of wireless applications. Further, researchers can easily identify the suitability of a specific feeding technique in various wireless applications such as Wi-MAX, Wi-Fi, WLAN, and 5G/mm-wave. The advantages of each feeding technique and geometry of DRs followed by their specific uses have been given in detail so one can directly go to or select the appropriate CP generation technique.

This book began with a comprehensive introduction and recent advancements in the field of dielectric resonator antennas. It delivered a complete overview of DRAs through the literature surveys and provided all the necessary details regarding the generation of circularly polarized fields in DRA. Comprehensive research reviews based on the feeding techniques were included along with their advantages and disadvantages to achieve circular polarization radiation in DRA. The different feeding techniques were described as follows: in Chapter 3, the microstrip line coupling technique was discussed, which has a potential advantage for the generation of circular polarization by implementing different configurations of feeding by using DRA as single, multiple, modified, multiple-input and multiple-output, array, and filtering structures. This coupling technique provides great flexibility for CP designing. Chapter 4 contained a modified version of microstrip line coupling (i.e., the conformal metallic strip coupling technique). A coplanar waveguide technique (Chapter 5) was discussed next, which has not yet been fully explored for CP applications and is a potential candidate for future CP designs. In Chapter 6, the aperture coupling technique was investigated, which is the most widely used feeding method and provides a greater amount of flexibility. Chapter 7 presented one of the simplest probe coupling techniques, which generally involves the modification of DRA geometry. A new technique in the sectorization approach is investigated in Chapter 8, in which the sectorization method has been implemented for the first time in DRA technology for circular polarization analysis and the corresponding physics were explained. The property of degenerate modes in DRA geometry was also investigated for CP generation in this chapter. The proposed approach of modification is simple to fabricate compared to the available geometry modification techniques that are generally complicated due to which fabrication difficulty increases.

This book concludes with this chapter, which provided a short survey on the current state of the art in circularly polarized DRAs and focused on the SIW technique for mm-wave DRA arrays, liquid DRAs mainly for polarization reconfigurability, and energy harvesting techniques. In each previous chapter, a design example was also included that explained the design methodology, provided a physical explanation for CP generation with proper simulation results and electric field distributions that were illustrated for better understanding of the CP process. Each chapter also provided an attempt to extend the use of each coupling technique to MIMO and arrays systems.

We conclude that DRAs show great flexibility in terms of shapes and excitation techniques to easily achieve wideband circularly polarized fields, and that research on circularly polarized DRAs can be extended for future applications. This book will be useful to the scientific community for research in future 5G/mm-wave applications for selecting different feeding techniques and the geometry of dielectric resonators. As per the current trend of circularly polarized dielectric resonator antennas in mm-wave applications, DRAs cannot be ignored. It is our intent that the feeding discussions in each chapter work as a push for current antenna designers to find the gap before introducing a new CP antenna model.

References

[1] Leung, K. W., and S. K. Mok, "Circularly Polarised Dielectric Resonator Antenna Excited by Perturbed Annular Slot with Backing Cavity," *IET Electronics Letters*, Vol. 37, No. 15, 2001, pp. 934–936.

[2] Chowdhury, R., R. Kumar, and R. K. Chaudhary, "A Coaxial Probe Fed Wideband Circularly Polarized Antenna Using Unequal and Adjacent-Slided Rectangular Dielectric Reso-

nators for WLAN Applications," *International Journal of RF and Microwave Computer-Aided Engineering,* Vol. 28, No. 4, 2018.

[3] Kumar, R., and R. K. Chaudhary, "A Wideband Circularly Polarized Cubic Dielectric Resonator Antenna Excited With Modified Microstrip Feed," *IEEE Antennas and Wireless Propagation Letters,* Vol. 15, 2016, pp. 1285–1288.

[4] Xu, F., and K. Wu, "Guided-Wave and Leakage Characteristics of Substrate Integrated Waveguide," *IEEE Transactions on Microwave Theory and Techniques,* Vol. 53, No. 1, 2005, pp. 66–73.

[5] Chu, H., and Yong-Xin Guo, "A Novel Approach for Millimeter-Wave Dielectric Resonator Antenna Array Designs by using the Substrate Integrated Technology," *IEEE Transactions on Antennas and Propagation* Vol. 65, No. 2, 2016, pp. 909–914.

[6] Bozzi, M., A. Georgiadis, and K. Wu, "Review of Substrate-Integrated Waveguide Circuits and Antennas," *IET Microwaves, Antennas and Propagation,* Vol. 05, No. 8, 2011, pp. 909–920.

[7] Wahab, W., D. Busuioc and S.Naeini, "Millimeter-Wave High Radiation Efficiency Planar Waveguide Series-Fed Dielectric Resonator Antenna (DRA) Array: Analysis, Design, and Measurements," *IEEE Transactions on Antennas and Propagation,* Vol. 59, No. 8, 2011, pp. 2834–2843.

[8] Sun, Y., and K. W. Leung, "Circularly Polarized Substrate-Integrated Cylindrical Dielectric Resonator Antenna Array for 60 GHz Applications," *IEEE Antennas and Wireless Propagation Letters,* Vol. 17, No. 8, 2018, pp. 1401–1405.

[9] Chu, H., and Y. Guo, "A Novel Approach for Millimeter-Wave Dielectric Resonator Antenna Array Designs by using the Substrate Integrated Technology," *IEEE Transactions on Antennas and Propagation,* Vol. 65, No. 2, 2016, pp. 909–914.

[10] Yang, M., Y. Pan, Y. Sun, and K. W. Leung, "Wideband Circularly Polarized Substrate-Integrated Embedded Dielectric Resonator Antenna for Millimeter-Wave Applications," *IEEE Transactions on Antennas and Propagation,* Vol. 68, No. 2, 2019, pp. 1145–1150.

[11] Lu, L., Y. Jiao, Z. Weng, L. Zhang, C. Cui, and R. Wang, "High-Efficiency Circularly Polarized Dielectric Resonator Antenna Array Fed by The Cavity-Backed SIW," *IEEE Antennas and Wireless Propagation Letters,* Vol. 17, No. 7, 2018, pp. 1145–1148.

[12] Chen, Z., and H. Wong, "Liquid Dielectric Resonator Antenna with Circular Polarization Reconfigurability, "*IEEE Transactions on Antennas and Propagation,* Vol. 66, No. 1, 2017, pp. 444–449.

[13] Song, C., E. Bennett, J. Xiao, A. Alieldin, K. Luk, and Y. Huang, "Metasurfaced, Broadband, and Circularly Polarized Liquid Antennas Using a Simple Structure," *IEEE Transactions on Antennas and Propagation,* Vol. 67, No. 7, 2019, pp. 4907–4913.

[14] Liu, S., D. Yang, Y. Chen, S. Huang, and Y. Xiang, "Broadband Dual Circularly Polarized Dielectric Resonator Antenna for Ambient Electromagnetic Energy Harvesting," *IEEE Transactions on Antennas and Propagation,* Vol. 68, No. 6, 2020, pp. 4961–4966.

[15] Singhwal, S. S., B. K. Kanaujia, A. Singh, and Jugul Kishor, "Novel Circularly Polarized Dielectric Resonator Antenna for Microwave Image Sensing Application," *Microwave and Optical Technology Letters,* Vol. 61, No. 7, 2019, pp. 1821–1827.

[16] Rana, B., and S. K. Parui, "Microstrip Line Fed Wideband Circularly-Polarized Dielectric Resonator Antenna Array for Microwave Image Sensing," *IEEE Sensors Letters*, Vol. 1, No. 3, 2017.

[17] Illahi, U., J. Iqbal, M. I. Sulaiman, M. M. Alam, M. M. Su'Ud, M. Jamaluddin, and M. Yasin, "Design of New Circularly Polarized Wearable Dielectric Resonator Antenna for Off-Body Communication in WBAN Applications," *IEEE Access,* Vol. 7, 2019, pp. 150573–150582.

About the Authors

Raghvendra Kumar Chaudhary is an associate professor of the Department of Electronics Engineering, Indian Institute of Technology (ISM) Dhanbad, India. He received a B.Tech. from UIET Kanpur, India, in 2007, an M.Tech. from IIT(BHU) Varanasi, India, in 2009, and a Ph.D. from IIT Kanpur, India, in 2014. He has researched and developed a metamaterial antenna and dielectric resonator antenna (DRA), and developing a circularly polarized compact antenna is one of his major areas of contribution. He has published over 125 papers in top scientific journals. Dr. Chaudhary has guided 5 Ph.D. students, and currently, 11 Ph.D. students are working under him. He was a recipient of the Young Engineers Award (2020) from the Indian National Academy of Engineering (INAE), India, the Young Scientist Award (2020) from the Institution of Electronics and Telecommunication Engineers (IETE), India, and the Young Engineers Award (2019–2020) from the Institution of Engineers, India (IEI). He has received many Best Paper awards from different national and international conferences, such as IEEE APACE Malaysia, PIERS Singapore, and ATMS India. He has served as a chair for the IEEE Student Branch of Uttar Pradesh Section, in 2012–2013. Presently, he serves as a counsellor of IEEE Student Branch and Advisor of AP/MTT Societies Student Branch Joint Chapter of IIT (ISM) Dhanbad. Dr. Chaudhary serves as an associate editor for IET *Microwave Antennas & Propagation* (U.K.), *IEEE Access* (U.S.), and *Microwave and Optical Technology Letters* (U.S.). He is a Senior Member of IEEE and URSI, and frequent reviewer for international journals such as *IEEE Transactions on Antennas and Propagation* and *IEEE Antennas and Wireless Propagation Letters.* He has also been a featured interview in *IET Electronics Letters* (U.K.). His current research interests include MIMO and cognitive radio, reconfigurable metamaterial/metasurfaces for antenna and absorber design, circularly polarized DRA, and frequency reconfigurable active array antenna with beamsteering.

Rajkishor Kumar was born in West-Champaran (Bettiah), Bihar, India. He received his Ph.D. in RF and microwave engineering from the Department of Electronics Engineering, Indian Institute of Technology (Indian School of Mines), Dhanbad, India, in 2019, and his B. E. in electronics and communication engineering from Nagpur University, Nagpur, India, in 2012. He is currently working as an assistant professor, senior grade, at the School of Electronics Engineering (SENSE) in the Department of Communication Engineering, Vellore Institute of Technology (VIT) University, Vellore, Tamil Nadu. He has authored/coauthored more than 34 refereed journal and conference papers and received a fellowship from MHRD, Government of India during his Ph.D. studies. He was the recipient of the International Travel Grant from DST (SERB) International Travel Support (ITS) Scheme, India. He is a reviewer of journals such as *IEEE Antenna & Wireless Propagation Letters, IEEE Access, IET Microwaves, Antennas & Propagation, IET Electronics Letter, IET The Journal of Engineering, Microwave Optical Technology Letters, AEU-International Journal of Electronics and Communications, ETRI Journal, Wiley Journals,* and *International Journal of RF and Microwave Computer-Aided Engineering.* His current research interests include RF and microwave antennas, RF energy harvesting and transfer systems, and antennas in biomedical applications.

Rakesh Chowdhury was born in West Bengal, India, in 1991. He received his B.Tech. in electronics and communication engineering from WBUT University, Kolkata, in 2013. He received his M.Tech. and Ph.D. from the Indian Institute of Technology (Indian School of Mines) Dhanbad, India, in 2015 and 2021 respectively. He received a certificate of honor for excelling in academics at graduation from BIET, SURI (WBUT University) He also received a research publication award in 2018 from the Electronics Department from IIT(ISM), Dhanbad. He has authored or coauthored more than 20 research articles in international journals/conference proceedings. He is student member of IEEE and frequent reviewer for many journals, such as *IEEE Antenna and Wireless Propagation Letters, IEEE Access, International Journal of RF and Microwave Computer-Aided Engineering,* and *Journal of Alloys and Compounds.* His current research interests include dielectric resonator antennas, high-gain antennas, electromagnetic bandgap (EBG) structures, and metamaterials.

Index

Recent Titles in the Artech House Antennas and Propagation Library

Christos Christodoulou, Series Editor

Adaptive Array Measurements in Communications, M. A. Halim

Antenna-on-Chip: Design, Challenges, and Opportunities, Hammad M. Cheema, Fatima Khalid, and Atif Shamim

Antenna Design for Cognitive Radio, Youssef Tawk, Joseph Costantine, and Christos Christodoulou

Antenna Design with Fiber Optics, A. Kumar

Antenna Engineering Using Physical Optics: Practical CAD Techniques and Software, Leo Diaz and Thomas Milligan

Antenna Measurement Techniques, Gary E. Evans.

Antennas and Propagation for Body-Centric Wireless Communications, Second Edition, Peter S. Hall and Yang Hao, editors

Antennas and Site Engineering for Mobile Radio Networks, Bruno Delorme

Antennas for Small Mobile Terminals, Kyohei Fujimoto and Koichi Ito, editors

Analysis, Design and Measurement of Small and Low-profile Antennas, Kazuhiro Hirasawa

Analysis of Radome-Enclosed Antennas, Second Edition, Dennis J. Kozakoff

Applications of Geographic Information Systems for Wireless Network Planning, Francisco Saez de Adana, Josefa Gómez Pérez, Abdelhamid Tayebi Tayebi, and Juan Casado Ballesteros

The Art and Science of Ultrawideband Antennas, Second Edition, Hans G. Schantz

AWAS for Windows Version 2.0: Analysis of Wire Antennas and Scatterers, Antonije R. Djordjević, et al.

Broadband Microstrip Antennas, Girsh Kumar and K. P. Ray

Broadband Patch Antennas, Jean-François Zürcher and Fred E. Gardiol

Circularly Polarized Dielectric Resonator Antennas, Raghvendra Kumar Chaudhary, Rajkishor Kumar, and Rakesh Chowdhury

Design and Applications of Active Integrated Antennas, Mohammad S. Sharawi and Oualid Hammi

Designer Notes for Microwave Antennas, Richard C. Johnson

Dielectric Resonator Antenna Handbook, Aldo Petosa

Electromagnetics and Antenna Technology, Alan J. Fenn

Electromagnetics, Microwave Circuit and Antenna Design for Communications Engineering, Peter Russer

Engineering Applications of the Modulated Scatterer Technique, Jean-Charles Bolomey and Fred E. Gardiol

Fixed & Mobile Terminal Antennas, Akhileswar Kumar

Foundations of Antenna Engineering: A Unified Approach for Line-of -Sight and Multipath, Per-Simon Kildal

Four-Arm Spiral Antennas, Joseph A. Mosko

Frequency-Agile Antennas for Wireless Communications, Aldo Petosa

Fresnel Zones in Wireless Links, Zone Plate Lenses and Antennas, Hristo D. Hristov

Handbook of Antennas for EMC, Second Edition, Thereza Macnamara with John McAuley

Handbook of Reflector Antennas and Feed Systems, Volume I: Theory and Design of Reflectors, Satish Sharma, Sudhakar Rao, and Lotfollah Shafai, editors

Handbook of Reflector Antennas and Feed Systems, Volume II: Feed Systems, Lotfollah Shafai, Satish Sharma, and Sudhakar Rao, editors

Handbook of Reflector Antennas and Feed Systems, Volume III: Applications of Reflectors, Sudhakar Rao, Lotfollah Shafai, and Satish Sharma, editors

Interference Suppression Techniques for Microwave Antennas and Transmitters, Ernest Freeman

Introduction to Antenna Analysis Using EM Simulators, Hiroaki Kogure, Yoshie Kogure, and James C. Rautio

LONRS: Low-Noise Receiving Systems Performance and Analysis Toolkit, Charles T. Stelzried, Macgregor S. Reid, and Arthur J. Freiley

Measurement of Mobile Antenna Systems, Second Edition, Hiroyuki Arai

Microstrip Lines and Slotlines, Third Edition, Ramesh Garg Inder J. Bahl, Maurizio Bozzi

Microstrip Antenna Design, K.C Gupta and Abdelaziz Benalla

Microwave and Millimeter-Wave Remote Sensing for Security Applications, Jeffrey A. Nanzer

Microwave Cavity Antennas, Akhileswar Kumar and Hristo D. Hristov

Millimeter Wave Microstrip and Printed Circuit Antennas, Prakash Bhartia

Mobile Antenna Systems Handbook, Third Edition, Kyohei Fujimoto, editor

Moment Methods in Antennas and Scattering, Robert C. Hansen

Multiband Integrated Antennas for 4G Terminals, David A. Sánchez-Hernández, editor

Near-Field Antenna Measurements, Dan Slater

Noise Temperature Theory and Applications for Deep Space Communications Antenna Systems, Tom Y. Otoshi

Phased Array Antenna Handbook, Third Edition, Robert J. Mailloux

Phased Array Antennas, Arthur A. Oliner

Phased Array Antennas with Optimized Element Patterns, Sergei P. Skobelev

Plasma Antennas, Second Edition, Theodore Anderson

Polarization in Electromagnetic Systems, Second Edition, Warren Stutzman

Practical Antenna Design for Wireless Products. Henry Lau

Practical Microstrip and Printed Antenna Design, Anil Pandey

Practical Phased Array Antenna Systems, Eli Brookner

Practical Simulation of Radar Antennas and Radomes, Herbert L. Hirsch

Printed MIMO Antenna Engineering, Mohammad S. Sharawi

Radiowave Propagation and Antennas for Personal Communications, Third Edition, Kazimierz Siwiak

Reconfigurable Antenna Design and Analysis, Mohammod Ali

Reflectarray Antennas: Analysis, Design, Fabrication and Measurement, Jafar Shaker, Mohammad Reza Chaharmir, and Jonathan Ethier

Reflector Lens Antennas: Analysis and Design Using Personal Computers, Carlyle J. Sletten, editor

RF Coaxial Slot Radiators: Modeling, Measurements, and Applications, Kok Yeow You

Shipboard Antennas, Second Edition, Preston Law

Significant Phased Array Papers, Robert C. Hansen

Smart Antenna Engineering, Ahmed El Zooghby

Solid Dielectric Horn Antennas, Carlos Salema, Carlos Fernandes, and Rama Kant Jha

Ultrawideband Antennas for Microwave Imaging Systems, Tayeb A. Denidni and Gijo Augustin

Ultrawideband Short-Pulse Radio Systems, V. I. Koshelev, Yu. I. Buyanov, and V. P. Belichenko

Waveguide Components for Antenna Feed Systems: Theory and CAD, Jaroslaw Uher, Jens Bornemann, and Uwe Rosenberg

For further information on these and other Artech House titles, including previously considered out-of-print books now available through our In-Print-Forever® (IPF®) program, contact:

Artech House
685 Canton Street
Norwood, MA 02062
Phone: 781-769-9750
Fax: 781-769-6334
e-mail: artech@artechhouse.com

Artech House
16 Sussex Street
London SW1V HRW UK
Phone: +44 (0)20 7596-8750
Fax: +44 (0)20 7630 0166
e-mail: artech-uk@artechhouse.com

Find us on the World Wide Web at: www.artechhouse.com